Lecture Notes in Mathematics   2246

More information about this subseries at http://www.springer.com/series/3114

Fondazione C.I.M.E., Firenze

C.I.M.E. stands for *Centro Internazionale Matematico Estivo*, that is, International Mathematical Summer Centre. Conceived in the early fifties, it was born in 1954 in Florence, Italy, and welcomed by the world mathematical community: it continues successfully, year for year, to this day.

Many mathematicians from all over the world have been involved in a way or another in C.I.M.E.'s activities over the years. The main purpose and mode of functioning of the Centre may be summarised as follows: every year, during the summer, sessions on different themes from pure and applied mathematics are offered by application to mathematicians from all countries. A Session is generally based on three or four main courses given by specialists of international renown, plus a certain number of seminars, and is held in an attractive rural location in Italy.

The aim of a C.I.M.E. session is to bring to the attention of younger researchers the origins, development, and perspectives of some very active branch of mathematical research. The topics of the courses are generally of international resonance. The full immersion atmosphere of the courses and the daily exchange among participants are thus an initiation to international collaboration in mathematical research.

C.I.M.E. Director (2002 – 2014)
Pietro Zecca
Dipartimento di Energetica "S. Stecco"
Università di Firenze
Via S. Marta, 3
50139 Florence
Italy
e-mail: zecca@unifi.it

C.I.M.E. Director (2015 –)
Elvira Mascolo
Dipartimento di Matematica "U. Dini"
Università di Firenze
viale G.B. Morgagni 67/A
50134 Florence
Italy
e-mail: mascolo@math.unifi.it

C.I.M.E. Secretary
Paolo Salani
Dipartimento di Matematica "U. Dini"
Università di Firenze
viale G.B. Morgagni 67/A
50134 Florence
Italy
e-mail: salani@math.unifi.it

CIME activity is carried out with the collaboration and financial support of INdAM (Istituto Nazionale di Alta Matematica)

For more information see CIME's homepage: **http://www.cime.unifi.it**

Sławomir Dinew • Sebastien Picard •
Andrei Teleman • Alberto Verjovsky

Complex Non-Kähler Geometry

Cetraro, Italy 2018

Daniele Angella • Leandro Arosio •
Eleonora Di Nezza

Editors

FONDAZIONE
CIME
ROBERTO CONTI
CENTRO INTERNAZIONALE MATEMATICO ESTIVO
INTERNATIONAL MATHEMATICAL SUMMER CENTER

Authors
Sławomir Dinew
Department of Mathematics and Computer Science
Jagiellonian University
Krakow, Poland

Sebastien Picard
Harvard University
Cambridge
MA, USA

Andrei Teleman
Aix-Marseille Université, CNRS
Marseille, France

Alberto Verjovsky
Instituto de Matematicas, Unidad Cuernavaca
Universidad Nacional Autónoma
Mexico, Morelos, Mexico

Editors
Daniele Angella (iD)
Dipartimento di Matematica e Informatica "Ulisse Dini"
Università di Firenze
Firenze, Italy

Leandro Arosio
Dipartimento di Matematica
Università di Roma Tor Vergata
Roma, Italy

Eleonora Di Nezza
Sorbonne Université
Paris, France

ISSN 0075-8434 ISSN 1617-9692 (electronic)
Lecture Notes in Mathematics
C.I.M.E. Foundation Subseries
ISBN 978-3-030-25882-5 ISBN 978-3-030-25883-2 (eBook)
https://doi.org/10.1007/978-3-030-25883-2

Mathematics Subject Classification (2010): Primary: 53C55, 32W20; Secondary: 32Q57, 32U05, 53C44, 57R17, 32U15, 32Q15

This Springer imprint is published by the registered company Springer Nature Switzerland AG.
The registered company address is: Gewerbestrasse 11, 6330 Cham, Switzerland

Preface

The study and the construction of complex non-Kähler manifolds is a demanding and interesting mathematical problem, with applications to Theoretical Physics. On the one side, in the words of [dBT13], "a sort of chemical analysis of symplectic and holomorphic contribution can be successfully performed in order to better understand the role of the different components of the theory" of Kähler Geometry, which "represents a perfect synthesis of the Symplectic and the Holomorphic worlds." On the other side, deeply studied problems in Kähler geometry have often a non-Kähler counterpart that is naturally motivated by Theoretical Physics, as, for example, the Hull-Strominger system [Str86, Hul86, GF16] that extends the Calabi–Yau theorem to non-Kähler manifolds. When studying such problems, one has to deal with the lack of examples and techniques or at least with choosing the right context where doing so.

The CIME Summer School on "Complex non-Kähler Geometry", held in Cetraro (Italy), on July 9–13, 2018, was aimed at introducing young students and researchers to this field, by presenting a vast range of techniques from different aspects of the theory: from complex analysis to complex differential geometry, from holomorphic dynamics to geometric analysis.

In complex dimension one, the classification of holomorphic curves is the "crown jewel of differential geometry" [Gro00]. In higher dimension, even in complex dimension two, the classification of compact complex manifolds is still an open problem. The classification of compact complex surfaces was started by the Italian school of algebraic geometry and then continued by Kunihiko Kodaira in the 1950s, (see e.g., [BHPV04, Nak84]). The incomplete steps regard minimal compact complex surfaces with Kodaira dimension $-\infty$ and with first Betti number equal to one and positive second Betti number $b_2 > 0$: Masahide Kato gave a fascinating construction of complex surfaces in this class [Kat78], characterized by containing b_2 rational curves [DOT03]. Therefore, the problem of completing the classification of compact complex surfaces reduces to the construction of rational curves. **Andrei Teleman** gave important contributions in this direction [Tel05, Tel10]. His course on "Non-Kählerian Compact Complex Surfaces" presents the classical theory on Enriques-Kodaira classification and non-Kählerian surfaces, class VII surfaces, and the construction and the classification of Kato surfaces. He also pointed out the main ideas of his gauge theoretical methods to prove the existence of cycles of curves on class VII surfaces with small second Betti number, as well as of his recent results obtained with Dloussky [DT18] towards classification up to biholomorphism.

A construction of a compact complex surface in class VII, with a complicated fractal subset, can be done starting from a Hénon map to obtain a compact dynamical system. This idea by Hubbard et al. [HPV00] has been presented in his talk on "Compact complex surfaces from Hénon mappings, and a surprising surgery."

In higher dimension (complex dimension $n \geq 2$), examples of non-Kähler (not even symplectic) complex manifolds are given by LVM manifolds that have been initially introduced and studied by López de Medrano and Verjovsky [LMV97]. This construction was then generalized by Meersseman [Mee97, Mee00] and Bosio [Bos01], and these manifolds are known as LVMB manifolds. The essential idea (discovered by André Haefliger) is that one can obtain non-algebraic complex manifolds as the space of leaves of holomorphic foliations of complex algebraic manifolds. In his course on "Intersection of quadrics in $\mathbb{C}^n$, moment-angle manifolds, complex manifolds and convex polytopes", **Alberto Verjovsky** explains in details how to make use of such a general principle: we start with a linear action of $\mathbb{C}^m$ in $\mathbb{C}^n$ for $n > 2m$ or, in other words, with a configuration of n vectors in $\mathbb{C}^m$. If the configuration is "admissible" (more precisely, if it satisfies the "weak hyperbolicity" and "Siegel conditions"), the holomorphic (singular) foliation $\mathcal{F}$ in the projective space $\mathbb{CP}^{n-1}$ is the one given by the orbits of the linear action we start with.

Such examples confirm a conjecture by Bogomolov [Bog96] on the realizability of any compact complex manifold by a transverse embedding into a projective manifold equipped with an algebraic foliation. In his talk "Algebraic embeddings of complex and almost complex structures" Jean-Pierre Demailly explained a recent work joint with Gaussier [DG17] in the direction of the Bogomolov conjecture: they show an embedding theorem for compact almost complex manifolds as subvarieties transverse to an algebraic distribution in a complex affine algebraic manifold.

Another way to study and try to understand complex non-Kähler manifolds is to look for "canonical" metrics, where the word "canonical" may refer to some cohomological or curvature properties, e.g., having constant curvature. The story begins with the Calabi conjecture [Cal54] in Kähler geometry and its 40-year-old solution by Yau [Yau77, Yau78]. Together with [Aub76] and the recent works [CDS14, CDS15, Tia15] on the Yau-Tian-Donaldson conjecture, they give a complete description of Kähler-Einstein metrics. Kähler metrics being Ricci-flat are also characterized by the restricted Riemannian holonomy being in $SU(n)$, and (both elliptic and parabolic) complex Monge–Ampère equations are useful tools for their investigation. There are different ways to generalize the notion of Kähler Ricci-flat metrics to Hermitian non-Kähler manifolds. One possibility is to consider the so-called Calabi-Yau manifolds with torsion. They are characterized by the Bismut connection having restricted holonomy in $SU(n)$, namely, by a Bismut-Ricci-flat metric, and they are motivated by heterotic string theory [Hul86, Str86, LY05].

The course "The Anomaly flow and Hull-Strominger system" by **Sébastien Picard** focused on the problem of constructing canonical Hermitian metrics on complex non-Kähler manifolds. In particular, the lectures focused on the "Anomaly flow," a tool that Sébastien Picard, in joint work with Phong et al. [PPZ18a, PPZ18b, PPZ19], recently introduced and investigated to study this non-Kähler Calabi-Yau geometry. The simplest case, the "Anomaly flow with zero slope," preserves the conformally balanced condition and deforms conformally balanced metrics towards astheno-Kähler metrics, so that stationary points are Kähler. In this sense, it can be understood as a deformation path connecting non-Kähler to Kähler geometry.

Another possible way to generalize the Calabi conjecture to non-Kähler manifolds has been studied by Tosatti and Weinkove [TW10]. More precisely, they prove that each element in the first Bott-Chern class can be represented as the first Chern-Ricci form of a Hermitian metric. The result is obtained as a corollary of uniform estimates for a complex Monge–Ampère-type equation on a compact Hermitian manifold. Their work has been presented by Eleonora Di Nezza in the talk "The Calabi conjecture on compact Hermitian manifolds."

Geometric flows, such as the Kähler-Ricci flow or the Anomaly flow, give another way to study and construct canonical metrics on a compact Kähler or Hermitian manifold. The basic idea is very simple: we pick a random initial metric, we run the flow, and we hope that such flow of metrics will converge to a *special* one. This is what happens, for example, on a compact Kähler manifold with negative or zero first Chern class [Cao85]. The problem is that, in general, it may happen that the time of existence of the flow is finite and that the limit metric is singular. One then needs to restart the flow with a singular data. This is the reason why studying geometric flows in a weak sense is of major importance in this respect. "Pluripotential Theory on Hermitian Manifolds" was instead the subject of the course of **Sławomir Dinew**. The aim of the course was to introduce the Hermitian version of complex Monge–Ampère equations, mostly trying to generalize the analytic techniques coming from the Kählerien counterpart. First of all, he defined several "Kähler-type" conditions (such as "balanced," "Gauduchon," "astheno-Kähler," and "pluriclosed" metrics), giving a variety of examples. If we fix one of these conditions, a "canonical" metric on a compact Hermitian manifold has to satisfy such a condition and to have some special curvature property. Once again, the problem boils down to a complex Monge–Ampère equation. He then introduced the first notions of pluripotential theory on compact Hermitian manifolds in order to treat (even in contexts presenting singularities) such equations. This last part mostly relies on his works [Din16, DK12].

Recently, Guedj et al. [GLZ18] developed a parabolic pluripotential theory on compact Kähler manifolds, defining and studying the "weak" Kähler-Ricci flow and, more precisely, the weak solutions to degenerate parabolic complex Monge–Ampère equations. The results were announced in the talk by Vincent Guedj.

Acknowledgment The School has been proposed as part of the Fondazione CIME Summer School Program. The course directors would like to warmly thank CIME for the financial and organizational support. Many thanks in particular to the CIME Director Elvira Mascolo, to the CIME Scientific Secretary Paolo Salani, and to the secretary Rita. The activity of CIME is carried out also with the collaboration and financial support of INdAM, which we thank.

The School benefited from the financial support by MIUR SIR2014 Project RBSI14CFME "New methods in holomorphic iteration (NEWHOLITE)" at the Università di Roma Tor Vergata, by MIUR SIR2014 Project RBSI14DYEB "Analytic aspects in complex and hypercomplex geometry (AnHyC)" at the Università degli Studi di Firenze, and by the grant Jeunes Géomètres (Projet soutenu par un prix attribué par la Fondation Louis D. sur proposition de l'Académie des Sciences).

The course directors would like to thank Valentino Tosatti, with whom they discussed a first draft of the scientific program and for his suggestions and encouragment all along the way.

Our warm gratitude goes to the lecturers: Sławomir Dinew, Sébastien Picard, Andrei Teleman, and Alberto Verjovsky. Their wonderful notes completed their wonderful lectures. The enthusiasm with which they shared their ideas affected the audience, and the beautiful scenario they constructed during the lectures competed with the vista on the sea! The scientific program was excellently completed by a series of research talks by Jean-Pierre Demailly, Eleonora Di Nezza, Vincent Guedj, John Hubbard, who we warmly thank as well!

Thanks to Hotel San Michele in Cetraro and to its staff for the warm hospitality.

Thanks also to Ute McCrory at Springer for her assistance in preparing this volume. We hope this book will be an inspiring reference for many young students.

Last but not least, thanks so much to all the participants. Your passion made sense of the School, and the atmosphere you contributed to create, both scientifically and personally, became a place of growth for everyone.

Firenze, Italy
Roma, Italy
Paris, France
May 17, 2019

Daniele Angella
Leandro Arosio
Eleonora Di Nezza

References

[Aub76] Th. Aubin, Équations du type Monge–Ampère sur les variétés käh-leriennes compactes. C. R. Acad. Sci. Paris Sér. A-B **283**(3), Aiii, A119–A121 (1976)

[BHPV04] W.P. Barth, K. Hulek, C.A.M. Peters, A. Van de Ven, *Compact Complex Surfaces* (Springer, Berlin, 2004)

[Bog96] F. Bogomolov, Complex manifolds and algebraic foliations. Publ. RIMS-1084, Kyoto, June 1996 (unpublished)

[Bos01] F. Bosio, Variétés complexes compactes: une généralisation de la construction de Meersseman et López de Medrano-Verjovsky. Ann. Inst. Fourier **51**(5), 1259–1297 (2001)

[dBT13] P. de Bartolomeis, A. Tomassini, Exotic deformations of Calabi-Yau manifolds. Ann. Inst. Fourier **63**(2), 391–415 (2013)

[Cal54] E. Calabi, The space of Kähler metrics, in *Proceedings of the International Congress of Mathematicians (Amsterdam, 1954)*, vol. 2 (Noordhoff, Groningen, 1954), pp. 206–207

[Cao85] H.D. Cao, Deformation of Kähler metrics to Kähler-Einstein metrics on compact Kähler manifolds, Invent. Math. **81**(2), 359–372 (1985)

[CDS14] X.X. Chen, S. Donaldson, S. Sun, Kähler-Einstein metrics and stability. Int. Math. Res. Not. IMRN **2014**(8), 2119–2125 (2014)

[CDS15] X.X. Chen, S. Donaldson, S. Sun, Kähler-Einstein metrics on Fano manifolds. I: approximation of metrics with cone singularities. J. Am. Math. Soc. **28**(1), 183–197 (2015). Kähler-Einstein metrics on Fano manifolds. II: Limits with cone angle less than 2π. ibid., 199–234. Kähler-Einstein metrics on Fano manifolds. III: Limits as cone angle approaches 2π and completion of the main proof. ibid., 235–278

[DG17] J.-P. Demailly, H. Gaussier, Algebraic embeddings of smooth almost complex structures. J. Eur. Math. Soc. **19**(11), 3391–3419 (2017)

[Din16] S. Dinew, Pluripotential theory on compact Hermitian manifolds. Ann. Fac. Sci. Toulouse Math. **25**(1), 91–139 (2016)

[DK12] S. Dinew, S. Kołodziej, Pluripotential estimates on compact Hermitian Manifolds. Adv. Geom. Anal. 6–86 (2012). Adv. Lect. Math. (21)

[DOT03] G. Dloussky, K. Oeljeklaus, M. Toma, Class VII$_0$ surfaces with b_2 curves. Tohoku Math. J. **55**(2), 283–309 (2003)

[DT18] G. Dloussky, A. Teleman, Smooth deformations of singular contractions of class VII surfaces. https://arxiv.org/abs/1803.07631

[GF16] M. Garcia-Fernandez, Lectures on the Strominger system, in *Travaux Mathématiques. Vol. XXIV*, vol. 24 (Faculty of Science, Technology and Communication, University of Luxembourg, Luxembourg, 2016), pp. 7–61

[Gro00] M. Gromov, Spaces and questions, in *GAFA 2000 (Tel Aviv, 1999)*. Geometric and Functional Analysis, vol. 2000, Special Volume, Part I (2000), pp. 118–161

[GLZ18] V. Guedj, H.C. Lu, A. Zeriahi, Pluripotential Kahler-Ricci flows. Geom. Topol. (to appear). https://arXiv.org/abs/1810.02121

[HPV00] J. Hubbard, P. Papadopol, V. Veselov, A compactification of Hénon mappings in $\mathbf{C}^2$ as dynamical systems. Acta Math. **184**(2), 203–270 (2000)

[Hul86] C.M. Hull, Superstring compactifications with torsion and spacetime supersymmetry, in *Superunification and Extra Dimensions (Torino, 1985)* (World Sci. Publishing, Singapore, 1986), pp. 347–375

[Kat78] Ma. Kato, Compact complex manifolds containing "global" spherical shells. I, in *Proceedings of the International Symposium on Algebraic Geometry (Kyoto Univ., Kyoto, 1977)* (Kinokuniya Book Store, Tokyo, 1978), pp. 45–84

[LY05] J. Li, S.-T. Yau, The existence of supersymmetric string theory with torsion. J. Differ. Geom. **70**(1), 143–181 (2005)

[LMV97] S. López de Medrano, A. Verjovsky, A new family of complex, compact non-symplectic manifolds. Bull. Braz. Math. Soc. **28**(2), 253–269 (1997)

[Mee97] L. Meersseman, Construction de variétés compactes complexes. C. R. Acad. Sci. Paris Sér. I Math. **325**(9), 1005–1008 (1997)

Mee00 L. Meersseman, A new geometric construction of compact complex manifolds in any dimension. Math. Ann. **317**(1), 79–115 (2000)

[Nak84] I. Nakamura, Classification of non-Kähler complex surfaces. Sugaku **36**(2), 110–124 (1984). Translated in Sugaku Expositions **2**(2), 209–229 (1989)

[PPZ18a] D.H. Phong, S. Picard, X.W. Zhang, Geometric flows and Strominger systems. Math. Z. **288**, 101–113 (2018)

[PPZ18b] D.H. Phong, S. Picard, X.W. Zhang, Anomaly flows. Commun. Anal. Geom. **26**(4), 955–1008 (2018)

[PPZ19] D.H. Phong, S. Picard, X.W. Zhang, A flow of conformally balanced metrics with Kähler fixed points. Math. Ann. **374**(3–4), 2005–2040 (2019)

[Str86] A. Strominger, Superstrings with torsion, Nuclear Phys. B **274**(2), 253–284 (1986)

[Tel05] A. Teleman, Donaldson theory on non-Kählerian surfaces and class VII surfaces with $b_2 = 1$. Invent. Math. **162**(3), 493–521 (2005)

[Tel10] A. Teleman, Instantons and curves on class VII surfaces. Ann. Math. **172**(3), 1749–1804 (2010)

[Tia15] G. Tian, K-stability and Kähler-Einstein metrics. Commun. Pure Appl. Math. **68**(7), 1085–1156 (2015). Corrigendum: K-stability and Kähler-Einstein metrics. ibid. no. 11, 2082–2083

[TW10] V. Tosatti, B. Weinkove, The complex Monge–Ampère equation on compact Hermitian manifolds. J. Am. Math. Soc. **23**(4), 1187–1195 (2010)

[Yau77] S.-T. Yau, Calabi's conjecture and some new results in algebraic geometry. Proc. Nat. Acad. Sci. U.S.A. **74**(5), 1798–1799 (1977)

[Yau78] S.-T. Yau, On the Ricci curvature of a compact Kähler manifold and the complex Monge–Ampère equation. I. Commun. Pure Appl. Math. **31**(3), 339–411 (1978)

Contents

List of Participants

Al-Abdallah Abdel Rahman	(Regina)	alabdaab@uregina.ca
Albanese Michael	(Stony Brook)	michael.albanese@stonybrook.com
Altavilla Amedeo	(Roma Tor Vergata)	amedeoaltavilla@gmail.com
Angella Daniele	(Firenze)	daniele.angella@unifi.it
Arosio Leandro	(Roma Tor Vergata)	arosio@mat.uniroma2.it
Bazzoni Giovanni	(UCM Madrid)	gbazzoni@ucm.es
Bianchi Fabrizio	(Imperial College)	fbianchi@imperial.ac.uk
Bisi Cinzia	(Ferrara)	bsicnz@unife.it
Bodian Mamadou Eramane	(Assane Seck)	m.bodian2966@zig.univ.sn
Bonechi Niccolò	(Firenze)	niccolo.bonechi1@stud.unifi.it
Clemente Gabriella	(Rutgers)	gabriella.clemente@rutgers.edu
de Fabritiis Chiara	(Marche)	fabritiis@dipmat.univpm.it
Deev Rodion	(New York)	rodion@cims.nyu.edu
Demailly Jean-Pierre	(Grenoble Alpes)	Jean-Pierre.Demailly@univ-grenoble-alpes.fr
Dinew Sławomir	(Jagiellonian)	slawomir.dinew@im.uj.edu.pl
Di Nezza Eleonora	(IHES)	dinezza@ihes.fr
Dloussky Georges	(Aix-Marseille)	georges.dloussky@univ-amu.fr
Eremin Boris	(Moscow)	Eremin.ba@phystech.edu
Gallup Nathaniel	(UC Davis)	npgallup@math.ucdavis.edu
Gori Anna	(Milano)	anna.gori@unimi.it
Guedj Vincent	(Toulouse)	vincent.guedj@math.univ-toulouse.fr

Guenancia Henri	(CNRS Toulouse)	henri.guenancia@math.univ-toulouse.fr
Hashimoto Yoshinori	(Firenze)	yoshinori.hashimoto@unifi.it
Hubbard John	(Cornell)	jhh8@cornell.edu
Korshunov Dmitrii	(HSE Moscow)	dkorshunov@hse.ru
Latorre Adela	(CUD Zaragoza)	adela@unizar.es
Le Meur Dimitri	(ENS Paris)	lemeur@clipper.ens.fr
Lopez de Medrano Santiago	(UNAM Mexico)	santiago@matem.unam.mx
Lukzen Elena	(SISSA Trieste)	elukzen@sissa.it
Maccheroni Roberta	(Parma)	roberta.maccheroni@unife.it
Manero Víctor	(Zaragoza)	vmanero@unizar.es
Marini Stefano	(Parma)	marinistefano86@gmail.com
McCleerey Nicholas	(Northwestern)	njm2@math.northwestern.edu
Mongodi Samuele	(Politecnico Milano)	samuele.mongodi@polimi.it
Myga Szymon	(Jagiellonian)	szymon.myga@uj.edu.pl
Otiman Alexandra	(MPI Bonn)	alexandra_otiman@yahoo.com
Pediconi Francesco	(Firenze)	francesco.pediconi@unifi.it
Picard Sébastien	(Columbia)	picard@math.columbia.edu
Pontecorvo Massimiliano	(Roma Tre)	max@mat.uniroma3.it
Preda Ovidiu	(IMAR Bucarest)	ovidiu.preda@imar.ro
Pujia Mattia	(Torino)	mattia.pujia@unito.it
Raissy Jasmin	(Toulouse)	jraissy@math.univ-toulouse.fr
Rashkovskii Alexander	(Stavanger)	alexander.rashkovskii@uis.no
Reppekus Josias	(Roma Tor Vergata)	hife@gmx.de
Rogov Vasilii	(HSE Moscow)	vasirog@gmail.com
Rossi Federico Alberto	(Milano Bicocca)	federico.rossi@unimib.it
Ruggiero Matteo	(Paris Diderot)	matteo.ruggiero@imj-prg.fr
Sarfatti Giulia	(Firenze)	giulia.sarfatti@unifi.it
Schemken Tobias	(Bochum)	tobias.schemken@rub.de
Sferruzza Tommaso	(Firenze)	tommaso.sferruzza@stud.unifi.it
Shen Xi Sisi	(Northwestern)	xss@math.northwestern.edu
Smith Kevin	(Columbia)	kjs@math.columbia.edu
Sroka Marcin	(Jagiellonian)	Oton123@poczta.fm
Stanciu Miron	(Bucarest)	mirostnc@gmail.com

Stelzig Jonas	(WWU Münster)	jonas.stelzig@wwu.de
Swinson Jenny	(King's London)	j.swinson.16@ucl.ac.uk
Tardini Nicoletta	(Firenze)	nicoletta.tardini@gmail.com
Teleman Andrei	(Aix-Marseille)	andrei.teleman@univ-amu.fr
Tô Tat Dat	(Toulouse)	Tat-Dat.To@math.univ-toulouse.fr
Tong Freid	(Columbia)	tong@math.columbia.edu
Ugolini Riccardo	(Ljubljana)	riccardo.ugolini@imfm.si
Ustinovskiy Yury	(Princeton)	yuryu@math.princeton.edu
Verjovsky Alberto	(UNAM Mexico)	alberto@matcuer.unam.mx
Vlacci Fabio	(Firenze)	vlacci@math.unifi.it
Zeriahi Ahmed	(Toulouse)	zeriahi@math.univ-toulouse.fr
Zheng Tao	(Grenoble Alpes)	zhengtao08@amss.ac.cn

Chapter 1
Lectures on Pluripotential Theory on Compact Hermitian Manifolds

Sławomir Dinew

Abstract The note is an extended version of lectures pluripotential theory in the setting of compact Hermitian manifolds given by the author in July 2018 at Cetraro.

1.1 Introduction

Let (X, J) be a compact complex and connected manifold with J denoting the fixed (integrable) almost complex structure. Unless otherwise stated by n we shall always denote the complex dimension of X. We begin with the basic fact in complex geometry which follows easily from patching local data in coordinate charts:

Theorem 1.1 (X, J) *admits a Hermitian metric.*

If g is such a Hermitian metric then in local coordinates we write

$$g = (g)_{j\bar{k}} := \sum_{j,k=1}^{n} g_{j\bar{k}} dz_j \otimes d\bar{z}_k,$$

where the coefficients $g_{j\bar{k}}$ are smooth local complex valued functions, such that pointwise $g_{j\bar{k}}(z)$ is a positive definite Hermitian symmetric matrix.

Given a Hermitian metric g on X we identify it with the positive definite (**not necessarily closed!**) $(1, 1)$ form ω defined by

$$\forall X, Y \in T_z X \quad \omega(z)(X, Y) := g(z)(JX, Y). \tag{1.1}$$

This form is often called the Kähler form of g in the literature, but we shall not use this terminology in order to avoid confusion with the Kähler condition.

S. Dinew (✉)

Institute of Mathematics, Jagiellonian University, Kraków, Poland

e-mail: slawomir.dinew@im.uj.edu.pl

© Springer Nature Switzerland AG 2019

D. Angella et al. (eds.), *Complex Non-Kähler Geometry*, Lecture Notes in Mathematics 2246, https://doi.org/10.1007/978-3-030-25883-2_1

1

The very existence of a Hermitian metric has profound implications on the geometry and analysis of X. For example there is a natural **volume form** given locally by $det(g_{j\bar{k}})dV$. (Note that the formula differs from its Riemmanian counterpart as there is no square root over the determinant!)

Exercise 1.2 Show that this locally defined volume form is global in the sense that it does not depend on the choice of the coordinate chart.

Just as in the Riemmanian geometry the metric allows one to compute length, to measure angles etc.

The construction based on gluing local data implies in fact that Hermitian metrics exist in abundance. Hence a very natural question appears:

Question 1.3 Are there Hermitian metrics that are better than the others?

While the question is far too vague it raises various problems related to geometry and analysis.

One of the classical "good" Hermitian metrics are the Kähler ones.

Definition 1.4 Let (X, J) be a Hermitian manifold equipped with a Hermitian form ω. If $d\omega = 0$ the metric is called Kähler. A complex manifold X is called a Kähler manifold if it admits a Kähler metric.

There are many reasons why kählerness is a natural condition both geometrically and analytically: its Levi-Civita connection coincides with the *Chern connection*, one has $\nabla J = 0$, we have the so-called Kähler identities relating the canonical operators, the $i\partial\bar{\partial}$-lemma holds and so on. Analytically kählerness means that (in a suitable coordinates) the metric is Euclidean up to terms of order 2. Also locally there exist *potentials* for the $(1, 1)$-Kähler form associated to the metric.

Note that each Kähler metric defines a de Rham cohomology class in $H^{1,1}(X, \mathbb{R})$. The $i\partial\bar{\partial}$-lemma (see [Dem]) which holds on Kähler manifolds allows the following relation between two *cohomologous* Kähler metrics:

Theorem 1.5 *Let X be a Kähler manifold. If ω_1, ω_2 are two Kähler metrics representing the same de Rham cohomology class then there exists a smooth real valued function φ such that*

$$\omega_2 = \omega_1 + i\partial\bar{\partial}\varphi.$$

The theorem says that the only way to perturb a Kähler metric within its cohomology class (and to remain Kähler!) is by adding $i\partial\bar{\partial}$ of a smooth function.

Of course such a perturbation by any function defines a closed $(1, 1)$—form and it will be Kähler form if it is additionally positive definite.

Definition 1.6 A smooth real valued function φ is called *admissible* (or smooth strictly ω-plurisubharmonic) if

$$\omega + i\partial\bar{\partial}\varphi > 0.$$

We shall call these functions ω-psh and we shall often use the notation ω_φ for $\omega + i\partial\bar\partial\varphi$.

Returning to the general Hermitian setting a question arises: how to perturb a given Hermitian metric hoping to get metrics with better properties? Since there are no obvious cohomological constraints in general the answer pretty much depends on one's background: people from conformal geometry may for example wish to perturb the Hermitian form ω by a multiplicative factor e^ρ with ρ being some smooth function.

People coming from Kähler geometry may in turn hope that $\omega + i\partial\bar\partial\varphi$ for a suitably chosen function φ might still be a good choice even though there is no cohomology class to be preserved.

In the lectures we shall pursue this second approach (which does not mean that the first one is not worth a try too!). Relying on the existent theory in the Kähler setting we shall investigate what remains true in the general Hermitian setting, what are the new phenomena to cope with and so on.

Pluripotential theory in the setting of compact Kähler manifolds has proven to be a very effective tool in the study of degeneration of metrics in geometrically motivated problems (see [Kol98, Kol03, EGZ09, KT08], which is by far incomplete list of the literature on the subject). Usually in such a setting *singular* Kähler metrics do appear as limits of smooth ones. Then pluripotential theory provides a natural background for defining the *singular* volume forms associated to such metrics. More importantly it provides useful information on the behavior of the Kähler potentials exactly along the singularity locus, which is hard to achieve by standard PDE techniques. On the other hand the theory does not rely on strong geometric assumptions, as most of the results are either local in nature or are modelled on local ones. It is therefore natural to expect that at least some of the methods and applications carry through in the more general Hermitian setting.

Of course there is inevitably some price to pay. Computations on general Hermitian manifolds are messier. We lack many important tools from the Kähler setting. Arguably the most important difference for us however will be the lack of invariance of the total volume $\int_X (\omega + dd^c u)^n$ for an admissible function u (which is easily seen after two applications of Stokes theorem). As one will soon verify this leads to troublesome additional terms involving $d\omega$ and/or $d\omega \wedge d^c \omega$ and controlling these in a suitable sense is the main technical difficulty in the whole theory.

The interest towards Hermitian versions of the complex Monge-Ampère equation has grown rapidly in the recent years. The first steps were laid down by the French school most notably by Cherrier [Che87] and Delanoë [De81]. In these papers the Authors followed Aubin and Yau's arguments [Y78] to get existence of smooth solutions of the Monge-Ampère equation in the case of smooth data. The Authors were successful only in particular cases (that is under geometric assumptions on the background metric). The main problem to overcome were the a priori estimates needed to establish the closedness part in the continuity method. The renewed interest towards Hermitian Monge-Ampère equations came with the breakthroughs by Guan-Li [GL10] and especially Tosatti-Weinkove [TW10a, TW10b]. Guan and Li were able to solve the equation assuming geometric conditions different than these from [Che87] and [De81], while the missing uniform estimate was finally established without any assumptions in [TW10b]. Parallel to these recent advances foundations of the corresponding pluripotential theory were laid down (see [DK12, BL11] and [KN1, KN2, KN3]). The theory is still in an infant state, and the techniques are technical modifications of their Kählerian counterparts.

In order to motivate the construction of such a theory we shall list a couple of arguments relying on (Kählerian) pluripotential reasonings and try to investigate what happens in the Hermitian realm. One of our first discoveries is that there is a condition strictly more general than kählerness that yields almost the same pluripotential theory. It was studied by Guan and Li [GL10]. The Authors assumed that $dd^c\omega = 0$ and $dd^c(\omega^2) = 0$. Under this condition almost every pluripotential argument from the Kähler setting carries through verbatim. It should be emphasized that in Hermitian geometry there are many other conditions imitating kählerness. These are motivated by various geometric considerations. Some of these conditions have consequences that are relevant to pluripotential theory.

The notes are organized as follows. We start with some basic notation and motivate the theory by listing some applications of Kähler pluripotential theory. Section 1.4 is devoted to some explicit constructions of *bad* plurisubharmonic functions and Hermitian metrics. Hopefully these examples shed more light on what kind of singular behavior can be expected within the theory. Later we define some of the "close-to-Kähler" conditions which can be found in the literature. In the next section we describe some explicit examples of Hermitian manifolds and "canonical" metrics on them. This list is of course only a glimpse into the vast world of Hermitian geometry. The existence of suitable adapted coordinates (due to [GL10]) is shown in Sect. 1.7. Such a coordinate system will turn out to be very useful in the proof of higher order a priori estimates for the Dirichlet problem later on. The main pluripotential tools are discussed in Sect. 1.8. In particular we show that the complex Monge-Ampère operator is well defined on bounded ω-plurisubharmonic functions and it shares the convergence properties known from the Kähler case. Section 1.9 is devoted to the most important tool in the whole theory—the comparison principle. As explained it differs considerably from the one known in the Kähler setting unless the form ω satisfies some restrictive additional conditions. In the next section the solution of the Dirichlet problem is presented in detail. For the openness part we follow [TW10a], while for the C^2 estimates we borrow the main idea from [GL10]. The uniform estimate is taken from [DK12]. Then we solve the Monge-Ampère equation with right hand side being an L^p function with $p > 1$ following [KN1].

These are expanded lecture notes of the course that I taught during a C.I.M.E workshop "Complex non-Kähler geometry" in Cetraro 9-13.07.2018. The lectures are based on the manuscript [D16] and more recent developments. It is a great pleasure to thank the organizers of this event for the invitation.

1.2   Notation

Throughout the paper X will denote a compact, complex and connected manifold. Unless otherwise specified n will always be the complex dimension of X.

As usual d will denote the exterior differentiation operator, while ∂ and $\bar{\partial}$ will be the $(1, 0)$ and $(0, 1)$ part of it under the standard splitting. In some arguments involving integration by parts it is more convenient to use the operator $d^c := i(\bar{\partial} - \partial)$, so that $dd^c = 2i\partial\bar{\partial}$. These will be used interchangeably. We shall also make use of the standard notation ω_u standing for $\omega + dd^c u$.

δ_{ij} will denote the Kronecker delta symbol. We shall make use of Einstein summation convention unless otherwise stated.

Throughout the note we shall use the common practice of denoting constants independent of the relevant quantities by C. In particular these constants may vary line-to-line. If special distinction between the constants is needed in some arguments these will be further distinguished by $\tilde{C}$, $\bar{C}$, C_i and so on.

A special constant that controls the geometry of (X, ω) (see below for details) is denoted by B—it is the infimum over all positive numbers b satisfying

$$- b\omega^2 \leq ni\partial\bar{\partial}\omega \leq b\omega^2 \text{ and} \tag{1.2}$$

$$-b\omega^3 \leq n^2 i\partial\omega \wedge \bar{\partial}\omega \leq b\omega^3.$$

It should be emphasized that this constant measures how far our metric is from satisfying a special condition studied by Guan and Li [GL10]. Of course if ω is Kähler then $B = 0$.

1.3   Why Pluripotential Theory?

In this section we shall briefly list some applications of the pluripotential theory on Hermitian manifolds. As the theory is still developing it is expected that this list will grow rapidly in the near future.

To begin with we recall that a basic example of a local plurisubharmonic function is $log(\|F(z)\|)$ with F being a local holomorphic mapping. Thus the theory is tightly linked to complex analysis. More globally let L be a holomorphic line bundle over (X, ω) with σ a (holomorphic) section (for analysts: it will be a collection of nowhere vanishing holomorphic functions $g_{\alpha\beta}$ defined on the intersections of coordinate charts $U_{\alpha\beta} := U_\alpha \cap U_\beta$ which satisfy the relations $g_{\alpha\alpha} = Id$ and $g_{\alpha\beta}g_{\beta\gamma}g_{\gamma\alpha} = Id$). If $\|\cdot\|$ is a smooth norm on the space of sections (collections of holomorphic functions f_α on U_α such that on intersections $f_\alpha = g_{\alpha\beta} f_\beta$), then

$$u(z) := log\|\sigma(z)\|$$

is a global function which is smooth except on the divisor $\{\sigma = 0\}$ and $i\partial\bar{\partial}log\|\sigma\| \geq -C\omega$, for some constant C dependent only on the choice of the norm on $X \setminus \{\sigma = 0\}$. It can be proven that this inequality extends past $\{\sigma = 0\}$ in the distributional sense—u is *quasiplurisubharmonic*.

This simple observation has important consequences: pluripotential theory might be useful for constructing analytic objects on the manifold as $-\infty$-values of suitably constructed functions. Unfortunately for a general ω-psh function u its $-\infty$ locus is much more complicated and hard to deal with as we shall see in the next section.

Reversing a bit the discussion above a natural question is whether one can construct ω-psh functions with *prescribed* singularities. One way of doing so is by solving suitable Monge-Ampère equations. Such an approach was initiated in the paper [TW12], which was motivated by the fundamental paper [Dem93]. Precisely the Authors goal was to construct ω-plurisubharmonic functions with prescribed logarithmic singularities at a collection of isolated points. Such singular quasiplurisubharmonic functions can be applied as weights in various Ohsawa-Takegoshi type L^2 extension problems or $\bar{\partial}$ problems.

The construction based on Demailly's idea in [Dem93] is by solving a family of Monge-Ampère equations with right hand sides converging to Dirac delta measures. More specifically in the Kähler case a family of Monge-Ampère equations

$$\begin{cases} \phi_\epsilon \in C^\infty(X), sup_X\phi_\epsilon = 0 \\ \omega + dd^c\phi_\epsilon > 0 \\ (\omega + dd^c\phi_\epsilon)^n = \chi_\epsilon\omega^n \end{cases} \tag{1.3}$$

is considered, where for each $\epsilon > 0$ χ_ϵ is a smooth strictly positive function with suitably normalized total integral. Moreover it is required that χ_ϵ converge weakly to a combination $\sum c_j\delta_j$ of weighted Dirac delta measures as ϵ tends to zero. Then the weak limit of the solutions (which exist by the Calabi-Yau theorem [Y78]) is the required function.

In the Hermitian setting such a technique requires a modification of the approximating equations:

$$\begin{cases} \phi_\epsilon \in C^\infty(X), sup_X\phi_\epsilon = 0 \\ \omega + dd^c\phi_\epsilon > 0 \\ (\omega + dd^c\phi_\epsilon)^n = e^{c_\epsilon}\chi_\epsilon\omega^n, \end{cases} \tag{1.4}$$

where c_ϵ is some constant (the equations are then solvable by [TW10b]). Successful repetition of the argument relies crucially on controlling total volumes, that is on the uniform control of the constants c_ϵ. This is why the results in [TW12] are complete only in dimension 2 and 3.

It is worth pointing out that construction of ω-psh functions with non-isolated analytic singularities is substantially harder, partially due to a lack of reasonable Monge-Ampère theory for such functions.

The Monge-Ampère equation is also related to the *Ricci curvature* in the Kähler setting through the following construction:

Given a Kähler metric ω_0 and a representative α of the first Chern class on a manifold X the Calabi problem boils down to finding a metric ω cohomologous to ω_0, such that $Ric(\omega) = \alpha$. By the dd^c lemma any such ω can be written as $\omega_0 + dd^c\phi$ for some smooth potential ϕ. Furthermore $Ric(\omega_0) = \alpha + dd^c h$, where the Ricci potential h is a function uniquely defined modulo an additive constant (which can be fixed if we assume the normalization $\int_X e^h \omega_0^n = \int_X \omega^n$). Recall that in the Kähler setting one has $Ric(\omega) = -dd^c log((\omega)^n)$ with ω^n denoting n-th wedge product of ω (modulo the identification of the coefficient of the volume form with the volume form itself). Hence $Ric(\omega) = \alpha$ is equivalent to

$$Ric(\omega_0 + dd^c\phi) = Ric(\omega_0) - dd^c h \Leftrightarrow -dd^c log \frac{(\omega_0 + dd^c\phi)^n}{(\omega_0)^n} = -dd^c h$$

$$\Leftrightarrow (\omega_0 + dd^c\phi)^n = e^{h+c}\omega_0^n$$

for some constant c. Exploiting the kählerness of ω_0 and integration by parts one easily sees that under our normalization $c = 0$ and we end up with the standard Monge-Ampère equation

$$(\omega_0 + dd^c\phi)^n = e^h\omega_0^n \tag{1.5}$$

with prescribed right hand side.

This equation for smooth h and ω_0 was solved in the celebrated paper of Yau [Y78]. In modern Kähler geometry it is of crucial importance to understand the behavior of the potential ϕ (or the form $\omega_0 + dd^c\phi$ itself) if we drop the smoothness assumptions on h and/or the strict positivity of ω_0. Such a situation occurs if we work on *mildly singular* Kähler varieties (see for example [EGZ09]) or when one tries to understand the limiting behavior of the Kähler-Ricci flow (see [KT08] and references therein). It is exactly the setting where pluripotential theory can be applied an indeed in such settings the uniform estimate for the potential ϕ (a starting point for the regularity analysis) is usually obtained in this way (compare [EGZ09, KT08]).

Returning to the Hermitian background the picture described above has to be modified. The obvious obstacles are that a Hermitian metric ω_0 need not define a cohomology class and the dd^c lemma may fail. On the bright side the first Chern class can still be reasonably defined in the *Bott-Chern* cohomology, that is the cohomology given by

$$H_{BC}^{p,q} = \frac{ker\{d : C_{p,q}(X) \to C_{p,q+1}(X) \oplus C_{p+1,q}(X)\}}{Im\{dd^c C_{p-1,q-1}(X)\}}, \tag{1.6}$$

where $C_{p,q}(X)$ denotes the space of smooth (p, q)-forms.

Given a Hermitian metric ω_0 its *first Chern form* can be defined analogously to the Kähler setting by

$$Ric^{BC}(\omega_0) := -dd^c log(\omega_0^n).$$

It turns out that the first Ricci forms represent the first Bott-Chern cohomology class $c_1^{BC}(X)$ in the Bott-Chern cohomology. Hence a natural question arises whether any form α in $c_1^{BC}(X)$ is representable as the Ricci form of some metric $\omega_0 + dd^c\phi(X)$. A computation analogous to the one above shows that such a ϕ has to satisfy the equation

$$(\omega_0 + dd^c\phi)^n = e^{h+c}\omega_0^n, \tag{1.7}$$

with a function h as above and some constant $c > 0$. Contrary to the Kähler case, however, the constant need not be equal to zero and thus the Hermitian Monge-Ampère equation has one more degree of freedom. As we shall see later this adds some technical difficulties into the solution of the equation.

The discussion above resulted in the fact that solutions to Hermitian Monge-Ampère equation prescribe the Ricci form in the Bott-Chern cohomology. Thus weakening of the smoothness assumptions on f and/or strict positivity of ω_0 is helpful in situations analogous to the ones in the Kähler setting above.

Arguably one of the most exciting problems in Hermitian geometry is the classification of class VII surfaces. To this end the conjectural picture reduces the problem to finding rational curves on such a surface. This is an extremely hard geometric problem. Essentially the only working tool in some special cases is a deep gauge theoretic argument of Teleman [T10].

It thus worth mentioning that another approach to construction of rational curves exploiting some *singularity magnifying* Monge-Ampère equations has been proposed by Siu [Siu09].

It is thus quite intriguing to investigate the relationships between Monge-Ampère equations and the existence of rational curves.

1.4 A Couple of Inspiring Examples

1.4.1 Local Theory

As we have already mentioned the functions

$$log(\|F(z)\|)$$

are plurisubharmonic for holomorphic mappings. Thus obviously analytic sets are locally contained in a $-\infty$-locus of some plurisubharmonic functions.

Is this the general picture? Let us begin with the following example:

Example 4.1 Let Ω be the disk centered at zero with radius $1/2$ in $\mathbb{C}$. Let $\{a_n\}_{n=1}^{\infty}$ be the set of all complex numbers in $\Omega \setminus \{0\}$ with both coefficients being rational (ordered in some fashion). Consider a sequence of real positive numbers b_n decreasing sufficiently rapidly to 0 such that

$$\sum_{n=1}^{\infty} b_n log|a_n| > -\infty.$$

Consider the function

$$u(z) := \sum_{n=1}^{\infty} b_n log|z - a_n|.$$

Obviously $u_m := \sum_{n=1}^{m} b_n log|z - a_n|$ are subharmonic and decrease towards u. Hence u is also subharmonic, $u(0) > -\infty$, yet $\{u = -\infty\}$ contains a dense subset of Ω!

Our next example taken from [Dem] is, in a sense, even more surprising—it shows that even if a plurisubharmonic function is nowhere equal to $-\infty$ it still may fail to be locally bounded from below:

Example 4.2 The function

$$v(z) := \sum_{k=1}^{\infty} \frac{1}{k^2} log(|z - 1/k| + e^{-k^3})$$

is everywhere finite but is not locally bounded from below at zero.

Exercise 4.3 Is it possible, using a countable collection of such v's for every rational complex number to get a dense set of points such that a plurisubharmonic function is everywhere finite but unbounded from below near any point from the dense set?

These examples lead to the following definitions:

Definition 4.4 A set E is said to be pluripolar if it is locally contained in a $-\infty$ locus of a plurisubharmonic function. Given any plurisubharmonic function the set $\{u = -\infty\}$ is called the pole set of u.

Exercise 4.5 A pluripolar set is contained in a pole set of some function but need not be *equal* to a pole set. Construct an example in $\mathbb{C}^2$ of a pluripolar set which is not a pole set.

Definition 4.6 Given a plurisubharmonic function u the unbounded locus set of u is the set of points z, such that u is not bounded from below in every neighborhood of z.

In pluripotential theory there are different tools for measuring the pointwise singularities of plurisubharmonic functions. Among the basic ones (see [Dem]) is the *Lelong number*:

Definition 4.7 (Lelong Number) Let u be a plurisubharmonic function defined in a neighbourhood of a point $z_0 \in \mathbb{C}^n$. Then the limit $lim_{r \to 0^+}$ of the quantity

$$\int_{|z-z_0| \leq r} dd^c u \wedge (dd^c \log|z - z_0|)^{n-1} = \frac{1}{r^{2n-2}} \int_{|z-z_0| \leq r} dd^c u \wedge \beta^{n-1}$$

is called a Lelong number of the function u at z_0.

Note that unless u is unbounded near z_0 the Lelong number vanishes. This is however not a sufficient condition as the plurisubharmonic function $-log(-log|z|)$ near zero shows. Intuitively speaking the Lelong number measures whether u has *logarithmic* singularity at z_0—these are the heaviest singularities that plurisubharmonic functions could have.

The equality above (whose proof can be found in [Dem]) in particular implies that the quantity $\frac{1}{r^{2n-2}} \int_{|z-z_0| \leq r} dd^c u \wedge \beta^{n-1}$ (which is up to a universal multiplicative constant equal to $\frac{1}{r^{2n-2}} \int_{|z-z_0| \leq r} \Delta u$) is increasing with r. This implies that the set

$$E_c(u) := \{z | u \text{ has a Lelong number at least } c \text{ at } z\}$$

is small for any $c > 0$. More precisely for any $\varepsilon > 0$ it has zero $2n - 2 + \varepsilon$ Hausdorff measure.

It turns out however that more is true: a deep theorem of Siu [Siu] states that the set $E_c(u)$ are always analytic for $c > 0$:

Theorem 4.8 (Siu) *Let u be a plurisubharmonic function in a domain $\Omega \subset \mathbb{C}^n$. Then for any $c > 0$ the set $E_c(u)$ is an analytic subset of Ω.*

This result is one instance of appearance of analytic objects in pluripotential theory.

1.4.2 Kähler Versus Hermitian

Below we discuss an example where general Hermitian pluripotential theory behaves differently to its Kählerian counterpart.

A broad field where pluripotential theory applies is the study of *singular metrics* i.e. in the case where the background $(1, 1)$-form fails to be a metric. One such instance occurs if some of the eigenvalues are zero i.e. we deal simply with semipositive forms.

Suppose ω_j is a sequence of smooth Kähler forms converging smoothly to a limiting smooth semipositive form ω. The local example to keep in mind is

$$\omega_j = i dz_1 \wedge d\bar{z}_1 + \frac{i}{j} dz_2 \wedge d\bar{z}_2$$

in $\mathbb{C}^2$.

Geometrically these metrics *shrink* the z_2 direction, so the limiting space can be identified as a *metric space* with $\mathbb{C} \times \{0\}$ as the z_2-factor is collapsed (this is a very easy example of Gromov-Hausdorff convergence).

Recall that the Frobenius theorem (under mild additional assumptions) implies that the *kernel* of ω is an integrable distribution i.e. we get a foliation by holomorphic leaves. As a result we end up with a limiting space that has some sort of complex structure.

In the Hermitian case obviously there is no Frobenius type theorem for the limiting form. Can we thus extract a sort of complex structure in the limit? The following example shows that the answer is no in general:

Example 4.9 ([TW14]) Consider the standard Hopf surface X i.e. $\mathbb{C}^2 \setminus \{(0,0)\}$ modulo the action of the group generated by the contraction $(z_1, z_2) \rightarrow (\frac{1}{2}z_1, \frac{1}{2}z_2)$ equipped with the family of metrics

$$\omega(t) = \sum_{j,k=1}^{2} \frac{1}{|z_1|^2 + |z_2|^2}\left((1 - 2t)\delta_{jk} + 2t\frac{\bar{z}_j z_k}{|z_1|^2 + |z_2|^2}\right)i dz_j \wedge d\bar{z}_k.$$

for $t \in (0, \frac{1}{2})$. As t converges to $\frac{1}{2}$ the metrics converge to the nonnegative form

$$\omega(1/2) = \sum_{j,k=1}^{2} \frac{\bar{z}_j z_k}{(|z_1|^2 + |z_2|^2)^2} i dz_j \wedge d\bar{z}_k.$$

It is easy to see that the kernel distribution of $\omega(1/2)$ are the vectors $X = \sum_j X^j \frac{\partial}{\partial z_j}$ satisfying $\sum_{j=1}^{n} \bar{z}_j X_j = 0$ i.e. the complex tangent directions of the spheres in $\mathbb{C}^2$ centered at zero.

Exercise 4.10 If the distribution were integrable that would mean that the boundary of the unit sphere in $\mathbb{C}^2$ would contain locally a holomorphic curve. Show that this is impossible.

More careful analysis shows that $\omega(t)$ collapses, as t tends to $\frac{1}{2}$, the spheres centered at zero. It can be shown that the *limiting space* is in fact the radial direction modulo the group action i.e. a circle! For obvious reasons then the limiting space cannot admit a complex structure!

1.5   Kähler Type Conditions

Given a fixed Hermitian manifold X it is natural to search for the "best" metric that X admits. The reason is at least twofold: nice metrics usually significantly simplify computations and more importantly it is sometimes possible to deduce geometric or topological information from the existence of these.

Unlike the Kähler case there is a large number of mutually different "Kähler type" conditions. Below we list the most common ones. Our discussion is borrowed from [D16].

Definition 5.1 (Balanced Metric) Let (X, ω) be a n-dimensional Hermitian manifold. The form ω is said to be balanced if it satisfies

$$d(\omega^{n-1}) = 0.$$

Of course this definition differs from the Kähler condition only if $n \geq 3$. The motivation behind such a condition partially comes from string theory (see [AB95, FIUV09, FLY12] and the references therein). There are various constructions of explicit examples of non-Kähler, balanced manifolds in the literature. For example using *conifold transitions* Fu, Li and Yau in [FLY12] proved that such a metric exists on the connected sum $\sharp_k S^3 \times S^3$ of k copies of the product of two three-dimensional spheres. Another example is the *Iwasawa manifold* which will be given in the next section.

Balanced metrics impose some geometric restrictions on the underlying manifold (for example it follows from the Stokes theorem that no smooth one-codimensional complex subvariety can be homologous to zero) and hence not every manifold can be endowed with such a metric.

From potential theoretic point of view the most important property of such metrics is that the Laplacian of any admissible (or even merely smooth) function u on X integrates to zero. Namely if we choose the canonical Laplacian associated to the Chern connection on X then we get

$$\int_X (\Delta_\omega u)\omega^n = n \int_X i\partial\bar\partial u \wedge \omega^{n-1} = -n \int_X \bar\partial u \wedge \partial(\omega^{n-1}) = 0.$$

An interesting exercise, left to the Reader, is to check that in the intermediate cases between the balanced and Kähler conditions we do not get anything besides Kählerness:

Exercise 5.2 *Suppose* $1 < k < n - 1$. *If* ω *is a form such that*

$$d(\omega^{n-k}) = 0,$$

then $d\omega = 0$ *i.e.* ω *is Kähler.*

A second family that we consider are the so-called Gauduchon metrics [Ga].

Definition 5.3 (Gauduchon Metric) Let (X, ω) be a n-dimensional Hermitian manifold. The form ω is said to be Gauduchon if it satisfies

$$dd^c(\omega^{n-1}) = 0.$$

Unlike balanced ones, these exist on **any** compact Hermitian manifold. Moreover a theorem of Gauduchon [Ga] states that given any Hermitian form ω there exists a conformal factor e^{ϕ_ω} such that the new form $e^{\phi_\omega}\omega$ is Gauduchon. Gauduchon metrics are useful in many geometric contexts, for example the notion of a degree of a line bundle over a Gauduchon manifold is well defined via the formula

$$deg_\omega(L) = \int c_1(L) \wedge \omega^{n-1},$$

where $c_1(L)$ is the first Chern class of L. This is the starting point for a *stability theory* for vector bundles in the Hermitian setting (see [LT95]).

Yet another difference is that after the exchange of the power $n - 1$ to a lesser power we do get nontrivial new conditions. This is in fact how Astheno-Kähler metrics are defined.

Definition 5.4 (Astheno-Kähler Metric) Let (X, ω) be a n-dimensional Hermitian manifold ($n \geq 2$). The form ω is said to be Astheno-Kähler if it satisfies

$$dd^c(\omega^{n-2}) = 0.$$

This condition was used by Jost and Yau [JY93] in their study of harmonic maps from Hermitian manifolds to general Riemmanian manifolds.

Unlike the Gauduchon metrics Astheno-Kähler metrics impose some constraints on the underlying manifold. It can be shown that any holomorphic 1-form on such a manifold must be closed. Explicit examples of Astheno-Kähler but non-Kähler manifolds can be found in dimension 3 where they coincide with the *pluriclosed* metrics to be defined below. Another type of examples are the so-called *Calabi-Eckmann* manifolds. These are topologically products $S^{2n-1} \times S^{2m-1}$, ($m > 1$, $n > 1$) of odd dimensional spheres. Any such manifold admits families of complex structures which can be constructed using Sasakian geometry. In [Mi09] it was shown that a special choice of such a complex structure yields an Astheno-Kähler manifold. Since $H^2(S^{2n-1} \times S^{2m-1}) = 0$ such manifolds are never Kähler.

Much more information regarding Astheno-Kähler geometry can be found in [FT].

Finally the important notion of the aforementioned *pluriclosed* metrics is defined as follows:

Definition 5.5 (Pluriclosed Metric) Let (X, ω) be a n-dimensional Hermitian manifold. The form ω is said to be pluriclosed if it satisfies

$$dd^c\omega = 0.$$

The pluriclosed metrics are also known as SKT (strong Kähler with torsion) in the literature [FPS04]. Of course in dimension 2 this notion coincides with the Gauduchon condition, hence any complex surface admits pluriclosed metrics. In complex dimension 3 some nontrivial examples of non-Kähler *nilmanifolds* admitting pluriclosed metrics were constructed by Fino, Parton and Salamon in [FPS04].

As is easily verified, Gauduchon metrics also have the property that the Laplacian of a smooth function integrates to zero. This is not the case for Astheno-Kähler and pluriclosed metrics in general.

A strengthened version of the Gauduchon condition was considered by Popovici in [Pop13]:

Definition 5.6 (Strongly Gauduchon Metric) If (X, ω) is n-dimensional Hermitian manifold, the form ω is said to be strongly Gauduchon if $\partial(\omega^{n-1})$ is $\bar{\partial}$ exact.

Of course strongly Gauduchon implies Gauduchon and these notions coincide if the $\partial\bar{\partial}$-lemma holds on X (see [Pop13]) but in general the inclusion is strict. Note also that any balanced metric is strongly Gauduchon.

The strongly Gauduchon condition was introduced by Popovici in [Pop13] in connection with studies of deformation limits of projective or Kähler manifolds. We refer to [Pop13] for the geometric conditions imposed by this structure. In particular a necessary and sufficient condition of existence of such a metric on a manifold X is the nonexistence of a positive d-exact $(1, 1)$-current on X.

None of the conditions above actually guarantee the invariance of the total volume of the perturbed metric. More precisely the value $\int_X (\omega + dd^c u)^n$ does depend on u and this is the main source of troubles in pluripotential theory. Still a condition weaker that being Käher can be imposed so that the total volume remains invariant. This condition has been investigated by Guan and Li [GL10]:

Definition 5.7 A metric satisfies the condition imposed by Guan and Li if $dd^c \omega = 0$ and $dd^c(\omega^2) = 0$.

Observe that this is weaker than Kähler yet by twofold application of Stokes' theorem it can be shown that the total volume remains invariant. Let us also stress once again that the constant B from Sect. 1.2 measures how far our metric is from satisfying the above condition.

Remark 5.8 Recently Chiose [Chi] has shown that Guan and Li condition is equivalent to the constancy of the total volume $\int_X \omega^n$ for all ω differing by a dd^c of a quasiplurisubharmonic function.

Remark 5.9 Non Kähler metrics satisfying the above property do exist. A trivial example, taken from [TW10a], is simply the product of a compact complex curve equipped with a Kähler metric and a non-Kähler complex surface equipped with a Gauduchon metric.

We refer the interested reader to the article [Pop14], for more explicit examples and interactions between the notions above.

1.6 Explicit Examples of Non-Kähler Hermitian Manifolds

We begin this section by defining the most classical examples of non-Kähler manifolds—the Hopf manifolds. These were historically the first ones and were discovered by Hopf in 1948 [Ho48].

Definition 6.1 (Hopf Manifold) Let t be any nonzero complex number satisfying $|t| \neq 1$. Then it induces a $\mathbb{Z}$ action on $\mathbb{C}^n \setminus \{0\}$ by scaling i.e.

$$(k, w) \rightarrow t^k w,$$

for any $k \in \mathbb{Z}$, $w \in \mathbb{C}^n \setminus \{0\}$. The action is discrete and properly discontinuous, hence the quotient manifold $\mathbb{C}^n \setminus \{0\}/_{\mathbb{Z}}$ is a smooth manifold.

Remark 6.2 In the literature more general definitions are being considered. In particular some Authors define Hopf manifolds as above but with the $\mathbb{Z}$ action induced by any contracting-to-zero biholomorphic mapping of $\mathbb{C}^n \setminus \{0\}$ into itself.

It can be proved that the Hopf manifolds are all diffeomorphic to $\mathbb{S}^{2n-1} \times \mathbb{S}^1$, hence the first Betti numbers are odd—in particular these are never Kähler. Another obstruction is that $H^2(X, \mathbb{R})$ vanishes which also shows that X cannot be Kähler. In fact it can be proven that Hopf manifolds do not admit even balanced metrics.

On the bright side a Gauduchon metric is explicitly computable in the simplest case. Indeed, suppose that $n = 2$, then the metric

$$\omega = \frac{i dz \wedge d\bar{z} + i dw \wedge d\bar{w}}{|z|^2 + |w|^2}$$

is clearly invariant under the group action, hence descends onto the quotient manifold. Moreover it is easy to check that $dd^c \omega = 0$, so this metric is pluriclosed (or Gauduchon).

In the two dimensional case Hopf manifolds do belong to the special class of the so-called class VII surfaces, named after the original Kodaira classification list [Kod64, Kod66, Kod68a, Kod68b]. These are characterized by two conditions: the first Betti number $b_1(X)$ is equal to 1, while the Kodaira dimension $\kappa(X)$ is minus infinity. Class VII minimal surfaces are the only remaining class of two dimensional manifolds that is not fully classified yet. More precisely the classification was obtained by the works of Kato, Nakamura and most notably Teleman [Ka78, Na84, T10] in the cases when the second Betti number $b_2(X)$ is small. Classification is complete in the case $b_2(X) \leq 2$ (see [T10]). In the remaining cases a theorem of Dloussky-Oeljeklaus-Toma [DOT03] yields a classification provided one can find $b_2(X)$ rational curves (possibly singular) on X. Conjecturally this is always the case and indeed this holds in the classified cases $b_2(X) \leq 2$. Hence the classification problem boils down to the construction of rational curves.

Let us now present one of the simplest examples of a class VII manifold, called Inoue surface [In74] (in this case $b_2(X) = 0$).

Definition 6.3 (Inoue Surface) Let M be a 3×3 integer valued matrix with determinant equal to 1. Suppose that it has a positive eigenvalue α and two complex eigenvalues β and $\bar{\beta}$. Let also (a_1, a_2, a_3) and (b_1, b_2, b_3) be eigenvectors corresponding to α and β respectively. The Inoue surface is defined as the quotient $\mathbb{H} \times \mathbb{C}$, $\mathbb{H}$ being the upper half plane, by a group G generated by the following four automorphisms:

$$g_0(w, z) := (\alpha w, \beta z),$$

$$g_i(w, z) = (w + a_i, z + b_i) \quad i = 1, 2, 3.$$

Remark 6.4 It can be proven that the action is discrete and properly discontinuous, hence the quotient is a smooth manifold. An important property of G in this construction is that it is not an Abelian group but is a solvable one. There are two other classes of surfaces defined by Inoue, also being quotients of $\mathbb{H} \times \mathbb{C}$ by a solvable group.

On Inoue surfaces one can also find an explicit pluriclosed/Gauduchon metric:

Definition 6.5 (Tricerri Metric) Let $\omega(z, w) := \frac{i dw \wedge d\bar{w}}{Im^2(w)} + Im(w) i dz \wedge d\bar{z}$. This metric is invariant under the action of G and hence descends to the Inoue surface. It can be computed that $dd^c \omega = 0$.

Our last example is known as Iwasawa threefold. It is not Kähler for it admits a non-closed holomorphic 1-form:

Definition 6.6 (Iwasawa Manifold) Let

$$M := \{A \in GL_3(\mathbb{C}) | \ A = \begin{bmatrix} 1 & z_1 & z_3 \\ 0 & 1 & z_2 \\ 0 & 0 & 1 \end{bmatrix}, \ z_i \in \mathbb{C}, i = 1, 2, 3\}.$$

The Iwasawa threefold is defined as quotient of M by the lattice of such matrices with coefficients being Gaussian integers acting on M by a left multiplication.

It is easily observed that dz_1, dz_2 and $dz_3 - z_1 dz_2$ are invariant holomorphic one forms on M. As $d(dz_3 - z_1 dz_2) = -dz_1 \wedge dz_2$ is also invariant, it descends to a non-zero 2-form. Thus $dz_3 - z_1 dz_2$ is a non closed holomorphic one form on M. It can be shown that

$$i dz_1 \wedge d\bar{z}_1 + i dz_2 \wedge d\bar{z}_2 + i (dz_3 - z_1 dz_2) \wedge \overline{(dz_3 - z_1 dz_2)}$$

descends to a balanced (hence strongly Gauduchon) metric on the Iwasawa threefold.

1.7   Canonical Coordinates

In the Kähler setting many local computations are significantly simplified by the use of canonical coordinates. More specifically such coordinates not only diagonalize the metric at a given point (which we assume to be the center of the associated coordinate chart) but also yield vanishing of all third order derivative terms while the fourth order terms are the coefficients of the curvature tensor.

   Of course in the general Hermitian setting one cannot expect vanishing of all third order terms. Yet getting more information than pointwise diagonalization is crucial in some laborious computations. Hence a question appears whether some milder "interpolating" conditions on third order terms are achievable. As observed by Guan and Li [GL10] this is indeed possible:

Theorem 7.1 (Guan-Li) *Given a Hermitian manifold (X, ω) and a point $p \in X$ it is possible to choose coordinates near p, such that $g_{i\bar{j}}(p) = \delta_{ij}$ and for any pair i, k one has $\frac{\partial g_{i\bar{i}}}{\partial z_k}(p) = 0$.*

Proof Choose first local coordinates z_i around p (identified with 0 in the coordinate chart), such that at this point the metric is diagonalized. Then rechoose coordinates by adding some quadratic terms:

$$w_r = z_r + \sum_{m \neq r} \frac{\partial g_{r\bar{r}}}{\partial z_m} z_m z_r + \frac{1}{2} \frac{\partial g_{r\bar{r}}}{\partial z_r} z_r^2.$$

Observe that

$$\frac{\partial z_r}{\partial w_i} = \delta_{ri} \quad \text{at } p; \tag{1.8}$$

$$\frac{\partial^2 z_r}{\partial w_i \partial w_k} = -\sum_{m \neq r} \frac{\partial g_{r\bar{r}}}{\partial z_m} \left(\frac{\partial z_m}{\partial w_i} \frac{\partial z_r}{\partial w_k} + \frac{\partial z_m}{\partial w_k} \frac{\partial z_r}{\partial w_i} \right) - \frac{\partial g_{r\bar{r}}}{\partial z_r} \frac{\partial z_r}{\partial w_i} \frac{\partial z_r}{\partial w_k}. \tag{1.9}$$

Computing now $\tilde{g}_{i\bar{j}} := g(\frac{\partial}{\partial w_i}, \frac{\partial}{\partial \bar{w}_j})$, one gets

$$\frac{\partial \tilde{g}_{i\bar{j}}}{\partial w_k} = \sum_{r,s=1}^{n} g_{r\bar{s}} \frac{\partial^2 z_r}{\partial w_i \partial w_k} \frac{\partial \bar{z}_s}{\partial \bar{w}_j}$$

$$+ \sum_{r,s,p=1}^{n} \frac{\partial g_{r\bar{s}}}{\partial z_p} \frac{\partial z_p}{\partial w_k} \frac{\partial z_r}{\partial w_i} \frac{\partial \bar{z}_s}{\partial \bar{w}_j}.$$

Plugging now (1.8) and (1.9) into the formula above we get

$$\frac{\partial \tilde{g}_{i\bar{i}}}{\partial w_k} = \sum_{r=1}^{n} (-\sum_{m \neq r} -\frac{\partial g_{r\bar{r}}}{\partial z_m}(\delta_{mi}\delta_{rk} + \delta_{mk}\delta_{ri})\delta_{ri} - \frac{\partial g_{r\bar{r}}}{\partial z_r}\delta_{ri}\delta_{rk})$$

$$+ \sum_{r,s,p=1}^{n} \frac{\partial g_{r\bar{s}}}{\partial z_p}\delta_{pk}\delta_{ri}\delta_{si} = 0.$$

$\square$

1.8 Basic Notions of Pluripotential Theory: Currents and Capacities

In this section we shall define all the basic tools in Hermitian pluripotential theory. A good reference for classical plurisubharmonic functions is [Hö2]. The pluripotential theory in the local setting was developed by Bedford and Taylor in [BT82]. For Kählerian counterparts of the discussed notions we refer to [Kol03, GZ05].

1.8.1 Some Linear Algebra

Given a $(1, 1)$-form $\alpha = \alpha_{jk}idz_j \wedge d\bar{z}_k$ it is easy to see that α is *real* ($\alpha = \bar{\alpha}$) iff the coefficients pointwise form a Hermitian matrix. Hence the following definition is natural:

Definition 8.1 Let ω be a real $(1, 1)$-form. Then ω is said to be positive if the coefficients $\omega_{j\bar{k}}$ form pointwise a nonnegative Hermitian matrix.

Exercise 8.2 Let μ be any smooth $(1, 0)$-form. Show that $i\mu \wedge \bar{\mu}$ is positive. Show that any constant coefficient positive $(1, 1)$-form in $\mathbb{C}^n$ can be written as a sum of at most n forms of the type $i\mu \wedge \bar{\mu}$.

By duality any $(n - 1, n - 1)$ real form is representable by a $n \times n$ matrix of its coefficients and once again one can define positivity through the positivity of the Hermitian matrix.

In intermediate degrees the coefficient matrix is substantially larger. One may still use its positivity properties for a definition:

Definition 8.3 Let τ be a (p, p)-form in $\mathbb{C}^n$, where $1 < p < n - 1$. We say that τ is strictly positive if the coefficient matrix is pointwise a nonnegative Hermitian matrix.

Way subtler notion of positivity (which is however more useful!) can be given through an action on simple positive forms:

Definition 8.4 A (p, p)-form is said to be simple positive if it can be written as $\Pi_{j=1}^{p}(i\mu_j \wedge \overline{\mu}_j)$ for some $(1, 0)$-forms μ_j. A $(n - p, n - p)$-form γ is said to be positive if for any simple positive (p, p)-form η one has $\gamma \wedge \eta \geq 0$.

Exercise 8.5 Inspect the differences between positivity and strict positivity in the first nontrivial case i.e. when $p = 2$ and $n = 4$.

1.8.2   Currents

Below we recall the notion of a *current* which generalizes in a sense the notion of an analytic subvariety. First we define the space of test forms.

Definition 8.6 Let $\mathcal{D}_{p,q}(\Omega)$ denote the space of smooth (p, q)-forms with compact support in Ω equipped with the Schwartz topology (i.e. a sequence α_j converges to α if the coefficients converge in C^∞ and the union of the supports of α_j is compact). Elements of $\mathcal{D}_{p,q}(\Omega)$ are called test forms.

Exercise 8.7 Let $\chi : \mathbb{C} \to \mathbb{R}$ be any smooth function with compact support. Consider the forms $\alpha_j(z) := \chi(z + j)idz \wedge d\bar{z}$. Do α_j converge to 0 in the Schwartz topology?

Given the space of test forms we define its dual—the space of currents:

Definition 8.8 A current of bidegree $(n - p, n - q)$ (or of bidimension (p, q)) is a continuous linear functional on the space $\mathcal{D}_{p,q}(\Omega)$.

Currents have all the standard properties of linear functionals: they can be added, multiplied by a scalar etc. A special feature of currents is that they can be *differentiated*. Formally if D denotes any partial derivative then

$$DT(\alpha) := \varepsilon T(D\alpha)$$

with $\varepsilon \in \{-1, 1\}$ depending on the bidegree so that the sign is consistent with the standard Stokes formula.

Exercise 8.9 Determine the sign of ε in terms of p and q.

Exercise 8.10 Let the Dirac delta measure δ_z act on a $(1, 1)$ form $f(z)idz \wedge d\bar{z}$ in $\mathbb{C}$ by

$$\delta_z(f) = f(z).$$

Determine whether the following operators are currents of dimension $(1, 1)$

$$a) \sum_{j=1}^{\infty} \delta_j;$$

$$b) \sum_{j=1}^{\infty} \frac{\partial^j}{\partial^j z} \delta_j;$$

$$c) \sum_{j=1}^{\infty} \frac{\partial^j}{\partial^j z} \delta_0.$$

A current T which is equal to its conjugate $\overline{T}$ is called *real* (this is only possible if $p = q$). A very special role in pluripotential theory is played by *positive currents*:

Definition 8.11 A real current T of bidimension (p, p) is said to be positive if for any simple positive test form γ one has

$$T(\gamma) \geq 0.$$

Exercise 8.12 Determine which of the currents from the previous exercise are positive.

A crucial fact that shall be used repeatedly is that positive currents have *coefficients* that are particularly nice:

Theorem 8.13 (Riesz Theorem) *Let T be a current of bidimension (p, p). It can be written uniquely as*

$$T = \sum_{|J|=n-p, |K|=n-p}' T_{JK} dz_{j_1} \wedge \cdots dz_{j_{n-p}} \wedge d\bar{z}_{k_1} \wedge \cdots d\bar{z}_{k_{n-p}},$$

where $'$ denotes summation over increasing multiindices and T_{JK} are distributions (currents of bidimension $(0, 0)$). If T is positive then T_{JK} are complex valued measures.

Exercise 8.14 Riesz theorem states that a distribution satisfying $T(\varphi) \geq 0$ for any nonnegative test function φ has to be a (positive) measure. Deduce from this that a positive $(1, 1)$-current has (complex valued) measures $\mu_{j\bar{k}}$ as coefficients. Furthermore $\mu_{j\bar{j}}$ is a real measure, whereas $\mu_{j\bar{k}}(A) = \bar{\mu}_{k\bar{j}}(A)$ for any Borel set A.

1.8.3 Plurisubharmonic Functions

We begin this Section by recalling the definition of the basic object of study: the ω-plurisubharmonic functions:

Definition 8.15 The ω-plurisubharmonic functions are the elements of the function class

$$PSH_\omega(X) := \{u \in L^1(X, \omega) : dd^c u \geq -\omega, \ u \in C^\uparrow(X)\},$$

where $C^\uparrow(X)$ denotes the space of upper semicontinuous functions and the inequality is understood in the weak sense of currents.

We call the functions that belong to $PSH_\omega(X)$ either ω-plurisubharmonic or ω-psh for short. Recall the handy notation $\omega_u := \omega + dd^c u$.

Note that the definition coincides with the usual one in the Kähler setting. In particular ω-psh functions are locally standard plurisubharmonic functions plus some smooth function.

Let now U be a coordinate chart in a compact complex Hermitian manifold (X, ω). Shrinking U a bit if necessary one can find two smooth local strictly plurisubharmonic functions ρ_1 and ρ_2 such that $\Omega_1 := dd^c \rho_1 \leq \omega$, while $\omega \leq \Omega_2 := dd^c \rho_2$. This simple observation has powerful consequences: as ω-psh functions are (locally) Ω_2-plurisubharmonic all local properties of ω-psh functions are essentially the same as in the Kähler setting.

We note that *all* functions u in $PSH_\omega(X)$, normalized by the condition $sup_X u = 0$ are **uniformly** integrable. This follows from classical results in potential theory (see [Kol98]). We provide a proof following quite closely the one in [GZ05], where the Authors treat the Kähler case.

Proposition 8.16 *Let* $u \in PSH_\omega(X)$ *be a function satisfying* $sup_X u = 0$. *Then there exists a constant* C *dependent only on* X, ω *such that*

$$\int_X |u| \omega^n \leq C.$$

Proof Consider a double cover of X by coordinate balls $B_s^1 \subset\subset B_s^2 \subset X$, $s = 1, \cdots, N$. In each B_s^2 there exists a strictly plurisubharmonic potential ρ_s satisfying the following properties:

$$\begin{cases} \rho_s|_{\partial B_s^2} = 0 \\ inf_{B_s^2} \rho_s \geq -C \\ dd^c \rho_s = \omega_{2,s} \geq \omega, \end{cases}$$

where C is a constant dependent only on the covering and ω. Note that plurisubharmonicity coupled with the first condition above yields the inequality $\rho_s \leq 0$ on B_s^2.

Suppose now that there exists a sequence $u_j \in PSH_\omega(X), sup_X u_j = 0$ satisfying $lim_{j\to\infty} \int_X |u_j|\omega^n = \infty$. After choosing subsequence (which for the sake of brevity we still denote by u_j) we may assume that

$$\int_X |u_j|\omega^n \geq 2^j \tag{1.10}$$

and moreover a sequence of points x_j where u_j attains maximum is contained in some fixed ball B_s^1.

Note that $\rho_s + u_j$ is an ordinary plurisubharmonic function in B_s^2 and by the submean value property one has

$$\rho_s(x_j) = \rho_s(x_j) + u_j(x_j) \leq C \int_{B_s^2} \rho_s(z) + u_j(z)dV \leq C \int_{B_s^2} u_j(z)dV + C, \tag{1.11}$$

where dV is the Lebesgue measure in the local coordinate chart, while C denotes constants dependent only on B_s^1 and B_s^2. Thus (1.11) implies that for some constant C one has

$$\int_{B_s^2} |u_j(z)|dV \leq C. \tag{1.12}$$

Consider the function $v := \sum_{j=1}^\infty \frac{u_j}{2^j}$. By classical potential theory this is again an ω-psh function or constantly $-\infty$. By (1.12), however, the integral of v over B_s^2 is finite, thus it is a true ω-psh function. By the same reasoning we easily obtain that $v \in L^1(B_t^1)$ for any $t \in 1, \cdots, N$ and hence $v \in L^1(X)$. This contradicts (1.10), and thus the existence of a uniform bound is established. $\square$

Exercise 8.17 *In the Kähler case a much neater argument can be used to establish this fact. In fact suppose that $\int_X \omega^n = 1$ and let the function $G_\omega(z, w)$ satisfy $(\omega_z + dd_z^c G(z, w)) \wedge \omega_z^{n-1} = \delta_w$ with δ_w being the Dirac delta measure (G is thus the Green function with respect to ω). Then*

$$u(w) = \int_X u(z)(\omega_z + dd_z^c G(z, w)) \wedge \omega_z^{n-1}$$

and integration by parts finishes the proof.

Check for which classes of special Hermitian metrics this Green type arguments works without any adjustment assuming that the Green function exists.

In fact a much stronger result is true: ω-psh functions are uniformly exponentially integrable. To prove this we need the following ingredient (see Lemma 4.4 in [Hö1])

Lemma 8.18 *Let u be a negative plurisubharmonic function in $B_R(0) \subset \mathbb{C}^n$ and λ be any positive number. Assume further that $u(0) \geq -1$. Then*

$$\int_{B_r(0)} e^{-\lambda u(z)} dV(z) \leq C$$

for any $r \leq Re^{-\lambda/2}$ and some constant C dependent only on n, λ and R.

Assuming this fact the following theorem holds:

Theorem 8.19 *Let (X, ω) be a compact Hermitian manifold. Then there exists positive constants α and C, dependent only on X, ω such that for any ω-psh function ϕ, $\sup_X \phi = 0$ one has*

$$\int_X e^{-\alpha\phi} \omega^n \leq C.$$

Proof This result in the Kähler case can be found in [T89]. If $2r$ is the injectivity radius of (X, ω) we fix a $r/4$—net of points $\{x_1, \cdots x_N\}$ i.e. a collection of points such that the geodesic balls $B_{r/4}(x_i)$ cover X and N is the smallest cardinality of such configuration. Using the uniform integrability of ω-psh functions we have for any ω-psh function ϕ normalized so that $\sup_X \phi = 0$ the inequalities

$$sup_{B_{r/4}(x_i)}\phi \geq -\frac{C}{\int_{B_{r/4}(x_i)} \omega^n}.$$

If ρ_i is a Kähler potential of a metric $\Omega > \omega$ in $B_{2r}(x_i)$ then ρ_i is uniformly bounded in $B_{\frac{3r}{4}}(x_i)$ and hence the local plurisubharmonic function $\rho_i + \phi$ satisfies

$$(\rho_i + \phi)(y_i) := sup_{B_{r/4}(x_i)}\rho_i + \phi \geq -\frac{C}{\int_{B_{r/4}(x_i)} \omega^n},$$

$$\rho_i + \phi \leq C \text{ in } B_{\frac{3r}{4}}(x_i)$$

for some constant C under control. Taking $\alpha := C + \frac{min_i \int_{B_{r/4}(x_i)} \omega^n}{C+1}$ by Lemma 8.18 we get

$$\int_{B_{r/2}(y_i)} e^{-\alpha(\rho_i+\phi-C)} dV \leq C.$$

It remains to observe that y_i are (by definition) in $B_{r/4}(x_i)$, hence $B_{r/4}(x_i) \subset B_{r/2}(y_i)$ and adding the integrals above we get the statement. $\square$

Remark 8.20 Note that we used a similar localization argument as in the proof of uniform integrability. As the exponent α depends on the geometry of (X, ω) we needed a more delicate reasoning. In fact in Kähler geometry the supremum of all such α for fixed (X, ω) is a very important invariant (known as the α-invariant) and its computation is a subtle problem. It is yet to be seen whether its Hermitian analog has interesting geometric applications.

1.8.4   The Monge-Ampère Measure

By construction $\omega + dd^c u$ is a positive $(1, 1)$-current i.e. a differential form with distributional coefficients. This raises a serious problem in defining $(\omega + dd^c u) \wedge (\omega + dd^c u)$—we would have to multiply distributions to get the coefficients!

We will follow Bedford and Taylor's idea [BT82, BT76] to construct this product. First, by Riesz theorem a positive current has *measure* coefficients i.e. each of the distributional coefficients is a complex valued measure.

The crucial observation in Bedford-Taylor theory is that for a locally bounded plurisubharmonic function u the current $u(dd^c u)$ also has measure coefficients. Note that this may still be the case for *some* unbounded functions but in general there is no reason why the product of an integrable function and a measure may still be a measure.

Theorem 8.21 (Bedford-Taylor) *The inductively constructed currents*

$$(dd^c u)^k := dd^c (u(dd^c u)^{k-1})$$

are well defined, closed and positive. Furthermore if u_j is a decreasing sequence of locally bounded plurisubharmonic functions with u as a limit then

$$(dd^c u_j)^k \to (dd^c u)^k$$

as currents.

Proof Recall that $T_j \to T$ as currents if for any test form ψ one has $lim_{j \to \infty} T_j(\psi) = T(\psi)$.

We begin with the following basic observation: if T_j is a sequence of currents converging to T then $dd^c T_j \to dd^c T$ i.e. distributional differentiation is *continuous* with respect to convergence of currents. The proof hinges on the fact that, by definition, $dd^c T_j(\psi) = T_j(dd^c \psi)$ and is left as an easy exercise.

Note that once we know that $(dd^c u)^{k-1}$ is a positive current it has measure coefficients, and hence $u(dd^c u)^{k-1}$ is well defined. Thus it remains to show that $(dd^c u)^k$ is positive.

Pick $\tilde{u}_j$ a local sequence of smooth plurisubharmonic functions decreasing towards u (such a sequence can be constructed using a standard mollification with a smoothing kernel). Note that $\tilde{u}_j (dd^c u)^{k-1} \to u(dd^c u)^{k-1}$ since $(dd^c u)^{k-1}$ has measure coefficients and we can use dominated convergence theorem.

But on the other hand $dd^c(\tilde{u}_j(dd^cu)^{k-1})$ is equal to $dd^c\tilde{u}_j \wedge (dd^cu)^{k-1}$ by definition once $\tilde{u}_j$ is smooth and $(dd^cu)^{k-1}$ is closed. Finally $dd^c\tilde{u}_j \wedge (dd^cu)^{k-1}$ is positive since it is a product of a positive current and a $(1,1)$-positive form. Passing to the limit we obtain the positivity of $(dd^cu)^k$.

It remains to prove the claimed convergence for decreasing sequences. As the result is purely local it suffices to prove it in any small ball $B_{r/2}(z_0) \subset B_r(z_0)$. Furthermore if $\rho(z) := A(|z-z_0|^2 - r^2)$ for a large enough A taking the maximums of u_j's and u with ρ we get new functions U_j, U agreeing with the old ones on $B_{r/2}(z_0)$ and smoothly approaching zero near $\partial B_r(z_0)$. Then it suffices to show that $U_j(dd^cU_j)^{k-1} \to U(dd^cU)^{k-1}$ on $B_r(z_0)$.

We show this convergence in two steps. First of all we show that any cluster point of $U_j(dd^cU_j)^{k-1}$ in the weak topology of currents has to be bounded from above by $U(dd^cU)^{k-1}$ as a current (note that $U_j(dd^cU_j)^{k-1}$ is locally of finite mass, hence cluster points exist by the Banach-Alaouglu theorem).

To this end suppose (relabelling if necessary) that the whole sequence weakly converges. Fix ψ a smooth positive closed test form of the appropriate bidegree. Fix also $j_0 \in \mathbb{N}$. Note that

$$lim_{j\to\infty}U_j(dd^cU_j)^{k-1} \wedge \psi \le U_{j_0}(dd^cU_j)^{k-1} \wedge \psi \le g_{j_0}(dd^cU_j)^{k-1} \wedge \psi,$$

where g_{j_0} is any continuous function majorizing U_{j_0}. But then the right hand side converges as measures to $g_{j_0}(dd^cU)^{k-1} \wedge \psi$. As U_{j_0} is a decreasing limit of such continuous functions g_{j_0} we obtain

$$lim_{j\to\infty}U_j(dd^cU_j)^{k-1} \wedge \psi \le U_{j_0}(dd^cU)^{k-1} \wedge \psi.$$

Passing now with j_0 to ∞ we obtain the claimed bound.

Next we will show that the inverse inequality holds for the total masses of the currents involved. Note that

$$\int_{B_r(z_0)} U(dd^cU)^{k-1} \wedge \psi$$

$$\le \int_{B_r(z_0)} U_j(dd^cU)^{k-1} \wedge \psi = \int_{B_r(z_0)} U dd^cU_j \wedge (dd^cU)^{k-2} \wedge \psi$$

$$\le \int_{B_r(z_0)} U_j dd^cU_j \wedge (dd^cU)^{k-2} \wedge \psi \cdots \le \int_{B_r(z_0)} U_j(dd^cU_j)^{k-1} \wedge \psi.$$

Taking the limit as $j \to \infty$ concludes the proof.

□

Exercise 8.22 *In the proof above we modified u_j's so that they are smooth near the boundary and vanish on the boundary. Where is this used? Also where the continuity of g_{j_0} was used?*

Returning to the Hermitian setting we argue in a local chart where a Kähler form $\Omega = dd^c\eta$ can be found so that $\Omega > \omega$.

Then formally in that chart

$$(\omega+dd^c u)^k = (dd^c(\eta+u)-(\Omega-\omega))^k = \sum_{j=0}^{k} \binom{k}{j}(-1)^{k-j}(dd^c(\eta+u))^j \wedge (\Omega-\omega)^{k-j}$$

and all the terms on the right hand side are well defined by the Bedford-Taylor construction.

To get the convergence for decreasing sequences we write

$$dd^c u_j + \omega = dd^c(u_j + \eta) - T, \quad T = (\Omega - \omega).$$

Then by the Newton expansion again

$$(dd^c u_j + \omega)^k = (dd^c u_j + \Omega)^k - k(dd^c u_j + \Omega)^{k-1} \wedge T + \ldots \pm T^k. \qquad (1.13)$$

By the convergence theorem for local psh functions all the terms on the right converge as currents, and the sum of their limits is

$$(dd^c u + \Omega)^k - k(dd^c u + \Omega)^{k-1} \wedge T + \ldots \pm T^k = (dd^c u + \omega)^k.$$

This allows the use of some local results from pluripotential theory developed by Bedford and Taylor in [BT82]. In particular the Monge-Ampère operator

$$\omega_u^n := \omega_u \wedge \cdots \wedge \omega_u$$

is well defined for bounded ω-psh functions.

Having the convergence for monotonely decreasing sequences it is natural to ask whether continuity of the Monge-Ampère operator holds for any sequence u_j of plurisubharmonic functions converging weakly towards $u \in PSH \cap L_{loc}^\infty$. The answer (a bit surprisingly) is *no* as shown by Cegrell [CE83]:

Example 8.23 There is a plurisubharmonic function $u \in L_{loc}^\infty$ in $\mathbb{C}^2$ and a sequence $u_j \in PSH \cap L_{loc}^\infty$ such that $u_j \to u$ weakly (in fact the convergence is even in L^p for any $p \in [1, \infty)$) but

$$(dd^c u_j)^2 \nrightarrow (dd^c u)^2.$$

Proof Let $f(z), g(z)$ be non negative subharmonic functions of one complex variable. Then $(f(z_1) + g(z_2))^2$ is a plurisubharmonic function in (z_1, z_2) with Monge-Ampère measure equal to

$$8(f + g)^2 dd^c f \wedge dd^c g + 8(f + g)[dg \wedge d^c g \wedge dd^c f + df \wedge d^c f \wedge dd^c g].$$

Pick now $g(z) = |z|^2$, say. It will be sufficient to show that the convergence $f_j \Delta f_j \to f \Delta f$ need not hold if f_j are bounded subharmonic functions converging weakly to a bounded subharmonic function f (we leave the details of this and the previous computation as an exercise).

To show the latter fact consider a compactly supported probability measure μ such that the potential

$$\tilde{f}(z) := \int log(|z - w|) d\mu(w)$$

is bounded from below—the (normalized) Lebesgue measure restricted to the unit disc would do. Let $a := inf \tilde{f}$ and define

$$f(z) := \tilde{f}(z) - a + 1.$$

By definition $f \geq 1$ everywhere. Consider now an approximation of $d\mu$ by combinations of discrete Dirac delta measures $d\mu_j := \sum_{i=1}^{i(j)} a_i^{(j)} \delta_{b_i^{(j)}}$ for some $a_i^{(j)} \geq 0, b_i^{(j)} \in supp\mu$. Then $\tilde{f}_j(z) := \int log(|z - w|) d\mu_j(w)$ converge weakly to $\tilde{f}$ yet they are clearly unbounded from below. Take then $f_j(z) = max\{\tilde{f}_j(z) - a + 1, 0\}$. As $\tilde{f}_j$'s converge to $\tilde{f} \geq a$ in L^p it is easy to see that f_j's converge to f in L^p too. Note however that $f \Delta f \geq \Delta f$ has total mass at least one while $f_j \Delta f_j = 0$ as measures since $\tilde{f}_j$'s are harmonic off their singularities.   □

Next we prove three very important inequalities known as Chern-Levine-Nirenberg (CLN) inequalities in the literature (see [CLN69]):

Theorem 8.24

(i) *(Local version) Let $u_1, \cdots u_n \in PSH(\Omega) \cap L^\infty(\Omega)$. Then for any two open relatively compact subsets $K \Subset L \Subset \Omega$ there is a constant $C = C(K, L, \Omega)$ such that*

$$\int_K dd^c u_1 \wedge \cdots \wedge dd^c u_n \leq C(K, L, \Omega) \Pi_{j=1}^n ||u_j||_{L^\infty(L)};$$

(ii) *(Local integral version) Let K, L and u_j be as above. Then there is a constant $C = C(K, L, \Omega)$ such that for any plurisubharmonic function v (not necessarily bounded!) normalized so that $sup_\Omega v \leq 0$ one has*

$$\int_K -v dd^c u_1 \wedge \cdots \wedge dd^c u_n \leq C(K, L, \Omega) ||v||_{L^1(L)} \Pi_{j=1}^n ||u_j||_{L^\infty(L)};$$

(iii) *(Global version) Let (X, ω) be a compact Hermitian manifold. Then there is a constant C dependent only on X, ω such that for any bounded ω-psh functions*

ϕ_j, $j = 1, \cdots, n$ and a non-positive ω-psh function ψ one has

$$\int_X -\psi(\omega + dd^c\phi_1) \wedge \cdots \wedge (\omega + dd^c\phi_n)$$

$$\leq C[\|\psi\|_{L^1(X,\omega)} + 1]\Pi^n_{j=1}(\|\phi_j\|_{L^\infty(X)} + 1).$$

Proof The first statement follows from (iterated) multiplication by a cut-off function and integration by parts. We leave the details as an exercise to the Reader.

Next we claim that (ii) implies (iii). To this end we again use localization in charts. In each chart of a triple cover $U_i \Subset V_i \Subset W_i$ we find potentials ρ_i of a local Kähler metric $\Omega > \omega$. Then the integral over U_i of

$$-\psi(\omega + dd^c\phi_1) \wedge \cdots \wedge (\omega + dd^c\phi_n)$$

is majorized by $\int_{U_i} -\psi(dd^c(\rho_i + \phi_1)) \wedge \cdots \wedge (dd^c(\rho_i + \phi_n))$. Applying the local integral version of the CLN inequalities this is bounded by

$$C[\|\psi\|_{L^1(V_i,\omega)} + 1]\Pi^n_{j=1}(\|\phi_j + \rho_i\|_{L^\infty(V_i)})$$

and the claim follows, since ρ_i is uniformly under control in V_i

It remains to prove (ii). As the result is local we shall argue in a fixed ball $B_{r/2}(z_0) \Subset B_r(z_0)$, such that $B_{2r}(z_0)$ is relatively compact in L. Just as in the proof of Theorem 8.21 we may modify u_j's close to the boundary so that they agree on $B_{r/2}(z_0)$ with the original u_j's and vanish smoothly on $\partial B_r(z_0)$ in the sense that they are all equal to $A(|z - z_0|^2 - r^2)$ for a constant A under control.

We also assume that v is smooth, the general case follows then by approximation. Hence

$$\int_{B_{r/2}(z_0)} -vdd^cu_1 \wedge \cdots \wedge dd^cu_n \leq \int_{B_r(z_0)} -vdd^cu_1 \wedge dd^cu_2 \wedge \cdots \wedge dd^cu_n.$$

At this stage we claim that in the last integral we can exchange each dd^cu_j factor by $Add^c|z|^2$. Indeed

$$\int_{B_r(z_0)} -vdd^cu_1 \wedge dd^cu_2 \wedge \cdots \wedge dd^c[u_n - A(|z - z_0|^2 - r^2)]$$

$$= \int_{B_r(z_0)} -[u_n - A(|z - z_0|^2 - r^2)]dd^cu_1 \wedge dd^cu_2 \wedge \cdots \wedge dd^cv$$

$$\leq C \int_{B_r(z_0)} dd^cu_1 \wedge dd^cu_2 \wedge \cdots \wedge dd^cv,$$

as is seen after two integrations by part (and we use the fact the last factor is constantly zero near the boundary).

Arguing as in (i) the last integral is controlled by

$$C \int_{B_{\frac{3}{2}r}(z_0)} (dd^c|z|^2)^{n-1} \wedge dd^c v,$$

and after yet another multiplication by a cut-off function and integration by parts we end up with $\int_{B_{2r}(z_0)} -v d\lambda \leq ||v||_{L^1(L)}$.

Arguing then in the same manner with the remaining factors $dd^c u_j$ we end up with

$$\int_{B_{r/2}(z_0)} -v dd^c u_1 \wedge \cdots \wedge dd^c u_n \leq C||v||_{L^1(L)} + \int_{B_r(z_0)} -v(dd^c|z|^2)^n,$$

which yields the claim. □

Exercise 8.25 Check carefully the localization argument. It is obvious provided that u_j's are uniformly negative on $B_r(z_0)$. Can we claim such a uniform bound?

1.8.5 Bedford-Taylor Capacities

In [BT76] Bedford and Taylor introduced a new *capacity* which has proven to be an extremely useful tool in pluripotential theory. Below we recall the definition:

Definition 8.26 Let Ω be a domain in $\mathbb{C}^n$. Given a Borel subset $E \subset \Omega$ its capacity is given by

$$cap(E, \Omega) := sup\{\int_E (dd^c u)^n | u \in PSH(\Omega), 0 \leq u \leq 1\}.$$

Exercise 8.27 Let K be a compact subset of Ω. Show that $cap(K, \Omega)$ is finite. Is $cap(\Omega, \Omega)$ finite for Ω—a ball in $\mathbb{C}^n$?

This notion was transplanted in [Kol03] to the setting of compact Kähler manifolds. The same construction can be applied in the Hermitian case.

If (X, ω) is Hermitian the Monge-Ampère capacity associated to (X, ω) is the function defined on Borel sets by

$$Cap_\omega(E) := sup\{\int_E (\omega + dd^c u)^n / u \in PSH(X, \omega) \text{ and } 0 \leq u \leq 1\}.$$

Exercise 8.28 In the Kähler case $Cap_\omega(E) \leq sup \int_X (\omega + dd^c u)^n = \int_X \omega^n$, hence the so-defined quantity is bounded. This reasoning fails in the Hermitian case. Nevertheless show, using Theorem 8.24 and integration by parts, that $Cap_\omega(K)$ is finite for any Borel subset K of the Hermitian manifold (X, ω).

We refer the reader to [Kol03, GZ05] for the basic properties of this capacity in the Kähler setting. In the Hermitian case one can repeat much of the Kählerian picture. Below we list some basic properties of cap_ω that will be useful later on:

Proposition 8.29

(i) If $E_1 \subset E_2 \subset X$ then $cap_\omega(E_1) \leq cap_\omega(E_2)$,

(ii) If U is open then $cap_\omega(U) = sup\{cap_\omega(K)|\ K - compact,\ K \subset U\}$,

(iii) If $U_j \nearrow U$, $U_j - open$ then $cap_\omega(U) = lim_{j \to \infty} cap_\omega(U_j)$.

Proof The first property follows from the very definition of cap_ω. To prove the second fix $\varepsilon > 0$ and a competitor u for the supremum, such that

$$cap_\omega(U) \leq \int_U \omega_u^n + \varepsilon.$$

Since ω_u^n is a regular Borel measure by inner regularity there is a compact set $K \subset U$ satisfying

$$\int_U \omega_u^n \leq \int_K \omega_u^n + \varepsilon \leq cap_\omega(K) + \varepsilon.$$

Coupling the above facts and letting ε converge to zero we end up with $cap_\omega(U) \leq sup\{cap_\omega(K)|\ K - compact,\ K \subset U\}$, and the reverse inequality follows from the first property.

Finally the third one can be proved as follows. Fix once more $\varepsilon > 0$ and a compact set $K \subset U$, such that

$$cap_\omega(U) \leq cap_\omega(K) + \varepsilon.$$

Observe that for j large enough $K \subset U_j$ and hence

$$cap_\omega(K) \leq cap_\omega(U_j) \leq lim_{j \to \infty} cap_\omega(U_j).$$

As a result we obtain

$$cap_\omega(U) \leq lim_{j \to \infty} cap_\omega(U_j),$$

while the reverse inequality is obvious.   □

For ω-Kähler the patched local Bedford-Taylor capacity was studied in [Kol03]. That is for a fixed double covering $B_s^1 \subset\subset B_s^2 \subset X$ of coordinate balls, we define the capacity cap'_ω of a Borel set E by

$$cap'_\omega(E) := \sum_{s=1}^n cap(E \cap B_s^1, B_s^2),$$

with $cap(E \cap B_s^1, B_s^2)$ denoting the classical Bedford-Taylor capacity [BT82]. It was shown in [Kol03] that cap_ω and cap_ω' are equicontinuous in the Kähler case. Our next result shows this equicontinuity in the Hermitian setting:

Proposition 8.30 *Let (X, ω) be a compact Hermitian manifold with a fixed double cover $B_s^1 \Subset B_s^2$ of coordinate charts. Then there exist a constant $C > 0$ dependent only on X, ω and the double cover such that for any Borel set E*

$$C^{-1} cap_\omega'(E) \le cap_\omega(E) \le C cap_\omega'(E).$$

Proof From the continuity properties of capacities it suffices to prove the result for compact sets.

To show the left inequality we fix a compact set $K \subset X$ and let $K_s := K \cap \overline{B_s^1}$. It suffices to prove that $cap_\omega(K) \ge cap_\omega(K_s) \ge C cap(K_s, B_s^2)$ for every fixed s.

To this end fix a smooth function χ on X with values in $[-1, 0]$, such that $\chi \equiv 0$ off B_s^2, while $\chi|_{B_s^1} = -1$. Then there is a small constant $\delta < \frac{1}{2}$ such that $2\delta\chi$ is ω-psh. Note that this δ depends only on X, ω and the covering.

Fix any $\varepsilon > 0$. Then from the definition there is a local plurisubharmonic function u on B_2^s, such that $0 \le u \le 1$, and

$$\int_{K_s} (dd^c u)^n \ge cap(K_s, B_s^2) - \varepsilon.$$

Consider now the function

$$\varphi := \begin{cases} max\{2\delta\chi(z), \delta(u(z) - 1)\} + 1 & z \in B_s^2 \\ 2\delta\chi + 1 & z \in X \setminus B_s^2. \end{cases} \tag{1.14}$$

By construction φ is a global ω-psh function equal to $\delta(u - 1) + 1$ on B_s^1. Note that $1 \ge \varphi \ge -2\delta + 1 > 0$. Then

$$cap_\omega(K) \ge \int_K (\omega + dd^c\varphi)^n \ge \int_{K_s} (\omega + dd^c\varphi)^n \ge \int_{K_s} (\omega + dd^c\delta u)^n.$$

Exploiting the positivity of $dd^c u$ this can be further estimated from below by

$$\int_{K_s} (dd^c\delta u)^n = \delta^n \int_{K_s} (dd^c u)^n \ge \delta^n [cap(K_s, B_s^2) - \varepsilon].$$

As ε was arbitrary passing to zero yields the left of the claimed inequalities.

In order to prove the right one note that in each chart B_s^2 there is a bounded strictly plurisubharmonic function ρ_s, such that $dd^c\rho_s \ge \omega$. Normalizing by adding a constant we may assume that for any s we have $0 \le \rho_s \le C$, with $C > 0$ a constant dependent only on the covering and the geometry of the manifold.

Again fix an ε and φ which is an ω-psh competitor in the definition of cap_ω, (we assume $0 \le \varphi \le 1$) such that

$$\int_K (\omega + dd^c\varphi)^n \ge cap_\omega(K) - \varepsilon.$$

The point is that the local function $u_s := \frac{\varphi + \rho_s}{C+1}$ is plurisubharmonic and satisfies $0 \le u_s \le 1$. Indeed, the inequalities are clear. As

$$dd^c\varphi + dd^c\rho_s \ge dd^c\varphi + \omega \ge 0$$

u_s is plurisubharmonic on B_s^2.

But then

$$cap_\omega(K) - \varepsilon \le \int_K (\omega + dd^c\varphi)^n \le \sum_s \int_{K_s} (\omega + dd^c\varphi)^n.$$

Exploiting the definition of ρ_s this string of inequalities continues as

$$\sum_s \int_{K_s} (dd^c(\rho_s + \varphi))^n = (C+1)^n \int_{K_s} (dd^c u_s)^n \le (C+1)^n \sum_s cap(K_s, B_s^2),$$

as claimed.   □

Before we proceed further we recall a basic local solvability result of Kolodziej (se [Kol05] for a much more general version):

Theorem 8.31 *Let B be a ball in $\mathbb{C}^n$ and the function f be L^p integrable on B for some $p > 1$. Then the Dirichlet problem*

$$\begin{cases} v \in PSH(B) \cap C(\overline{B}) \\ (dd^c v)^n = f & (1.15) \\ v|_{\partial B} = 0 \end{cases}$$

admits unique solution v. Moreover there is a constant $c > 0$ dependent only on n, p and the diameter of the ball B, such that

$$sup_{\overline{B}}(-v) \le c\|f\|_{L^p(B)}^{\frac{1}{n}}.$$

Theorem 8.31 yields immediately a comparison between the volume and capacity of a set:

Proposition 8.32 *Let K be a compact subset of a ball $B \subset \mathbb{C}^n$. Then for any $q > 1$ there is a constant C dependent only on n, q and the diameter of the ball such that $V(K) \le Ccap(K, B)^q$ with $V(K)$ denoting the Lebesgue measure of K.*

Proof If $V(K) = 0$ the inequality trivially holds. Suppose that $V(K) > 0$ and consider the function $f := \frac{\chi_K}{V(K)^{1-\frac{1}{q}}}$, where χ_K is the characteristic function of the set K. Obviously $\int_B f^p = 1$ for $p > 1$ such that $\frac{1}{p} + \frac{1}{q} = 1$. Using Theorem 8.31 one finds a continuous solution v to the problem

$$
\begin{cases}
v \in PSH(B) \cap C(\overline{B}) \\
(dd^c v)^n = f \\
v|_{\partial B} = 0,
\end{cases}
$$

for which $-c \leq v \leq 0$ for some uniform $c > 0$. But then $u := \frac{v}{c} + 1$ is a competitor in the definition of capacity. Hence

$$
cap(K, B) \geq \int_K (dd^c u)^n = \frac{1}{c^n} \int_K (dd^c v)^n = \frac{1}{c^n} \int_K \frac{\chi_K}{V(K)^{1-\frac{1}{q}}} = \frac{V(K)^{\frac{1}{q}}}{c^n},
$$

which yields the proof. □

Remark 8.33 It can be proven using subtler tools that the volume is controlled by the capacity in an even stronger way. See [Kol05] for details.

The local results above can be used to prove a volume-capacity estimate on a compact Hermitian manifold:

Theorem 8.34 *Let $p > 1$ and f be a non negative function belonging to $L^p(\omega^n)$. Then for any compact $K \subset X$ one has*

$$
\int_K f \omega^n \leq C(p, X) \|f\|_p cap_\omega(K)^2,
$$

where $C(p, X)$ is a constant dependent only on p and (X, ω).

Proof As $\int_K f \omega^n \leq \|f\|_p (\int_K \omega^n)^{\frac{p-1}{p}}$ it suffices to prove that

$$
\int_K \omega^n \leq C(q, X) cap_\omega(K)^q
$$

for any $q > 1$. To this end consider the double covering $B_s^1 \Subset B_s^2$ as in Proposition 8.30. We assume that the number of the balls is fixed, and their radii are under control. Recall that $K_s := K \cap B_s^1$.
Then

$$
\int_K \omega^n \leq \sum_s \int_{K \cap B_s^1} \omega^n \leq C \sum_s \int_{K_s} dV,
$$

as ω^n on compact subsets of coordinate charts differs from the Lebesgue measure by a bounded function. Invoking now Proposition 8.32 we have

$$\int_K \omega^n \leq C \sum_s cap(K_s, B_s^2)^q \leq C(\sum_s cap(K_s, B_s^2))^q.$$

But by Proposition 8.30 the latter quantity is controlled by $Ccap_\omega(K)^q$, as claimed.

□

As yet another consequence of psh-like property of ω-psh functions one gets the capacity estimate of sublevel sets of those functions.

Proposition 8.35 *Let* $u \in PSH_\omega(X), sup_X u = 0$. *Then there exists an independent constant* C *such that for any* $s > 1$ $cap_\omega(\{u < -t\}) \leq \frac{C}{t}$.

Proof We shall use the double covering introduced in Proposition 8.16. Fix a function $v \in PSH_\omega(X), 0 \leq v \leq 1$. Then we obtain

$$\int_{\{u<-t\}} \omega_v^n \leq \frac{1}{t} \int_X -u\omega_v^n.$$

Now by the generalized Chern-Levine-Nirenberg inequalities (Theorem 8.24) one obtains that the last quantity can be estimated by

$$C\frac{1}{t}\|u\|_{L^1(\omega^n)} \leq \frac{C}{t}$$

which completes the proof.

□

We finish this Section with a lemma which shall be used throughout the note. It follows from the proof of the comparison principle by Bedford and Taylor in [BT76].

Lemma 8.36 *Let* u, v *be bounded* $PSH_\omega(X)$ *functions and* T *a (positive but non necessarily closed) current of the form* $\omega_{u_1} \wedge \cdots \wedge \omega_{u_{n-1}}$ *for bounded functions* u_i *belonging to* $PSH_\omega(X)$. *Then*

$$\int_{\{u<v\}} dd^c(u-v) \wedge T \geq \int_{\{u<v\}} d^c(u-v) \wedge dT.$$

Proof Suppose first that u, v and the boundary of the set $\{u < v\}$ are smooth. If ρ is a smooth defining function of $\{u < v\}$, then $u - v = \alpha\rho$ for some positive function α on the closure of $\{u < v\}$.

Given any smooth positive $(n-1, n-1)$ form θ we thus get the equality

$$\int_{\partial\{u<v\}} d^c(u-v) \wedge \theta = \int_{\partial\{u<v\}} \alpha d^c\rho \wedge \theta.$$

On the other hand if σ denotes the surface area element on $\partial\{u < v\}$ induced by ω then $\sigma = \frac{*d\rho}{||d\rho||}$, where $*$ stands for the Hodge star operator with respect to ω.

Now if $d^c\rho \wedge \theta = f\,d\sigma$ for some function f we end up with the equality

$$\alpha d\rho \wedge d^c\rho \wedge \theta = \alpha f\,d\rho \wedge \frac{*d\rho}{||d\rho||}.$$

But $d\rho \wedge d^c\rho \wedge \theta \geq 0$, which yields that $\alpha f \geq 0$ and thus

$$\int_{\{u<v\}} (dd^c(u-v) \wedge \theta - d^c(u-v) \wedge d\theta) = \int_{\partial\{u<v\}} d^c(u-v) \wedge \theta$$

$$= \int_{\partial\{u<v\}} \alpha f\,d\sigma \geq 0.$$

The case of a current T of the given form is done by approximation of each u_j by a decreasing sequence of smooth ω-psh functions.

Finally if either u, v or $\partial\{u < v\}$ is not smooth we consider an approximating sequence of smooth ω-psh functions u^j, v^j. By the Sard theorem for almost every t the sets $\{u^j < v^j + t\}$ have smooth boundary. Thus we can apply the argument above to the pair $(u^j, v^j + t)$ and then let t to zero. Finally we let $j \to \infty$ and the desired inequality follows. $\square$

1.9 Comparison Principle in Hermitian Setting

In the Kähler setting the comparison principle says that for any u, $v \in PSH_\omega(X) \cap L^\infty(X)$ we have

$$\int_{\{u<v\}} \omega_v^n \leq \int_{\{u<v\}} \omega_u^n.$$

In a sense this integral inequality makes up for the lack of a classical maximum principle in pluripotential theory. It is a basic ingredients in many proofs—see [Kol05].

Such an inequality is in general impossible on Hermitian manifolds due to the following proposition:

Proposition 9.1 *A necessary condition for the comparison principle to hold is that*

$$\forall u \in PSH_\omega(X) \cap L^\infty(X) \quad \int_X (\omega + dd^c u)^n = \int_X \omega^n.$$

Proof Note that for any bounded ω-psh function u we can find a constant C such that $u - C < 0 < u + C$. Then applying the comparison principle to the pairs $(u - C, 0)$ and $(0, u + C)$ (the integration takes place over the whole of X) one gets that $\int_X \omega_u^n = \int_X \omega^n$, whence the result. □

It was recently proven by Chiose[1] [Chi] that such an invariance of total volume is in fact *equivalent* to the Guan-Li type conditions $dd^c\omega = 0$, $d\omega \wedge d^c\omega = 0$ imposed on ω.

Thus unless ω is of special type we have to allow some additional error terms into the inequality. The next theorem shows that such a result indeed holds. Below we present a weaker form of a comparison principle with "error terms" which will be useful in obtaining a priori estimates:

Theorem 9.2 ([DK12]) *Let ω be a Hermitian metric on a complex compact manifold X and let $u, v \in PSH_\omega(X) \cap L^\infty(X)$. Then there exists a polynomial P_n of degree $n - 1$ and zeroth degree coefficient equal to 0, such that*

$$\int_{\{u<v\}} \omega_v^n \leq \int_{\{u<v\}} \omega_u^n + P_n(BM) \sum_{k=0}^{n} \int_{\{u<v\}} \omega_u^k \wedge \omega^{n-k},$$

where B is defined by (1.2) and $M = sup_{\{u<v\}}(v - u)$. The coefficients of the polynomial are nonnegative and depend only on the dimension of X.

This claim says that provided the product of B and the supremum of $v - u$ is small enough the error terms are small. Of course these error terms are bounded anyway and can be incorporated in the coefficients of the polynomial P_n but here it is emphasized that P_n is independent of the functions u and v and also that the error terms involve lower order Hessians of ω_u. In general it is impossible to control these pointwise but it will turn out later that these can be controlled by ω_u^n in the integral sense over specific subdomains.

Proof Note that

$$\int_{\{u<v\}} \omega_v^n = \int_{\{u<v\}} \omega \wedge \omega_v^{n-1} + \int_{\{u<v\}} dd^c v \wedge \omega_v^{n-1} \leq \int_{\{u<v\}} \omega \wedge \omega_v^{n-1}$$
$$+ \int_{\{u<v\}} dd^c u \wedge \omega_v^{n-1} + \int_{\{u<v\}} d^c(v - u) \wedge d(\omega_v^{n-1}),$$

where we have used Lemma 8.36. Again by (1.2) we have

$$dd^c(\omega_v^{n-1}) \leq B[\omega^2 \wedge \omega_v^{n-2} + \omega^3 \wedge \omega_v^{n-3}].$$

---

[1]Professor Demailly informed me that he was aware of this equivalence long time ago.

Thus by Stokes' theorem

$$\int_{\{u<v\}} \omega_v^n \le \int_{\{u<v\}} \omega_u \wedge \omega_v^{n-1} - \int_{\{u<v\}} d(v-u) \wedge d^c(\omega_v^{n-1}) \le \int_{\{u<v\}} \omega_u \wedge \omega_v^{n-1}$$

$$+ \int_{\{u<v\}} (v-u) \wedge dd^c(\omega_v^{n-1}) \le \int_{\{u<v\}} \omega_u \wedge \omega_v^{n-1}$$

$$+ sup_{\{u<v\}}(v-u)B \int_{\{u<v\}} (\omega^2 \wedge \omega_v^{n-2} + \omega^3 \wedge \omega_v^{n-3}).$$

Repeating the above procedure of replacing ω_v by ω and ω_u in the end one obtains the statement.   □

In the computations above it is easy to see that the term $\int_{\{u<v\}} \omega_u^{n-1} \wedge \omega$ will never appear on the right hand side but we shall not use this fact. Also for small n the polynomials P_n are explicitly computable: in particular one can take $P_2(x) = 2x$, $P_3(x) = 2x^2 + 4x$. In general we can use the following (very) crude count: In the process we exchange a term $\int_{\{u<v\}} \omega_v^k \wedge \omega_u^l \wedge \omega^{k-l}$ for the term $\int_{\{u<v\}} \omega_v^{k-1} \wedge \omega_u^{l+1} \wedge \omega^{k-l}$ and $\int_{\{u<v\}} (v-u)dd^c(\omega_v^{k-1} \wedge \omega_u^l \wedge \omega^{k-l})$. The latter term splits into six pieces and each of them contains ω_v with power no higher than $k-1$. Of course there are special cases when some of these terms coincide or do not appear, but the upshot is that there will be at most 7^n terms in the very end. Thus one can take P_n as $P_n(x) = 7^n(x + x^2 + \cdots + x^{n-1})$.

Below we shall state a technical refined version of the above theorem. It works only for *special* sublevel domains but has the advantage that all the lower order Hessian terms are incorporated into the ω_u^n-term at the cost of enlarging the constant 1 in front of it. This inequality was proven by Cuong and Kolodziej in [KN1]:

Theorem 9.3 (Comparison Principle-Refined Version) *Let X, ω, u and v be as above. Take $0 < \varepsilon < 1$ and let $m(\varepsilon) = inf_X(u - (1-\varepsilon)v)$. Then for any small constant $0 < s < \frac{\varepsilon^3}{16B}$*

$$\int_{\{u<(1-\varepsilon)v+m(\varepsilon)+s\}} \omega_{(1-\varepsilon)v}^n \le \left(1 + n^2 14^n \frac{sB}{\varepsilon^n}\right) \int_{\{u<(1-\varepsilon)v+m(\varepsilon)+s\}} \omega_u^n$$

for some universal constant C dependent only on X, n and ω.

Observe that this comparison principle works only for sublevel sets very close to the empty set $\{u < (1-\varepsilon)v + m(\varepsilon)\}$. The bonus is that we control not only ω_v^n but also the (integrals of) lower order Hessians of ω_v.

Proof Observe that $(1-\varepsilon)v + m(\varepsilon) + s$ is ω-psh. Denote by

$$a_k = \int_{\{u<(1-\varepsilon)v+m(\varepsilon)+s\}} \omega_u^k \wedge \omega^{n-k}.$$

Plugging u and $(1 - \varepsilon)v + m(\varepsilon) + s$ in the first version of the comparison principle, we obtain that

$$sup_{\{u<(1-\varepsilon)v+m(\varepsilon)+s\}}\{(1 - \varepsilon)v + m(\varepsilon) + s - u\} \leq s.$$

Observe that from the assumptions made on s Bs is small, hence $P_n(Bs) \leq n7^n Bs$ (since $x^k \leq x$ for $k \geq 1$, $x \in (0, 1)$). Then it is enough to get rid of the lower order Hessians of ω_u.

Note that $\varepsilon\omega \leq \omega_{(1-\varepsilon)v+m(\varepsilon)+s}$ and hence

$$\varepsilon a_k \leq \int_{\{u<(1-\varepsilon)v+m(\varepsilon)+s\}} \omega_u^k \wedge \omega_{(1-\varepsilon)v} \wedge \omega^{n-k-1}.$$

Swapping now $(1 - \varepsilon)v + m(\varepsilon) + s$ with u as in the previous proof we get

$$\varepsilon a_k \leq a_{k+1} + sB(a_k + a_{k-1} + a_{k-2}) \tag{1.16}$$

(with the understanding that $a_{-1} = a_{-2} = 0$). Now we shall prove inductively that $a_k \leq \frac{2}{\varepsilon}a_{k+1}$. Indeed for $k = 0, 1$ this follows from inequality (1.16) and the assumption that $sB \leq \frac{\varepsilon^3}{16}$. Suppose now that the inequality is true for $k-2$ and $k-1$ then (1.16) results in

$$\varepsilon a_k \leq a_{k+1} + \frac{\varepsilon^3}{16}\left(a_k + \frac{2}{\varepsilon}a_k + \frac{4}{\varepsilon^2}a_k\right) \leq a_{k+1} + \frac{\varepsilon}{2}a_k,$$

which proves the claim.

Our inductive argument gives us the inequality $a_k \leq \frac{2^n}{\varepsilon^n}a_n$, so the integrals of lower order Hessians can be estimated by $\int_{\{u<(1-\varepsilon)v+m(\varepsilon)+s\}} \omega_u^n$ and the result follows.   □

Observe that when $B = 0$ (in particular when ω is Kähler) the theorem above gives us the standard comparison principle.

Finally we prove an analog of the comparison principles for the Laplacian with respect to a Gauduchon metric.

Proposition 9.4 *Let ω be a Gauduchon metric and let ϕ, $\psi \in PSH_\omega(X) \cap L^\infty(X)$. Then*

$$\int_{\{\phi<\psi\}} \omega_\psi \wedge \omega^{n-1} \leq \int_{\{\phi<\psi\}} \omega_\phi \wedge \omega^{n-1}.$$

Proof The claimed inequality can be rewritten as

$$\int_{\{\phi<\psi\}} (dd^c[\phi - \psi]) \wedge \omega^{n-1} \geq 0.$$

But this as an immediate consequence of the proof of Lemma 8.36 since the latter integral is at least

$$\int_{\{\phi<\psi\}} d^c[\phi - \psi]) \wedge d\omega^{n-1} = \int_{\{\phi<\psi\}} [\phi - \psi] \wedge dd^c\omega^{n-1} = 0$$

by the Gauduchon property of ω. Note that in the last integration by parts we used that $\phi - \psi = 0$ on the boundary.   □

1.10 The Complex Monge-Ampère Equation on Compact Hermitian Manifolds

In this section we shall discuss in detail the solvability of the Dirichlet problem for the complex Monge-Ampère equation in the Hermitian setting. First we shall prove existence and uniqueness in the smooth case. Although this is not part of pluripotential theory we include the detailed argument for the sake of completeness. Our goal will be the following theorem:

Theorem 10.1 *Let (X, ω) be a compact Hermitian manifold of complex dimension n. Let also f be any smooth strictly positive function on X. Then the following problem*

$$\begin{cases} u \in C^\infty(X), \quad \omega + dd^c u > 0, \\ \sup_X u = 0, \\ c \in \mathbb{R}, \\ (\omega + dd^c u)^n = e^c f \omega^n, \quad f \in C^\infty(X) \end{cases} \tag{1.17}$$

admits a unique solution (u, c). Furthermore there exist constants C_k, $k = 0, 1, 2, \cdots$ dependent only on X, ω and f, such that the C^k-th norm of the function u is bounded by C_k.

Note that we do not assume compatibility conditions on f (i.e. we do not assume that $\int_X f\omega^n = \int_X \omega^n$) but instead we introduce an additional constant c in the equation.

In the case when ω is Kähler the solvability of this equation was proved by Yau in his seminal paper [Y78]. The Hermitian case was studied by Cherrier [Che87], and later by Guan-Li, Tosatti-Weinkove [GL10, TW10a] up until the final resolution by Tosatti and Weinkove in [TW10b].

Our exposition is borrowed from [D16].

The method of proof will follow the classical continuity method approach. More precisely we consider the family of problems

$$(*)_t \quad \begin{cases} u_t \in PSH_\omega(X), \\ sup_X u_t = 0, \\ (\omega + dd^c u_t)^n = e^{c_t}(1 - t + tf)\omega^n \quad f \in C^\infty(X), f > 0, \end{cases} \qquad (1.18)$$

for $t \in [0, 1]$. Clearly the problem $(*)_0$ is solvable and it is enough to prove that the set

$$A := \{T \in [0, 1] | (*)_t \text{ is solvable for every } t \leq T\}$$

is open and closed in $[0, 1]$.

To this end we shall first prove uniqueness of the constant c and uniqueness of the solution u. Then we pass to the openness. The hard part of the argument is the closedness which is achieved by establishing a priori estimates for the solutions.

1.10.1 Uniqueness

In [TW10b] the authors proved that if u, v are smooth ω-psh functions and their Monge-Ampère measures satisfy $\omega_u^n = e^{c_1} f\omega^n$, $\omega_v^n = e^{c_2} f\omega^n$ for some smooth function f and some constants c_1 and c_2 then in fact $c_1 = c_2$ and u and v differ by a constant. This is the counterpart of the uniqueness of potentials in the Calabi conjecture from the Kähler case.

The equality $u = v$ is easy. Indeed, suppose that we already knew that $c_1 = c_2$. Then we have

$$0 = e^{c_1} f\omega^n - e^{c_1} f\omega^n = \omega_u^n - \omega_v^n = dd^c(u - v) \wedge \left(\sum_{k=0}^{n-1} \omega_u^k \wedge \omega_v^{n-1-k} \right).$$

This can be treated as a linear strictly elliptic equation with respect to $u - v$ for the coefficients of the form $\sum_{k=0}^{n-1} \omega_u^k \wedge \omega_v^{n-1-k}$ pointwise give strictly positive definite matrix. But then the strong maximum principle yields that $u - v$ must be a constant.

Now we show that $c_1 = c_2$. The proof is taken from [DK12] and is in the spirit of pluripotential theory. Suppose, to the contrary, that

$$\omega_u^n = e^{c_1} f\omega^n, \quad \omega_v^n = e^{c_2} f\omega^n$$

for some smooth u, v and $c_1 \neq c_2$. We can without loss of generality assume that $c_2 > c_1$.

Consider the Hermitian metric $\omega + dd^c u$. Since by the assumptions above it is smooth and strictly positive one finds a unique Gauduchon function ϕ_u, such that

$$inf_X \phi_u = 0, \ dd^c(e^{(n-1)\phi_u}(\omega + dd^c u)^{n-1}) = 0.$$

Then one can apply the comparison principle for the Laplacian with respect to the Gauduchon metric (Proposition 9.4) $e^{\phi_u}(\omega + dd^c u)$ which yields

$$\int_{\{u<v\}} e^{(n-1)\phi_u}(\omega + dd^c u)^{n-1} \wedge \omega_v \leq \int_{\{u<v\}} e^{(n-1)\phi_u} \omega_u^n.$$

Exchanging now v with $v + C$ (which does not affect the reasoning above) for big enough C one obtains

$$\int_X e^{(n-1)\phi_u}(\omega + dd^c u)^{n-1} \wedge \omega_v \leq \int_X e^{(n-1)\phi_u} \omega_u^n.$$

Note that the left hand side can be estimated from below using (pointwise) the AM-GM inequality:

$$\int_X e^{(n-1)\phi_u}(\omega + dd^c u)^{n-1} \wedge \omega_v \geq \int_X e^{(n-1)\phi_u + \frac{(c_2-c_1)}{n}} \omega_u^n.$$

Coupling the above estimates one obtains

$$1 < e^{\frac{(c_2-c_1)}{n}} \leq 1,$$

a contradiction.

1.10.2 Continuity Method: Openness

The openness part boils down to showing that if $(*)_T$ is solvable then the problem $(*)_t$ is also solvable for t close enough to T. This is achieved by applying the implicit function theorem between well chosen Banach spaces and linearization of the equation. Here the linearized operator is essentially the Laplacian, and we shall prove that this operator is bijective in our setting. The details are taken from [TW10a].

First of all we need the following classical fact:

Proposition 10.2 *Let ω be a Gauduchon metric on X and let Δ_ω be the Laplacian operator with respect to ω. Then, given any $f \in L^2(X, \omega)$ there is a unique $W^{2,2}$ function u which solves the problem*

$$\Delta_\omega u = f, \ \int_X v\omega^n = 0$$

if and only if $\int_X f\omega^n = 0$. *Furthermore if* $\alpha \in (0, 1)$ *and* $f \in \mathcal{C}^\alpha(X)$, *then* $u \in \mathcal{C}^{2,\alpha}(X)$.

Proof Uniqueness of normalized solutions follows from the ellipticity of Δ_ω. The formal computation

$$\int_X < \Delta_\omega u, g > \omega^n = \int_X gdd^c u \wedge \omega^{n-1} = \int_X udd^c(g\omega^{n-1})$$

$$= \int_X (udd^c g \wedge \omega^{n-1} + udg \wedge d^c(\omega^{n-1}) - ud^c g \wedge d(\omega^{n-1}))$$

$$= \int_X < u, \Delta_\omega^* g > \omega^n$$

shows that the adjoint operator Δ_ω^* is second order elliptic and moreover it contains no zero order term (note that we use the Gauduchon condition here!) thus it contains only constant functions in its kernel. On the other hand, again by classical elliptic theory the image of Δ_ω in L^2 is perpendicular to the kernel of Δ_ω^* which proves the first assertion. The second assertion is a consequence of the classical Schauder theory of linear elliptic equations.   □

Suppose now that at time T we have a smooth solution u to the problem $(*)_T$ (we skip the index T for the ease of notation). Let ϕ_u denote the Gauduchon function associated to ω_u. We normalize it by adding a constant if needed so that $\int_X e^{(n-1)\phi_u}(\omega+dd^c u)^n = 1$. We also fix a small positive constant $\alpha < 1$ (dependent on X, ω and n—the dependence will be important in the later stages when we prove higher order a priori estimates).

Consider the two Banach manifolds

$$B_1 := \{w \in \mathcal{C}^{2,\alpha}(X)| \int_X we^{(n-1)\phi_u}\omega_u^n = 0\}$$

and

$$B_2 := \{h \in \mathcal{C}^\alpha(X)| \int_X e^{h+(n-1)\phi_u}\omega_u^n = 1\}.$$

Consider the mapping $\mathcal{T}: B_1 \to B_2$ given by

$$\mathcal{T}(v) := log\frac{(\omega + dd^c u + dd^c v)^n}{(\omega + dd^c u)^n} - log \int_X e^{(n-1)\phi_u}(\omega + dd^c u + dd^c v)^n.$$

Observe that $\mathcal{T}(0) = 0$ and that any function v sufficiently close to 0 in $\mathcal{C}^{2,\alpha}$-norm is $(\omega + dd^c u)$—plurisubharmonic.

By the implicit function theorem the equation $\mathcal{T}(v) = h$ is solvable for any $h \in B_2$ sufficiently close in C^α norm to zero if the Frechet derivative

$$(D\mathcal{T}) : T_0 B_1 = B_1 \rightarrow T_0 B_2 = \{g \in C^\alpha(X)| \int_X g e^{(n-1)\phi_u} \omega_u^n = 0\}$$

is an invertible linear mapping.

But a computation shows that

$$(D\mathcal{T})(\eta) = \Delta_{\omega+dd^c u} \eta - n \int_X e^{(n-1)\phi_u} \omega_u^{n-1} \wedge dd^c \eta.$$

Note that the last summand is zero because $e^{\phi_u}(\omega + dd^c u)$ is Gauduchon. The question is thus whether $\Delta_{\omega+dd^c u} : B_1 \rightarrow T_0 B_2$ is a continuous bijective mapping.

By Proposition 10.2 (recall that $e^{\phi_u}(\omega + dd^c u)$ is Gauduchon metric!) the equation

$$\Delta_{e^{\phi_u}(\omega+dd^c u)}(\eta) = \tau$$

is solvable if and only if $\int_X \tau e^{n\phi_u}(\omega + dd^c u)^n = 0$ and the solution is unique up to an additive constant. Thus we can assume that $\int_X \eta e^{(n-1)\phi_u} \omega_u^n = 0$. Furthermore, if $\tau \in C^\alpha(X)$ then η belongs to $C^{2,\alpha}(X)$ and hence it belongs to B_1. Note that $\Delta_{e^{\phi_u}(\omega+dd^c u)}(\eta) = e^{-\phi_u} \Delta_{(\omega+dd^c u)}(\eta)$ thus $(D\mathcal{T})(\eta) = \tau$ is solvable if and only if $\int_X \tau e^{(n-1)\phi_u}(\omega + dd^c u)^n = 0$ i.e. exactly if τ belongs to $T_0 B_2$. This proves the surjectivity of $(D\mathcal{T})$ and injectivity follows from the normalization condition. Finally continuity of $(D\mathcal{T})$ follows from the Schauder $C^{2,\alpha}$ a priori estimates for the Laplace equation.

1.10.3 Continuity Method: Closedness—Higher Order Estimates

Before starting the proofs of a priori estimates let us stress that third and higher order ones follow from standard Schauder elliptic theory as long as $C^{2,\alpha}$ estimates are proven for some small positive $\alpha < 1$. Thus we are left with proving estimates up to order $2 + \alpha$.

By the complex version of the Evans-Krylov theory (see [TWWY14] for a nice overview) there is a constant

$$C = C(X, \omega, n, ||\Delta u||_{C^0}, ||u||_{C^0}, ||f||_{C^1})$$

and $0 < \alpha < 1$ dependent on the same quantities, such that if u solves Eq. (1.17) then

$$||u||_{C^{2,\alpha}} \leq C.$$

Thus what remains is to prove uniform bound for the Laplacian of u, and of u itself.

1.10.4 Continuity Method: Closedness—Second Order Estimate

The aim of this subsection is to prove the following estimate:

Theorem 10.3 ([GL10]) *If u is a solution to Eq. (1.17) then there exists a constant $C = C(X, \omega, n, ||\Delta f||_{C^0}, ||u||_{C^0})$, such that*

$$0 \leq n + \Delta u \leq C,$$

where the Laplacian is the ordinary Chern Laplacian with respect to the metric ω.

Once we have second order estimates the gradient estimate follows by interpolation. Our proof will differ slightly from the one in [GL10] but, of course, the main idea remains the same.

Proof Consider the function $A(u) := log(n + \Delta u) + h \circ u$, where h is an additional uniformly bounded strictly decreasing function that we shall choose later on. If we can prove that at the point z where A attains maximum we have that $n + \Delta u$ is bounded then we are done since at any other point x we have

$$log(n + \Delta u)(x) \leq A(z) - h(u(x)) \leq C.$$

Thus let us fix a point of maximum of A and identify it with zero in a local chart. We shall use ordinary partial derivatives in this chart—in particular $g_{i\bar{j},k}$ will denote $\frac{\partial g_{i\bar{j}}}{\partial z_k}$ and so on. Let us also denote by g' the metric $g_{i\bar{j}} + u_{i\bar{j}}$, while $g^{k\bar{l}}, g'^{k\bar{l}}$ will denote the inverse transposed matrices of g and g' respectively.

In order to simplify the computations let us assume that we have chosen coordinates diagonalizing the metric $g_{i\bar{j}}$ and $\frac{\partial^2 u}{\partial z_i \partial \bar{z}_j}$ and then rechoose the canonical coordinates so that additionally $g_{i\bar{i},k}(0) = 0$ for any i, k. Observe that the Hessian of u is still diagonal at zero. Moreover we can safely assume that $\Delta u(0) \geq 1$, say, for otherwise we are done.

Applying logarithm to both sides of Eq. (1.17) and differentiating twice at z we get

$$g'^{p\bar{r}}(g_{p\bar{r},k} + u_{p\bar{r}k}) = log(f)_k + g^{p\bar{r}}g_{p\bar{r},k}; \tag{1.19}$$

$$-g'^{p\bar{s}}g'^{h\bar{r}}(g_{h\bar{s},\bar{l}}+u_{h\bar{s}\bar{l}})(g_{p\bar{r},k}+u_{p\bar{r}k})+g'^{p\bar{r}}(g_{p\bar{r},k\bar{l}}+u_{p\bar{r}k\bar{l}})$$

$$=log(f)_{k\bar{l}}-g^{p\bar{s}}g^{h\bar{r}}g_{h\bar{s},\bar{l}}g_{p\bar{r},k}+g^{p\bar{r}}g_{p\bar{r},k\bar{l}}. \qquad (1.20)$$

Taking trace in the second equation we obtain

$$-g'^{p\bar{p}}g'^{r\bar{r}}|g_{r\bar{p},k}+u_{r\bar{p}k}|^2+g'^{r\bar{r}}(g_{r\bar{r},k\bar{k}}+u_{r\bar{r}k\bar{k}})=\Delta log(f)-|g_{p\bar{r},k}|^2+g_{r\bar{r},k\bar{k}}. \qquad (1.21)$$

Let us now investigate the function A at the point of maximum. From the vanishing of the first derivative of A we get the equalities

$$0=\frac{g_{,k}^{i\bar{j}}u_{i\bar{j}}+g^{i\bar{j}}u_{i\bar{j}k}}{\Delta u+n}+h'u_k=\frac{u_{i\bar{i}k}}{\Delta u+n}+h'u_k. \qquad (1.22)$$

(The first term in the first summand vanishes because we have chosen the special coordinates!) Taking now the trace of the Hessian of A at the point z with respect to g' we obtain the inequality

$$0\geq g'^{k\bar{k}}A_{k\bar{k}}=g'^{k\bar{k}}[\frac{(g^{i\bar{j}}u_{i\bar{j}})_{k\bar{k}}}{\Delta u+n}-\frac{|\sum_i u_{i\bar{i}k}|^2}{(\Delta u+n)^2}+h'u_{k\bar{k}}+h''|u_k|^2]. \qquad (1.23)$$

From Eq. (1.22) the second term can be exchanged by $-(h')^2g'^{k\bar{k}}|u_k|^2$, while the third one reads $h'(n-\sum_k g'^{k\bar{k}})$. In order to estimate the first term we observe that

$$(g^{i\bar{j}}u_{i\bar{j}})_{k\bar{k}}=g_{,k\bar{k}}^{i\bar{i}}u_{i\bar{i}}+u_{i\bar{i}k\bar{k}}+2Re(g_k^{i\bar{j}}u_{i\bar{j}\bar{k}}).$$

The fourth order term, after taking trace with $g'^{k\bar{k}}$ can be exchanged using Eq. (1.21).
Note that, exploiting the diagonality of g at z one has

$$g_{,k}^{i\bar{j}}=-g^{i\bar{s}}g^{l\bar{j}}g_{l\bar{s},k}=-g_{j\bar{i},k}.$$

Altogether the first term then reads

$$g'^{k\bar{k}}\frac{(g^{i\bar{j}}u_{i\bar{j}})_{,k\bar{k}}}{\Delta u+n}=g'^{k\bar{k}}\frac{g_{k\bar{k}}^{i\bar{i}}u_{i\bar{i}}}{\Delta u+n}-g'^{k\bar{k}}\frac{2Re(g_{j\bar{i},k}u_{i\bar{j}k})}{\Delta u+n}-g'^{k\bar{k}}\frac{g_{k\bar{k},i\bar{i}}-\Delta log f}{\Delta u+n}$$

$$-\frac{|g_{r\bar{k},i}|^2}{\Delta u+n}+\frac{g'^{r\bar{r}}g'^{k\bar{k}}|g_{r\bar{k},i}+u_{r\bar{k}i}|^2}{\Delta u+n}.$$

Note that the first summand above is controlled from below by $-C\sum_k g'^{k\bar{k}}$ with the constant C dependent on the sup norm of all second order derivatives of g. The same goes for all the terms in the third and fourth summand (the dependence of C on the relevant quantities is clear—note also that in a sense these terms are even "better" due to the Laplacian in the denominator).

Summing up our computations up to now inequality (1.23) results in

$$0 \geq [-h' - C]\sum_k g'^{k\bar{k}} - C + [h'' - (h')^2]\sum_k g'^{k\bar{k}}|u_k|^2 + \frac{g'^{r\bar{r}}g'^{k\bar{k}}|g_{r\bar{k},i} + u_{r\bar{k}i}|^2}{\Delta u + n}$$

$$- g'^{k\bar{k}}\frac{2Re(g_{j\bar{i},k}u_{i\bar{j}k})}{\Delta u + n}.$$

The last summand can be rewritten as follows:

$$g'^{k\bar{k}}\frac{2Re(g_{j\bar{i},k}u_{i\bar{j}k})}{\Delta u + n} = g'^{k\bar{k}}\frac{2Re(g_{j\bar{i},k}u_{i\bar{k}j})}{\Delta u + n} = g'^{k\bar{k}}\frac{2Re(g_{j\bar{i},k}(g_{i\bar{k},j} + u_{i\bar{k}j} - g_{i\bar{k},j}))}{\Delta u + n}$$

$$= g'^{k\bar{k}}\sum_{i \neq j}\sqrt{g'^{i\bar{i}}g'_{i\bar{i}}}\frac{2Re(g_{j\bar{i},k}g'_{i\bar{k},\bar{j}})}{\Delta u + n} - g'^{k\bar{k}}\frac{2Re(g_{j\bar{i},k}g_{i\bar{k},\bar{j}})}{\Delta u + n}.$$

(We sum only over indices $i \neq j$ for in the special coordinates $g_{i\bar{i},k} = 0$.) Applying Schwarz inequality the latter is bounded above by

$$g'^{k\bar{k}}\sum_{i \neq j}g'^{i\bar{i}}\frac{|g'_{i\bar{k},\bar{j}}|^2}{\Delta u + n} + g'^{k\bar{k}}\sum_{i \neq j}\frac{g'_{i\bar{i}}|g_{j\bar{i},k}|^2}{n + \Delta u} + C\sum_k g'^{k\bar{k}} \leq \sum_{i \neq j}g'^{k\bar{k}}g'^{i\bar{i}}\frac{|g'_{i\bar{k},\bar{j}}|^2}{\Delta u + n}$$

$$+ C\sum_k g'^{k\bar{k}},$$

where we have also used the elementary inequality $g'_{i\bar{i}} \leq \Delta u + n$.

Thus our main inequality reduces to

$$0 \geq [-h' - C]\sum_k g'^{k\bar{k}} - C + [h'' - (h')^2]\sum_k g'^{k\bar{k}}|u_k|^2 + \frac{g'^{r\bar{r}}g'^{k\bar{k}}|g'_{r\bar{k},k}|^2}{\Delta u + n}$$

The last term can be handled as follows

$$\frac{g'^{r\bar{r}}g'^{k\bar{k}}|g'_{r\bar{k},k}|^2}{\Delta u + n} = g'^{r\bar{r}}\frac{[(\sum_k g'^{k\bar{k}}|g'_{r\bar{k},k}|^2)(\sum_k g'_{k\bar{k}})]}{(\Delta u + n)^2} \geq g'^{r\bar{r}}\frac{|\sum_k(u_{r\bar{k}k} + g_{r\bar{k},k})|^2}{(\Delta u + n)^2}$$

$$= g'^{r\bar{r}}|h'u_r + \frac{\sum_k g_{r\bar{k},k}}{\Delta u + n}|^2,$$

where in the last equality we have made use of Eq. (1.22).

Expanding the squares and applying Schwarz inequality once more we end up with

$$\frac{g'^{r\bar{r}}g'^{k\bar{k}}|g'_{r\bar{k},k}|^2}{\Delta u + n} \geq g'^{r\bar{r}}((h')^2 + h')|u_r|^2 - |h'|g'^{r\bar{r}}\frac{|\sum_k g_{r\bar{k},k}|^2}{(\Delta u + n)^2},$$

and the last summand is estimated by $C\sum_r g'^{r\bar{r}}$.

Summing up our main inequality now reads

$$0 \geq [-h' - C]\sum_k g'^{k\bar{k}} - C + [h'' - h']\sum_k g'^{k\bar{k}}|u_k|^2.$$

So if we choose the function $h(t) = Ce^{-t}$ for a sufficiently large constant C, and assuming a bound on $osc_X u$ we end up with

$$0 \geq C\sum_k g'^{k\bar{k}} - C,$$

which shows that $g'^{k\bar{k}}$ are upper bounded and hence $g'_{k\bar{k}}$ are also lower bounded. From the equation we immediately get that $g'_{k\bar{k}}$ are upper bounded at the point z which establishes the desired estimate. □

Exercise 10.4 We used the special coordinates introduced in Sect. 1.7 in the computations. Check whether the usage is just simplifying the calculations or is it used in a substantial way.

1.10.5 Continuity Method: Closedness—Uniform Estimate

The last and historically the hardest step is to establish the uniform C^0 estimate. The uniform estimate was proven by Cherrier, Guan-Li and Tosatti-Weinkove [Che87, GL10, TW10a] under various additional assumptions on the metric ω. The general result with no assumptions on ω was first accomplished by Tosatti and Weinkove in [TW10b]. There the Authors used a version of Moser iteration to obtain the following bound:

$$Vol(\{u < inf_X u + \varepsilon\}) \geq \delta, \tag{1.24}$$

for some fixed constants ε and δ. Roughly speaking such an estimate tells us that there is some control from below on the volume of "small" sublevel sets. This coupled with suitable Sobolev inequality completes the proof, see [TW10b] for details.

Below we prove the uniform estimate using techniques from pluripotential theory taken from [DK12]. For different approaches we refer also to [BL11]. More specifically we shall prove the following result

Theorem 10.5 *Let u be a solution to Eq. (1.17). Then there exists a constant $C > 0$ dependent on $||f||_p$, p, X, ω, n, such that $inf_X u \geq -C$.*

In the proof we shall prove and exploit a similar bound to (1.24) but we shall use the capacity instead of the volume. Thus our goal is the inequality

$$cap_\omega(\{u < inf_X u + \varepsilon\}) \geq \delta.$$

Indeed suppose that such an inequality is already proven. Then exploiting Proposition 8.35 we immediately get a uniform bound of $inf_X u$ and we are done.

Let us first establish an additional capacity inequality which is modelled on an analogous argument from the Kähler setting:

Proposition 10.6 ([DK12, KN1]) *Let u be a ω-psh solution of the equation $\omega_u^n = f\omega^n$, where $f \in L^p(X, \omega)$ for some $p > 1$ and v be any bounded continuous ω-psh function satisfying $-C_0 \leq v \leq 0$. Take a constant $0 < \varepsilon < 1$ and let $0 < t << \varepsilon$, $0 < s << \varepsilon$ be two sufficiently small constants. Then there is a constant $C = C(n, X, \omega, p, \varepsilon, C_0)$, such that*

$$t^n cap_\omega(\{u < (1 - \varepsilon)v + inf_X[u - (1 - \varepsilon)v] + s\})$$
$$\leq C||f||_{L^p} cap_\omega(\{u < (1 - \varepsilon)v + inf_X[u - (1 - \varepsilon)v] + s + t\})^2.$$

Proof For notational simplicity we denote by $m(\varepsilon)$ the quantity

$$inf_X[u - (1 - \varepsilon)v]$$

and by $U(s, \varepsilon)$ the set $\{u < (1 - \varepsilon)v + m(\varepsilon) + s\}$. Throughout the proof we shall assume s and t are small enough, so that all technical requirements for the application of Theorem 9.3 are satisfied.

Pick any ω-psh function w such that $0 \leq w \leq 1$. As w is a competitor for the supremum in the definition of the capacity we need to bound from above the quantity $t^n \int_{\{u<(1-\varepsilon)v+m(\varepsilon)+s\}} \omega_w^n$.

To this end observe that the following inequality holds:

$$m(\varepsilon) - (C_0 + 1)t \leq inf_X[u - (1 - \varepsilon)((1 - t)v + tw)] \leq m(\varepsilon)$$

Thus we get the following string of set inclusions

$$U(s, \varepsilon) = \{u < (1 - \varepsilon)v + m(\varepsilon) + s\} \subset \{u < (1 - \varepsilon)((1 - t)v + tw) + m(\varepsilon) + s\}$$

$$\subset \{u < (1-\varepsilon)((1-t)v+tw)+\inf_X[u-(1-\varepsilon)((1-t)v+tw)]+s+(C_0+1)t\} = V$$

$$\subset \{u < (1 - \varepsilon)v + m(\varepsilon) + s + 2(C_0 + 1)t\} = U(s + 2(C_0 + 1)t, \varepsilon).$$

Note that $(1 - t)v + tw$ is a ω-psh function, and the set V is defined so that Theorem 9.3 can be applied for the pair $(u, (1 - t)v + tw)$ provided s and t are sufficiently small. Thus

$$((1 - \varepsilon)t)^n \int_{U(s,\varepsilon)} \omega_w^n \le ((1 - \varepsilon)t)^n \int_V \omega_w^n \le \int_V \omega_{(1-\varepsilon)((1-t)v+tw)}^n$$

$$\le C \int_V \omega_u^n \le C \int_{U(s+2(C_0+1)t,\varepsilon)} \omega_u^n,$$

where we have made use of Theorem 9.3 in the penultimate inequality. Note that the constant C depends on ε but is independent of u and v.

Continuing the string of inequalities we get

$$C \int_{U(s+2(C_0+1)t,\varepsilon)} \omega_u^n \le C||f||_{L^p} cap_\omega(U(s + 2(C_0 + 1)t, \varepsilon))^2,$$

where the last inequality follows from Theorem 8.34. Thus our claim follows after we exchange t with $2(C_0 + 1)t$.   □

Remark 10.7 Observe that we haven't made use of the continuity of v. This assumption will be used later to guarantee openness of the sets $U(s, \varepsilon)$.

Let us now explain how the above estimate implies that

$$cap_\omega(\{u < \inf_X u + \varepsilon\}) \ge \delta$$

for some ε and δ. In fact we shall prove the following more general statement:

Proposition 10.8 *There exists a small constant s_0, such that for any $s < s_0$ one has*
$s \le ||f||_{L^p}^{1/n} C cap_\omega(U(s, \varepsilon))^{\frac{1}{n}}$, *for a constant C dependent on n, ϵ, X, C_0, p and ω.*

In particular we get our desired bound by plugging $v = 0$ and taking any fixed positive $\varepsilon < 1$.

Proof Suppose s_0 is chosen so small that Proposition 10.6 applies for any $s, t \le s_0$. Define inductively s_i to be the supremum of all numbers between 0 and s_{i-1} such that

$$2cap_\omega(U(s, \varepsilon)) < cap_\omega(U(s_{i-1}, \varepsilon))\}.$$

Then s_i is clearly a decreasing sequence and any s_i is well defined for the sets
shrink to the empty set as s decreases to zero. Observe also that $U(s, \varepsilon)$ are
open sets and from the continuity of the capacity for increasing open sets (recall
Proposition 8.29) we get $2cap_\omega(U(s_{i+1}, \varepsilon)) \le cap_\omega(U(s_i, \varepsilon))$, while by definition
$lim_{s \to s_i^+} 2cap_\omega(U(s, \varepsilon)) \ge cap_\omega(U(s_{i-1}, \varepsilon))$.

Take now an s, such that $s_i \le s < s_{i-1}$. Then from Proposition 10.6 we get

$$(s_{i-1} - s)^n cap_\omega(U(s, \varepsilon)) \le Ccap_\omega(U(s_{i-1}, \varepsilon))^2.$$

Observe that since $s \ge s_i$ we have $2cap_\omega(U(s, \varepsilon)) \ge cap_\omega(U(s_{i-1}, \varepsilon))$.

Coupling these inequalities we obtain

$$(s_{i-1} - s)^n \le 4Ccap_\omega(U(s, \varepsilon)) \le 4C(\frac{1}{2})^{i-1} cap_\omega(U(s_0, \varepsilon)),$$

where the last inequality follows from iteration.

If we now let s to s_i, then take n-th roots and finally sum up the inequalities over
i we will obtain

$$s_0 = \sum_{i=1}^{\infty}(s_i - s_{i+1}) \le (4C)^{1/n} \sum_{j=0}^{\infty} (\frac{1}{2})^{j\frac{1}{n}} cap_\omega(U(s_0, \varepsilon))^{\frac{1}{n}},$$

which is the claimed result. $\square$

1.11   Weak Solutions for Degenerate Right Hand Side

We have already proved that solutions to the smooth Dirichlet problem for the
complex Monge-Ampère equation exist and are unique. Furthermore the argument
provides an *a priori* uniform bound for the solutions if the L^p norm of the right
hand side is under control for some $p > 1$.

In this section we shall discuss the solvability of the Dirichlet problem

$$\begin{cases} u \in PSH_\omega(X), sup_X u = 0 \\ (\omega + dd^c u)^n = e^c f\omega^n \quad f \in L^p(X, \omega), \ p > 1, \ f \ge 0. \end{cases} \tag{1.25}$$

The strategy is to use the smooth solvability to approximate the singular right hand
sides by smooth functions f_j in a suitable way, and then to extract a convergent
subsequence of solutions u_j. This approach leads to a problem, namely the behavior
of the constants c_j in such an approximation procedure. The technical heart of the
matter if we want to extract convergent subsequences is to show that these c_j's are
bounded from above and below **independently** of the supremum norms of f_j. This
was proven in [KN1]:

Theorem 11.1 *Let $X, \omega, f \neq 0$ and p be as above. Let also f_j be a sequence of smooth strictly positive functions convergent in L^p norm to f. Then the corresponding sequence of constants c_j associated to the problems*

$$(*)_i \begin{cases} u_i \in PSH_\omega(X), \\ \sup_X u_i = 0, \\ (\omega + dd^c u_i)^n = e^{c_i} f_i \omega^n \end{cases} \tag{1.26}$$

is uniformly bounded from above and below.

Proof Let us first give a lower bound for c_j's. For the sake of brevity we drop the index j in what follows. Recall that from the proof of Proposition 10.6 applied to $\varepsilon = \frac{1}{2}$, say, and $v = 0$ one has

$$t^n cap_\omega(\{u < inf_X u + s\}) \leq C cap_\omega(\{u < inf_X u + s + t\})^2$$

for all t, s smaller than a fixed constant ε_0. Taking $t = s$ and estimating the capacity on the right hand side by an uniform constant, which is legitimate since $cap_\omega(\{u < inf_X u + s + t\}) \leq cap_\omega(X)$, one gets the inequality

$$cap_\omega(\{u < inf_X u + s\}) \leq \frac{C}{s^n}.$$

On the other hand from Proposition 10.8 one has

$$s \leq (\tilde{C} e^c \|f\|_{L^p})^{1/n} cap_\omega(\{u < inf_X u + s_0\})^{\frac{1}{n}}.$$

Coupling these one obtains

$$s^2 \leq \bar{C} e^{c/n} \|f\|_{L^p}^{1/n},$$

for all $s \leq \varepsilon_0$. But then obviously c cannot decrease to minus infinity, hence we get a lower bound.

The upper bound is established as follows: since f_j converge to f in L^p, convergence also holds for $f_j^{1/n}$ towards $f^{1/n}$ in L^1 (we have to use the compactness of X here). Thus for j large enough

$$\int_X f_j^{1/n} \omega^n > \frac{\int_X f^{1/n} \omega^n}{2} > 0.$$

But from the AM-GM inequality one has $(\omega + dd^c u_j) \wedge \omega^{n-1} \geq (e^{c_j} f_j)^{1/n} \omega^n$ thus

$$e^{c_j/n} \leq \frac{2}{\int_X f^{1/n} \omega^n} \int_X (\omega + dd^c u_j) \wedge \omega^{n-1}.$$

if we multiply ω^{n-1} in the last integral by the Gauduchon function $e^{(n-1)\phi}$ (which is uniformly bounded) we get

$$e^{c_j/n} \leq \frac{2}{\int_X f^{1/n}\omega^n} e^{-(n-1)inf\phi} \int_X (\omega + dd^c u_j) \wedge e^{(n-1)\phi}\omega^{n-1}$$

$$= \frac{2}{\int_X f^{1/n}\omega^n} e^{-(n-1)inf\phi} \int_X e^{(n-1)\phi}\omega^n$$

by the Stokes theorem.   □

Now we are ready for the proof of the existence theorem:

Theorem 11.2 *The Dirichlet problem (1.25) admits a continuous solution.*

Proof It is enough to show that the sequence of solutions u_j of the problems (1.26) admits a Cauchy subsequence in the uniform topology. Indeed then one can extract a continuous limit. The Monge-Ampère operator is continuous with respect to uniform convergence, thus the limiting function solves the equation.

First of all we can assume that (after passing to a subsequence) the sequence of the constants c_j is convergent to some c. Let us still denote this subsequence by c_j.

Note that the family u_j is normalized by $sup_X u_j = 0$, hence it forms a relatively compact subset in the L^1-topology. Thus we can assume that the sequence u_j converges in L^1 to a ω-psh function u (take another subsequence if necessary).

Observe now that in the Dirichlet problems (1.26) the right hand sides are uniformly bounded in L^p for the chosen subsequence. By Theorem 10.5 we get that the sequence u_j is then uniformly bounded. Let then $C_0 > 0$ be a constant such that $u_j \geq -C_0$ for every j.

We shall argue by contradiction. To this end consider the quantities $S_{kj} := inf_X(u_k - u_j) \leq 0$. Since $sup_X(u_k - u_j) = -inf_X(u_j - u_k)$, it is enough to prove that the numbers S_{kj} converge to zero as k and j tend to infinity.

Suppose that this is not the case and let $1 > \varepsilon > 0$ be a constant such that $S_{kj} \leq -(C_0 + 3)\varepsilon$ for arbitrarily large $j \neq k$ (we can further decrease ε if needed). Then if $m_{kj}(\varepsilon)$ as usual denotes the infimum over X of the quantity $u_k - (1 - \varepsilon)u_j$ we obtain the inequality $m_{kj}(\varepsilon) \leq S_{kj}$.

As in the proof of Proposition 10.6 suppose that $s, t << \varepsilon$. Then we have a set inclusion

$$\{u_k < (1 - \varepsilon)u_j + m_{kj}(\varepsilon) + s + t\} \subset \{u_k < u_j + S_{kj} + \varepsilon C_0 + s + t\},$$

and the last set is in turn contained in

$$\{u_k < u_j - \varepsilon\} \subset \{|u_k - u_j| \geq \varepsilon\}$$

by our assumption on the constants S_{kj}.

From the proof of Proposition 10.6 we then know that for all t and s smaller than a (fixed) ε

$$t^n cap_\omega(\{u_k < (1-\varepsilon)u_j + m_{kj}(\varepsilon) + s\})$$

$$\leq C \int_{\{u_k < (1-\varepsilon)u_j + m_{kj}(\varepsilon)+s+t\}} \omega_{u_k}^n \leq C \int_{|u_k-u_j|\geq\varepsilon} \omega_{u_k}^n = C \int_{|u_k-u_j|\geq\varepsilon} e^{c_k} f_k \omega^n$$

$$\leq C \|e^{c_k} f_k\|_{L^p} (Vol(|u_k - u_j| \geq \varepsilon))^{p/(p-1)}.$$

The latter quantity converges to zero as $j, k \to \infty$, as u_k converge to u in L^1. But arguing analogously to the proof of Proposition 10.8 the capacity term on the left hand side cannot converge to zero when t and s are fixed, a contradiction. $\quad\square$

Note that the argument above yields existence, but not the uniqueness of the solutions.

Exercise 11.3 Recall the uniqueness argument in the smooth setting. What exactly *fails* if we try to apply it in the non-smooth setting as above?

Uniqueness and better regularity of the solutions can be nevertheless obtained, at least for *strictly positive* right hand side. This was proven in [KN3], where tools beyond the scope of these notes were used. We state the main results of [KN3] without proof:

Theorem 11.4 *Let u, v be ω-psh functions, such that $\omega_u^n = f\omega^n, \omega_v^n = g\omega^n$ and $sup_X u = sup_X v = 0$. Assume that $f \geq c > 0$, $\int_X g\omega^n > 0$ and both $f, g \in L^p(\omega^n)$ for some $p > 1$. Then for every $\alpha < \frac{1}{n+1}$ there is a constant C dependent on $n, \alpha, \|f\|_{L^p}, \|g\|_{L^p}$, such that*

$$sup_X|u - v| \leq C\|f - g\|_{L^p}^\alpha.$$

In particular the Dirichlet problem with L^p strictly positive right hand side admits a unique solution.

Theorem 11.5 *Let u be an ω-psh function, such that $\omega_u = f\omega^n$ for some $f \in L^p(\omega^n)$, $p > 1$. Assume that $f \geq c > 0$. Then u is Hölder continuous with any Hölder exponent α, such that*

$$\alpha < \frac{2(p-1)}{pn(n+1) + p - 1}.$$

It is unknown whether the strict positivity assumptions in the above theorems can be relaxed.

Acknowledgements The Author wishes to thank the referee for very careful reading and valuable suggestions. The Author was partially supported by NCN grant 2013/08/A/ST1/00312.

References

[AB95] L. Alessandrini, G. Bassanelli, Modifications of compact balanced manifolds. C. R. Acad. Sci. Paris **320**, 1517–1522 (1995)

[BT76] E. Bedford, B.A. Taylor, The Dirichlet problem for a complex Monge-Ampère equation. Invent. Math. **37**(1), 1–44 (1976)

[BT82] E. Bedford, B.A. Taylor, A new capacity for plurisubharmonic functions. Acta Math. **149**(1–2), 1–40 (1982)

[BL11] Z. Blocki, On the uniform estimate in the Calabi-Yau theorem, II. Sci. China Math. **54**(7), 1375–1377 (2011)

[CE83] U. Cegrell, Discontinuité de l'opérateur de Monge-Ampère complexe. C. R. Acad. Sci. Paris Sér. I Math. **296**(21), 869–871 (1983)

[Che87] P. Cherrier, Équations de Monge-Ampère sur les variétés hermitiennes compactes. (French) [Monge-Ampére equations on compact Hermitian manifolds]. Bull. Sci. Math. (2) **111**(4), 343–385 (1987)

[CLN69] S.S. Chern, H.I. Levine, L. Nirenberg, *Intrinsic Norms on a Complex Manifold*. Global Analysis (Papers in Honor of K. Kodaira) (Univ. Tokyo Press, Tokyo, 1969), pp. 119–139

[Chi] I. Chiose, On the invariance of the total Monge-Ampere volume of Hermitian metrics (2016). Preprint arXiv:1609.05945

[De81] P. Delanoë, Équations du type de Monge-Ampère sur les variétés riemanniennes compactes. II. (French) [Monge-Ampère equations on compact Riemannian manifolds. II]. J. Funct. Anal. **41**(3), 341–353 (1981)

[Dem93] J.P. Demailly, A numerical criterion for very ample line bundles. J. Differ. Geom. **37**(2), 323–374 (1993)

[Dem] J.P. Demailly, Complex Analytic and Differential geometry, self published e-book

[D16] S. Dinew, Pluripotential theory on compact Hermitian manifolds. Ann. Fac. Sci. Toulouse Math. (6) **25**(1), 91–139 (2016)

[DK12] S. Dinew, S. Kolodziej, Pluripotential estimates on compact Hermitian Manifolds, Advances in geometric analysis. Adv. Lect. Math. **21**, 69–86 (2012)

[DOT03] G. Dloussky, K. Oeljeklaus, M. Toma, Class VII_0 surfaces with b_2 curves. Tohoku Math. J. (2) **55**, 283–309 (2003)

[EGZ09] P. Eyssidieux, V. Guedj, A. Zeriahi, Singular Kähler-Einstein metrics. J. Am. Math. Soc. **22**, 607–639 (2009)

[FIUV09] M. Fernandez, S. Ivanov, L. Ugarte, R. Villacampa, Non Kaehler heterotic string compactifications with non-zero fluxes and constant dilation. Commun. Math. Phys. **288**, 677–697 (2009)

[FPS04] A. Fino, M. Parton, S. Salamon, Families of strong KT structures in six dimensions. Comment. Math. Helv. **79**, 317–340 (2004)

[FT] A. Fino, G. Tomassini, On Astheno-Kähler metrics. J. Lond. Math. Soc. **83**(2), 290–308 (2011)

[FLY12] Y. Fu, J. Li, S.T. Yau, Balanced metrics on non-Kähler Claabi-Yau threefolds. J. Diff. Geom. **90**, 81–130 (2012)

[Ga] P. Gauduchon, Le théorème de l'excentricitè nulle (French). C. R. Acad. Sci. Paris **285**, 387–390 (1977)

[GL10] B. Guan, Q. Li, Complex Monge-Ampère equations and totally real submanifolds. Adv. Math. **225**, 1185–1223 (2010)

[GZ05] V. Guedj, A. Zeriahi, Intrinsic capacities on compact Kähler manifolds. J. Geom. Anal. **15**(4), 607–639 (2005)

[Hö1] L. Hödrmander, *An Introduction to Complex Analysis in Several Variables* (Van Nostrand, Princeton, 1973)

[Ho48] H. Hopf, *Zur Topologie der komplexen Mannigfaltigkeiten*. Studies and Essays Presented to R. Courant on his 60th Birthday, January 8, 1948 (Interscience Publishers, Inc., New York, 1948), pp. 167–185

[Hö2] L. Hörmander, *Notions of Convexity. Progress in Mathematics*, vol. 127 (Birkhäuser Boston, Inc., Boston, 1994), viii+414 pp.

[In74] M. Inoue, On surfaces of Class VII_0. Invent. Math. **24**, 269–310 (1974)

[JY93] J. Jost, S.T. Yau, A non linear elliptic system for maps from Hermitian to Riemannian manifolds and rigidity theorems in Hermitian geoemtry. Acta Math. **170**, 221–254 (1993)

[Ka78] M. Kato, Compact complex manifolds containing "global" spherical shells. I, in *Proceedings of the International Symposium on Algebraic Geometry (Kyoto Univ., Kyoto, 1977)* (Kinokuniya Book Store, Tokyo, 1977), pp. 45–84

[Kod64] K. Kodaira, On the structure of compact complex analytic surfaces. I. Am. J. Math. **86**, 751–798 (1964)

[Kod66] K. Kodaira, On the structure of compact complex analytic surfaces. II. Am. J. Math. **88**, 682–721 (1966)

[Kod68a] K. Kodaira, On the structure of compact complex analytic surfaces. III. Am. J. Math. **90**, 55–83 (1968)

[Kod68b] K. Kodaira, On the structure of complex analytic surfaces. IV. Am. J. Math. **90**, 1048–1066 (1968)

[Kol98] S. Kolodziej, The complex Monge-Ampère equation. Acta Math. **180**(1), 69–117 (1998)

[Kol03] S. Kolodziej, The Monge-Ampère equation on compact Kähler manifolds. Indiana Univ. Math. J. **52**(3), 667–686 (2003)

[Kol05] S. Kolodziej, The complex Monge-Ampère equation and pluripotential theory. Mem. Am. Math. Soc. **178**(840), x+64 (2005)

[KN1] S. Kolodziej, N.C. Nguyen, Weak solutions to the complex Monge-Ampère equation on compact Hermitian manifolds. Contemp. Math. **644**, 141–158 (2015)

[KN2] S. Kolodziej, N.C. Nguyen, Stability and regularity of solutions of the Monge-Ampère equation on Hermitian manifolds. Adv. Math. **346**, 264–304 (2019)

[KN3] S. Kolodziej, N.C. Nguyen, Hölder continuous solutions of the Monge-Ampère equation on compact Hermitian manifolds. Ann. Inst. Fourier (2018). Preprint arXiv:1708.06516

[KT08] S. Kolodziej, G. Tian, A uniform L^∞ estimate for complex Monge-Ampère equations. Math. Ann. **342**(4), 773–787 (2008)

[LT95] M. Lübke, A. Teleman, *The Kobayashi-Hitchin Correspondence* (World Scientific Publishing Co., Singapore, 1995)

[Mi09] K. Mitsuo, Astheno-Kähler structures on Calabi-Eckman manifolds. Colloq. Math. **155**, 33–39 (2009)

[Na84] I. Nakamura, On surfaces of class VII_0 with curves. Invent. Math. **78**(3), 393–443 (1984)

[Pop13] D. Popovici, Deformation limits of projective manifolds: Hodge numbers and strongly Gauduchon metrics. Invent. Math. **194**(3), 515–534 (2013)

[Pop14] D. Popovici, Deformation openness and closedness of various classes of compact complex manifolds; examples. Ann. Sc. Norm. Super. Pisa Cl. Sci. (5) **13**(2), 255–305 (2014)

[Siu] Y.T. Siu, Analyticity of sets associated to Lelong numbers and the extension of closed positive currents. Invent. Math. **27**, 53–156 (1974)

[Siu09] Y.T. Siu, Dynamic multiplier ideal sheaves and the construction of rational curves in Fano manifolds (2009). Preprint arXiv:0902.2809

[T10] A. Teleman, Instantons and curves on class VII surfaces. Ann. Math. (2) **172**(3), 1749–1804 (2010)

[T89] G. Tian, On Kähler-Einstein metrics on certain Kähler manifolds with $C_1(M) > 0$. Invent. Math. **89**(2), 225–246 (1987)

[TW10a] V. Tosatti, B. Weinkove, Estimates for the complex Monge-Ampère equation on Hermitian and balanced manifolds. Asian J. Math. **14**(1), 19–40 (2010)

[TW10b] V. Tosatti, B. Weinkove, The complex Monge-Ampère equation on compact Hermitian manifolds. J. Am. Math. Soc. **23**(4), 1187–1195 (2010)

[TW12] V. Tosatti, B. Weinkove, Plurisubharmonic functions and nef classes on complex manifolds. Proc. Am. Math. Soc. **140**(11), 4003–4010 (2012)

[TWWY14] V. Tosatti, Y. Wang, B. Weinkove, X. Yang, $C^{2,\alpha}$ estimates for nonlinear elliptic equations in complex and almost complex geometry. Calc. Var. PDE **54**, 431–453 (2015)

[TW14] V. Tosatti, B. Weinkove, The Chern-Ricci flow on complex surfaces. Compos. Math. **149**(12), 2101–2138 (2013)

[Y78] S.T. Yau, On the Ricci curvature of a compact Kähler manifold and the complex Monge-Ampère equation. I. Commun. Pure Appl. Math. **31**(3), 339–411 (1978)

Chapter 2
Calabi–Yau Manifolds with Torsion and Geometric Flows

Sébastien Picard

Abstract The main theme of these lectures is the study of Hermitian metrics in non-Kähler complex geometry. We will specialize to a certain class of Hermitian metrics which generalize Kähler Ricci-flat metrics to the non-Kähler setting. These non-Kähler Calabi–Yau manifolds have their origins in theoretical physics, where they were introduced in the works of C. Hull and A. Strominger. We will introduce tools from geometric analysis, namely geometric flows, to study this non-Kähler Calabi–Yau geometry. More specifically, we will discuss the Anomaly flow, which is a version of the Ricci flow customized to this particular geometric setting. This flow was introduced in joint works with Duong Phong and Xiangwen Zhang. Section 2.1 contains a review of Hermitian metrics, connections, and curvature. Section 2.2 is dedicated to the geometry of Calabi–Yau manifolds equipped with a conformally balanced metric. Section 2.3 introduces the Anomaly flow in the simplest case of zero slope, where the flow can be understood as a deformation path connecting non-Kähler to Kähler geometry. Section 2.4 concerns the Anomaly flow with α' corrections, which is motivated from theoretical physics and canonical metrics in non-Kähler geometry.

2.1 Review of Hermitian Geometry

We start by reviewing non-Kähler metrics in complex geometry. In particular, we study unitary connections, torsion, and curvature associated to a Hermitian metric ω.

2.1.1 Hermitian Metrics

Let X be a complex manifold of dimension n. The manifold X is covered by holomorphic charts U_μ equipped with local holomorphic coordinates $(z^1, \ldots, z^n)$

S. Picard (✉)

Department of Mathematics, Harvard University, Cambridge, MA, USA

e-mail: spicard@math.harvard.edu

© Springer Nature Switzerland AG 2019

D. Angella et al. (eds.), *Complex Non-Kähler Geometry*, Lecture Notes in Mathematics 2246, https://doi.org/10.1007/978-3-030-25883-2_2

such that $X = \bigcup_\mu U_\mu$. The complexified tangent bundle of X will be denoted TX, which splits

$$TX = T^{1,0}X \oplus T^{0,1}X.$$

Using local coordinates, a tangent vector in $T^{1,0}X$ is a combination of

$$\left\{ \frac{\partial}{\partial z^1}, \cdots, \frac{\partial}{\partial z^n} \right\}$$

and a tangent vector in $T^{0,1}X$ is a combination of

$$\left\{ \frac{\partial}{\partial \bar{z}^1}, \cdots, \frac{\partial}{\partial \bar{z}^n} \right\}.$$

We will use the notation

$$\partial_k = \frac{\partial}{\partial z^k}, \quad \partial_{\bar{k}} = \frac{\partial}{\partial \bar{z}^k}.$$

Next, we will use $\Omega^{p,q}(X)$ to denote differential forms of (p, q) type. This means that in local coordinates, $\Omega^{p,q}(X)$ is generated by

$$dz^{i_1} \wedge \cdots \wedge dz^{i_p} \wedge d\bar{z}^{j_1} \wedge \cdots d\bar{z}^{j_q}.$$

We will use the following convention for the components $\Psi_{\bar{j}_1 \cdots \bar{j}_q i_1 \cdots i_p}$ of a differential form $\Psi \in \Omega^{p,q}(X)$

$$\Psi = \frac{1}{p!q!} \sum \Psi_{\bar{j}_1 \cdots \bar{j}_q i_1 \cdots i_p} dz^{i_p} \wedge \cdots dz^{i_1} \wedge d\bar{z}^{j_q} \wedge \cdots \wedge d\bar{z}^{j_1}. \qquad (2.1)$$

The exterior derivative d decomposes into

$$d = \partial + \bar{\partial},$$

where

$$\partial : \Omega^{p,q}(X) \to \Omega^{p+1,q}(X), \quad \bar{\partial} : \Omega^{p,q}(X) \to \Omega^{p,q+1}(X),$$

are the Dolbeault operators. A Hermitian metric g on X is a smooth section $(T^{1,0}X)^* \otimes (T^{0,1}X)^*$ such that in local coordinates

$$g = g_{\bar{k}j} \, dz^j \otimes d\bar{z}^k, \qquad (2.2)$$

where $g_{\bar{k}j}$ is a positive-definite Hermitian matrix at each point.

$$g_{\bar{k}j} > 0, \qquad \overline{g_{\bar{k}j}} = g_{\bar{j}k}.$$

In (2.2) we use the summation convention, which will be used throughout these notes, where we omit the summation sign for matching upper and lower indices. We use the notation $g^{j\bar{k}} = (g_{\bar{k}j})^{-1}$ for the inverse, meaning that

$$g^{i\bar{k}} g_{\bar{k}j} = \delta^i{}_j.$$

We can identify the metric g with a Hermitian form $\omega \in \Omega^{1,1}(X, \mathbf{R})$ via

$$\omega = i g_{\bar{k}j} \, dz^j \wedge d\bar{z}^k.$$

The metric g induces a metric on differential forms $\Omega^{p,q}(X)$, and we define the Hodge star operator $\star : \Omega^{p,q}(X) \to \Omega^{n-q,n-p}(X)$ by requiring

$$\alpha \wedge \star\bar{\beta} = g(\alpha, \beta) \, \frac{\omega^n}{n!}.$$

for all $\alpha, \beta \in \Omega^{p,q}(X)$.

A basic fact which will be often used in these notes is

Proposition 2.1 *Let X be a compact complex manifold with Hermitian metric g and $\partial X = \emptyset$. Let $f \in C^\infty(X, \mathbf{R})$. If*

$$g^{j\bar{k}} \partial_j \partial_{\bar{k}} f \geq 0,$$

everywhere on X, then f is a constant function.

Proof Let c denote the maximum value attained by f on X. The set

$$S = f^{-1}(c)$$

is closed. We claim that S is also open. Indeed, let $p \in S$. Let B be a ball in a local chart such that f attains a maximum in the center of B and satisfies $g^{j\bar{k}} \partial_j \partial_{\bar{k}} f \geq 0$ in B. By the Hopf strong maximum principle (e.g. Theorem 2.7 in [HL11]), we must have $f \equiv c$ in B. This shows that S is open, and hence $S = X$.   □

A Hermitian metric ω is Kähler if

$$d\omega = 0.$$

Kähler manifolds are of fundamental importance as they lie at the crossroads of both Riemannian geometry and algebraic geometry. In these notes, our goal is to generalize the Kähler condition while still retaining enough structure to develop an interesting theory.

There are many ways to generalize the Kähler condition. There is the notion of a pluriclosed metric, which satisfies

$$i\partial\bar\partial\omega = 0.$$

There are also astheno-Kähler metrics [JY93], which satisfy

$$i\partial\bar\partial\omega^{n-2} = 0.$$

It was shown by Gauduchon [GA77] that every compact complex manifold admits a Gauduchon metric, which satisfies

$$i\partial\bar\partial\omega^{n-1} = 0.$$

More generally, Fu-Wang-Wu [FWW13] introduced the notion of k-Gauduchon, for $1 \le k \le n-1$, which is defined by the condition

$$i\partial\bar\partial\omega^k \wedge \omega^{n-k-1} = 0.$$

All these notions generalize Kähler metrics in different ways. In these notes, we will mostly focus on another notion: we say a Hermitian metric ω is balanced if

$$d\omega^{n-1} = 0. \tag{2.3}$$

The special properties of balanced metrics were noticed early in the study of Hermitian geometry, arising for examples in articles of Gauduchon [GA75]. Balanced metrics were studied systematically by Michelsohn [MI82], and these metrics were rediscovered in theoretical physics in the development of heterotic string theory [HU186, ST86, LY05]. A main theme in Michelsohn's work is that balanced metrics are in some sense dual to the Kähler condition. For example, Kähler metrics are inherited by the ambient space (via pullback) while balanced metrics can be pushed forward [MI82].

Given a Hermitian metric ω, its torsion is defined by

$$T = i\partial\omega, \quad \bar T = -i\bar\partial\omega.$$

We see that a metric is Kähler if and only if its torsion vanishes. The components of the torsion are given by

$$T = \frac{1}{2}T_{\bar k jm}dz^m \wedge dz^j \wedge d\bar z^k, \quad \bar T = \frac{1}{2}\bar T_{k\bar j\bar m}d\bar z^m \wedge d\bar z^j \wedge dz^k.$$

Explicitly, we have

$$T_{\bar{k}jm} = \partial_j g_{\bar{k}m} - \partial_m g_{\bar{k}j}, \quad \bar{T}_{kj\bar{m}} = \partial_{\bar{j}} g_{\bar{m}k} - \partial_{\bar{m}} g_{\bar{j}k}. \tag{2.4}$$

We can raise indices using the metric, and we will write $T^k{}_{ij} = g^{k\bar{\ell}} T_{\bar{\ell}ij}$. We can also contract indices, and we will use the notation

$$T_j = g^{i\bar{k}} T_{\bar{k}ij}.$$

We will also use the 1-form τ defined by

$$\tau = T_k dz^k.$$

Taking norms, we have

$$|T|^2 = g^{m\bar{n}} g^{k\bar{\ell}} g^{j\bar{i}} T_{\bar{i}km} \bar{T}_{j\bar{\ell}\bar{n}}, \quad |\tau|^2 = g^{k\bar{\ell}} T_k \bar{T}_{\bar{\ell}}.$$

2.1.2 Connections

Let $E \to X$ be a complex vector bundle of rank r. The bundle E can be specified by an open cover $X = \bigcup_\mu U_\mu$ together with transition matrices $t_{\mu\nu} : U_\mu \cap U_\nu \to GL(r, \mathbf{C})$ satisfying

$$t_{\mu\mu}{}^\alpha{}_\beta = \delta^\alpha{}_\beta,$$

and

$$t_{\mu\nu}{}^\alpha{}_\beta t_{\nu\rho}{}^\beta{}_\gamma = t_{\mu\rho}{}^\alpha{}_\gamma \quad \text{on } U_\mu \cap U_\nu \cap U_\rho.$$

If all transition functions $t_{\mu\nu}$ are holomorphic, then E is a holomorphic bundle.

A section $s \in \Gamma(X, E)$ is given by local data $(U_\mu, s_\mu{}^\alpha)$, where

$$s = (s_\mu{}^1(z_\mu), \cdots, s_\mu{}^r(z_\mu)) \quad \text{on } U_\mu,$$

and $s_\mu : U_\mu \to \mathbf{C}^r$ is a smooth map which transforms via

$$(s_\mu)^\alpha = t_{\mu\nu}{}^\alpha{}_\beta s_\nu{}^\beta$$

on $U_\mu \cap U_\nu$. On a holomorphic bundle, we say s is holomorphic if the s_μ are holomorphic.

Let us illustrate this notation by considering the example of the holomorphic tangent bundle $T^{1,0}X$. Here the transition functions are

$$t_{\mu\nu}{}^i{}_k = \frac{\partial z_\mu{}^i}{\partial z_\nu{}^k},$$

which are defined on the intersection of coordinate patches $(U_\mu, z_\mu{}^i)$ and $(U_\nu, z_\nu{}^i)$. Sections of $T^{1,0}X$ are vector fields $V = V^i \partial_i \in \Gamma(X, T^{1,0}X)$, and on $U_\mu \cap U_\nu$,

$$V_\mu{}^k = \frac{\partial z_\mu{}^k}{\partial z_\nu{}^\ell} V_\nu{}^\ell.$$

Next, we recall that from a bundle E, we can induce bundles such as E^*, $\bar{E}$, and $\det E$. If the bundle E has transition matrices $t_{\mu\nu}$, then sections $\phi \in \Gamma(X, E^*)$ are given by data $(U_\mu, \phi_{\mu\alpha})$ which transform according to

$$(\phi_\mu)_\alpha = t_{\nu\mu}{}^\beta{}_\alpha \phi_{\nu\beta}.$$

Similarly, sections $s \in \Gamma(X, \bar{E})$ transform by

$$s_\mu{}^{\bar{\alpha}} = \overline{t_{\mu\nu}{}^\alpha{}_\beta}\, s_\nu{}^{\bar{\beta}},$$

and sections $\psi \in \Gamma(X, \det E)$ are given by local functions $\psi_\mu : U_\mu \to \mathbf{C}$ which transform by

$$\psi_\mu = (\det t_{\mu\nu})\, \psi_\nu.$$

To differentiate sections of a vector bundle, we use a connection ∇. Connections can be expressed locally as $\nabla = d + A_\mu$, where A_μ are local matrix-valued 1-forms $(A_\mu)_i{}^\alpha{}_\beta$ defined on U_μ. The local matrices $(A_\mu)_i$ satisfy the transformation law

$$(A_\mu)_i = t_{\mu\nu} (A_\nu)_i\, t_{\mu\nu}{}^{-1} - (\partial_i t_{\mu\nu}) t_{\mu\nu}{}^{-1}. \tag{2.5}$$

Here we omitted the indices for matrix multiplication. This transformation law is designed such that for any section $s \in \Gamma(X, E)$, its derivative $\nabla_i s$ is again a section. Explicitly, derivatives of s are given locally by

$$\nabla_i s^\alpha = \partial_i s^\alpha + A_i{}^\alpha{}_\beta s^\beta, \quad \nabla_{\bar{i}} s^\alpha = \partial_{\bar{i}} s^\alpha + A_{\bar{i}}{}^\alpha{}_\beta s^\beta.$$

with the notation

$$\nabla_i = \nabla_{\frac{\partial}{\partial z^i}}, \quad \nabla_{\bar{i}} = \nabla_{\frac{\partial}{\partial \bar{z}^i}}.$$

Given a connection on E, we can induce connections on E^*, $\bar{E}$, $\det E$, etc., by imposing the product rule. For example, the product rule $\partial_k(s_\alpha \phi^\alpha) = \nabla_k s_\alpha \phi^\alpha + s_\alpha \nabla_k \phi^\alpha$ leads to the definition

$$\nabla_k \phi_\alpha = \partial_k \phi_\alpha - \phi_\beta A_k{}^\beta{}_\alpha, \quad \nabla_{\bar{k}} \phi_\alpha = \partial_{\bar{k}} \phi_\alpha - \phi_\beta A_{\bar{k}}{}^\beta{}_\alpha$$

for sections $\phi \in \Gamma(X, E^*)$. Similarly, for a section $u \in \Gamma(X, \bar{E})$, the induced connection is defined by

$$\nabla_k u^{\bar{\alpha}} = \partial_k u^{\bar{\alpha}} + \overline{A_{\bar{k}}{}^\alpha{}_\beta} u^{\bar{\beta}}, \quad \nabla_{\bar{k}} u^{\bar{\alpha}} = \partial_{\bar{k}} u^{\bar{\alpha}} + \overline{A_k{}^\alpha{}_\beta} u^{\bar{\beta}},$$

and for a section $\psi \in \Gamma(X, \det E^*)$, the induced connection is

$$\nabla_i \psi = \partial_i \psi - A_i{}^\alpha{}_\alpha \psi, \quad \nabla_{\bar{i}} \psi = \partial_{\bar{i}} \psi - A_{\bar{i}}{}^\alpha{}_\alpha \psi. \tag{2.6}$$

As a final example, the induced connection on $\Gamma(X, E^* \otimes \bar{E}^*)$ is defined by

$$\nabla_k h_{\bar{\alpha}\beta} = \partial_k h_{\bar{\alpha}\beta} - \overline{A_{\bar{k}}{}^\gamma{}_\alpha} h_{\bar{\gamma}\beta} - A_k{}^\gamma{}_\beta h_{\bar{\alpha}\gamma}.$$

We now focus our attention on the holomorphic tangent bundle $T^{1,0}X$. Given a Hermitian metric $\omega = i g_{\bar{k}j} dz^j \wedge d\bar{z}^k$ on X, we say a connection ∇ on $T^{1,0}X$ is unitary with respect to ω if

$$\nabla_i g_{\bar{k}j} = 0.$$

On a Hermitian manifold (X, ω), the Chern connection is the unique unitary connection on $T^{1,0}X$ such that $A_{\bar{k}} = 0$. The Chern connection acts on sections $V \in \Gamma(X, T^{1,0}X)$ by

$$\nabla_k(V^i \partial_i) = (\nabla_k V^i)\partial_i, \quad \nabla_{\bar{k}}(V^i \partial_i) = (\nabla_{\bar{k}} V^i)\partial_i,$$

where

$$\nabla_k V^i = \partial_k V^i + \Gamma^i_{k\ell} V^\ell, \quad \nabla_{\bar{k}} V^i = \partial_{\bar{k}} V^i,$$

and

$$\Gamma^i_{k\ell} = g^{i\bar{p}} \partial_k g_{\bar{p}\ell}. \tag{2.7}$$

Due to its simplicity, the Chern connection is best suited for most computations. However, in non-Kähler geometry, there are other interesting connections on $T^{1,0}X$ to consider too. We start with the Levi-Civita connection, which acts on $V \in \Gamma(X, TX)$ by

$$\nabla^g_k(V^i \partial_i + V^{\bar{i}} \partial_{\bar{i}}) = (\nabla^g_k V^i)\partial_i + (\nabla^g_k V^{\bar{i}})\partial_{\bar{i}},$$

where

$$\nabla_k^g V^i = \partial_k V^i + \Gamma_{k\ell}^i V^\ell - \frac{T^i{}_{k\ell}}{2} V^\ell - \frac{g^{i\bar{j}}}{2} \bar{T}_{k\bar{j}\bar{\ell}} V^{\bar{\ell}},$$

$$\nabla_{\bar{k}}^g V^i = \partial_{\bar{k}} V^i + \frac{g^{i\bar{m}}}{2} \bar{T}_{\bar{\ell}\bar{k}\bar{m}} V^\ell,$$

and

$$\nabla_k^g V^{\bar{i}} = \overline{\nabla_{\bar{k}}^g V^i}, \quad \nabla_{\bar{k}}^g V^{\bar{i}} = \overline{\nabla_k^g V^i}.$$

To be clear, we note that here, and throughout these notes, $\Gamma_{k\ell}^i$ is reserved for the expression (2.7), which is not the Christoffel symbol of the Levi-Civita connection.

This well-known connection from Riemannian geometry preserves the metric $\nabla^g g = 0$ and has zero torsion tensor $\nabla_X^g Y - \nabla_Y^g X - [X, Y]$. For Kähler metrics, $T = 0$ and we see that the Levi-Civita connection coincides with the Chern connection.

However, for general Hermitian metrics, the tensor $T_{\bar{k}ij}$ is nonzero and the Levi-Civita connection does not preserve the decomposition $TX = T^{1,0}X \oplus T^{0,1}X$. In particular, it does not define a connection on the holomorphic bundle $T^{1,0}X$.

We can add a correction to ∇^g to obtain a new connection which does preserve $T^{1,0}X$. We define

$$\nabla^+ = \nabla^g + \frac{1}{2} g^{-1} H, \quad H = i(\bar{\partial} - \partial)\omega.$$

The new connection acts on $V \in \Gamma(X, T^{1,0}X)$ by $\nabla_k^+(V^i \partial_i) = (\nabla_k^+ V^i)\partial_i$ with components

$$\nabla_k^+ V^i = \partial_k V^i + (\Gamma_{k\ell}^i - T^i{}_{k\ell})V^\ell, \tag{2.8}$$

$$\nabla_{\bar{k}}^+ V^i = \partial_{\bar{k}} V^i + g^{i\bar{m}} \bar{T}_{\bar{\ell}\bar{k}\bar{m}} V^\ell.$$

We will call this connection the Strominger–Bismut connection [BI89, ST86]. It evidently preserves $T^{1,0}X$, and a straightforward computation shows that

$$\nabla^+ g_{\bar{k}j} = 0,$$

hence ∇^+ is a unitary connection. Furthermore, $\nabla^+ = \nabla^g + \frac{1}{2} g^{-1} H$ has the property that its torsion 3-form

$$\mathscr{T}(X, Y, Z) = g(\nabla_X^+ Y - \nabla_Y^+ X - [X, Y], Z)$$

is given by the skew-symmetric 3-form H.

Using the Chern connection ∇ and the Strominger–Bismut connection ∇^+, we can define a line of unitary connections which preserve the complex structure.

$$\nabla^{(\kappa)} = (1 - \kappa)\nabla + \kappa\nabla^+,$$

where $\kappa \in \mathbf{R}$ is a parameter. This family of connections is known as the Gauduchon line [GA97]. We note that this line collapses to a point when ω is Kähler.

There are other connections which play a role in theoretical physics which do not preserve the complex structure. One such example is the Hull connection [HU286, LE11, DS14], denoted by $\nabla^- = \nabla^g - \frac{1}{2}g^{-1}H$. Explicitly, this connection acts on $V \in \Gamma(X, TX)$ by

$$\nabla^-_k V^i = \partial_k V^i + \Gamma^i_{k\ell}V^\ell - g^{i\bar{j}}\bar{T}_{k\bar{j}\bar{\ell}}V^{\bar{\ell}}, \tag{2.9}$$

$$\nabla^-_{\bar{k}} V^i = \partial_{\bar{k}} V^i.$$

Although ∇^- does not preserve $T^{1,0}X$, a direct computation shows that $\nabla^- g = 0$.

Most computations in these notes will be done using the Chern connection, and from now on we reserve ∇ to denote the Chern connection. We will use superscripts e.g. ∇^+, to denote other connections.

Next, we review integration and adjoint operators in Hermitian geometry. The first identity is the divergence theorem for Hermitian metrics.

Lemma 2.1 *Let (X, ω) be a closed Hermitian manifold. The divergence theorem for the Chern connection ∇ is given by*

$$\int_X \nabla_i V^i \, \omega^n = \int_X T_i V^i \, \omega^n, \tag{2.10}$$

for any $V \in \Gamma(X, T^{1,0}X)$.

We see that the torsion components T_i play a role when integrating by parts. The proof is similar to the Kähler case, and is omitted.

Next, we recall the L^2 pairing of differential forms, given by $\langle \phi, \psi \rangle = \int_X g(\phi, \psi) \omega^n$, where $g(\phi, \psi)$ is the induced metric on $\phi, \psi \in \Omega^{p,q}(X)$. For example, for $\eta, \beta \in \Omega^{1,0}(X)$, we define

$$\langle \eta, \beta \rangle = \int_X g^{j\bar{k}}\eta_j\overline{\beta_k} \, \omega^n,$$

and for $\alpha, \chi \in \Omega^{1,1}(X)$,

$$\langle \alpha, \chi \rangle = \int_X g^{j\bar{k}}g^{\ell\bar{m}}\alpha_{\bar{k}\ell}\overline{\chi_{jm}} \, \omega^n.$$

The adjoint operators $\partial^\dagger : \Omega^{p,q}(X) \to \Omega^{p-1,q}(X)$ and $\bar{\partial}^\dagger : \Omega^{p,q}(X) \to \Omega^{p,q-1}(X)$ are defined by the property

$$\langle \partial\phi, \psi \rangle = \langle \phi, \partial^\dagger\psi \rangle, \quad \langle \bar{\partial}\phi, \psi \rangle = \langle \phi, \bar{\partial}^\dagger\psi \rangle.$$

We will also write $d^\dagger = \partial^\dagger + \bar{\partial}^\dagger$. We will need an explicit expressions for these adjoint operators in the following special case.

Lemma 2.2 *Let* (X, ω) *be a Hermitian manifold. The adjoint operators act on* $\alpha \in \Omega^{1,1}(X)$ *by*

$$(\partial^\dagger\alpha)_{\bar{k}} = -g^{p\bar{q}}\nabla_{\bar{q}}\alpha_{\bar{k}p} + g^{p\bar{q}}\bar{T}_{\bar{q}}\alpha_{\bar{k}p}. \tag{2.11}$$

$$(\bar{\partial}^\dagger\alpha)_k = g^{p\bar{q}}\nabla_p\alpha_{\bar{q}k} - g^{p\bar{q}}T_p\alpha_{\bar{q}k}. \tag{2.12}$$

Proof Let $\alpha \in \Omega^{1,1}(X)$ and $\beta \in \Omega^{0,1}(X)$. The components of $\partial\beta$ are

$$(\partial\beta)_{\bar{k}j} = \nabla_j\beta_{\bar{k}}.$$

The inner product $\langle \alpha, \partial\beta \rangle = \langle \partial^\dagger\alpha, \beta \rangle$ expands to

$$\int_X g^{j\bar{k}}g^{p\bar{q}}\alpha_{\bar{k}p}\overline{(\nabla_q\beta_{\bar{j}})}\,\omega^n = \int_X g^{j\bar{k}}(\partial^\dagger\alpha)_{\bar{k}}\overline{\beta_{\bar{j}}}\,\omega^n.$$

Applying the divergence theorem (2.10) to the left-hand side, we obtain (2.11). A similar computation leads to (2.12).   □

As a corollary, if we apply these identities to $\alpha = \omega = ig_{\bar{k}j}dz^j \wedge d\bar{z}^k$, we obtain

$$(\partial^\dagger\omega)_{\bar{k}} = i\bar{T}_{\bar{k}}, \quad (\bar{\partial}^\dagger\omega)_k = -iT_k. \tag{2.13}$$

and

$$d^\dagger\omega = i(\bar{\tau} - \tau).$$

2.1.3 Curvature

Let $E \to X$ be a complex vector bundle. The curvature of a connection $\nabla = d + A$ on E is a 2-form valued in the endomorphisms of E given by

$$F = dA + A \wedge A,$$

with components

$$F = \frac{1}{2} F_{kj}{}^{\alpha}{}_{\beta} dz^j \wedge dz^k + \frac{1}{2} F_{\bar{k}j}{}^{\alpha}{}_{\beta} d\bar{z}^j \wedge dz^k + F_{\bar{k}j}{}^{\alpha}{}_{\beta} \, dz^j \wedge d\bar{z}^k.$$

The curvature form of the Chern connection of a Hermitian metric ω will be denoted Rm. In this case, one can verify that the curvature form Rm is an endomorphism-valued $(1, 1)$ form

$$Rm = R_{\bar{k}j}{}^{p}{}_{q} dz^j \wedge d\bar{z}^k,$$

with components given by

$$R_{\bar{k}j}{}^{p}{}_{q} = -\partial_{\bar{k}} \Gamma_{jq}^{p} = -\partial_{\bar{k}} (g^{p\bar{s}} \partial_j g_{\bar{s}q}).$$

We may write this as

$$Rm = \bar{\partial}(g^{-1} \partial g), \tag{2.14}$$

which holds in a holomorphic frame on $T^{1,0}X$. We note that in general, when using unitary connections other than the Chern connection on $T^{1,0}X$, the curvature will have $(2, 0)$ and $(0, 2)$ components as well.

We can raise and lower indices of the curvature tensor using the metric $g_{\bar{k}j}$.

$$R_{\bar{k}j\bar{m}\ell} = g_{\bar{m}p} R_{\bar{k}j}{}^{p}{}_{\ell} = -\partial_{\bar{k}} \partial_j g_{\bar{m}\ell} + g^{s\bar{r}} \partial_{\bar{k}} g_{\bar{m}s} \partial_j g_{\bar{r}\ell}. \tag{2.15}$$

Lemma 2.3 *The curvature of the Chern connection on (X, ω) satisfies the following Bianchi identities*

$$R_{\bar{k}j\bar{m}\ell} = R_{\bar{m}j\bar{k}\ell} + \nabla_j \bar{T}_{\ell\bar{m}\bar{k}},$$

$$R_{\bar{k}j\bar{m}\ell} = R_{\bar{k}\ell\bar{m}j} + \nabla_{\bar{k}} T_{\bar{m}\ell j}.$$

Proof For example, we compute using the definition (2.15) and obtain

$$R_{\bar{k}j\bar{m}\ell} - R_{\bar{m}j\bar{k}\ell} = -\partial_{\bar{k}} \partial_j g_{\bar{m}\ell} + g^{s\bar{r}} \partial_{\bar{k}} g_{\bar{m}s} \partial_j g_{\bar{r}\ell} + \partial_{\bar{m}} \partial_j g_{\bar{k}\ell} - g^{s\bar{r}} \partial_{\bar{m}} g_{\bar{k}s} \partial_j g_{\bar{r}\ell}$$

$$= \partial_j (\partial_{\bar{m}} g_{\bar{k}\ell} - \partial_{\bar{k}} g_{\bar{m}\ell}) - g^{s\bar{r}} \partial_j g_{\bar{r}\ell} (\partial_{\bar{m}} g_{\bar{k}s} - \partial_{\bar{k}} g_{\bar{m}s})$$

$$= \partial_j \bar{T}_{\ell\bar{m}\bar{k}} - \Gamma_{j\ell}^{p} \bar{T}_{p\bar{m}\bar{k}}$$

$$= \nabla_j \bar{T}_{\ell\bar{m}\bar{k}}.$$

The other identity is derived in a similar way. $\qquad\square$

There are four notions of Ricci curvature for the Chern connection in Hermitian geometry, and we will use the notation

$$R_{\bar{k}j} = R_{\bar{k}j}{}^p{}_p, \quad \tilde{R}_{\bar{k}j} = R^p{}_{p\bar{k}j}, \quad R'_{\bar{k}j} = R_{\bar{k}p}{}^p{}_j, \quad R''_{\bar{k}j} = R^p{}_{j\bar{k}p}.$$

From the Bianchi identity, we see that these notions of Ricci curvature are all different. We will call $R_{\bar{k}j}$ the Chern–Ricci curvature, and it is also given by

$$R_{\bar{k}j} = -\partial_{\bar{k}}\partial_j \log \det g_{\bar{p}q}.$$

The Chern–Ricci form represents the first Chern class $[\frac{i}{2\pi}\mathrm{Ric}_\omega] = c_1(X)$ and is given by

$$\mathrm{Ric}_\omega = -\partial\bar{\partial} \log \det g_{\bar{p}q} = R_{\bar{k}j} dz^j \wedge d\bar{z}^k.$$

There are two notions of scalar curvature, denoted by

$$R = g^{\ell\bar{m}} g^{j\bar{k}} R_{\bar{k}j\bar{m}\ell} = R^p{}_p{}^j{}_j, \quad R' = g^{j\bar{m}} g^{\ell\bar{k}} R_{\bar{k}j\bar{m}\ell} = R^p{}_j{}^j{}_p.$$

2.1.4   *U(1) Principal Bundles*

2.1.4.1   Definitions

We denote the group of complex numbers with length equal to 1 by $U(1)$. A $U(1)$ principal bundle can be specified by an open cover $X = \bigcup_\mu U_\mu$ together with smooth maps

$$g_{\mu\nu} : U_\mu \cap U_\nu \to U(1),$$

such that

$$g_{\mu\mu} = 1, \quad g_{\mu\nu}^{-1} = g_{\nu\mu},$$

and

$$g_{\mu\nu} g_{\nu\rho} = g_{\mu\rho},$$

on an non-empty overlap $U_\mu \cap U_\nu \cap U_\rho$. In this section, we review how a connection on a line bundle defines a connection on a $U(1)$ principal bundle.

Let $L \to X$ be a smooth complex line bundle with data $(U_\mu \cap U_\nu, t_{\mu\nu})$, equipped with a connection $\nabla_A = d + A$ whose curvature is $F_A = dA$. We also consider the line bundle $L' \to X$ given by the data

$$(U_\mu \cap U_\nu, e^{i\tau_{\mu\nu}}), \quad \frac{t_{\mu\nu}}{|t_{\mu\nu}|} = e^{i\tau_{\mu\nu}}.$$

To compactify the fibers, we equip L with a metric h, which is locally given by (U_μ, h_μ) where h_μ are positive functions which transforms as

$$h_\mu = \frac{1}{|t_{\mu\nu}|^2} h_\nu.$$

The metric h provides an isomorphism of the line bundles L and L', where the connection $d + A$ on L becomes the connection $d + A'$ given by

$$A' = A - \frac{1}{2} d \log h,$$

on L'. It can be checked that this expression satisfies the transformation law for a connection (2.5), which in this case becomes

$$A'_\mu = A'_\nu - i d\tau_{\mu\nu}. \tag{2.16}$$

Thus we have induced a connection $d + A'$ on L' with curvature

$$dA' = F_A. \tag{2.17}$$

Let $\pi : P \to X$ be the $U(1)$ bundle determined by the data $(U_\mu \cap U_\nu, e^{i\tau_{\mu\nu}})$. Locally, points in P are given by $(z_\mu, e^{i\psi_\mu})$ with projection $\pi(z_\mu, e^{i\psi_\mu}) = z_\mu$, where the coordinates $e^{i\psi_\mu}$ on the fiber transform via

$$e^{i\psi_\mu} = e^{i\tau_{\mu\nu}} e^{i\psi_\nu}.$$

In other words, on $U_\mu \cap U_\nu$, there holds

$$\psi_\mu = \psi_\beta + \tau_{\mu\nu} + 2\pi k, \tag{2.18}$$

for an integer k. Combining this with the transformation law for the connection (2.16), it follows that

$$\theta = d\psi_\mu - i A'_\mu \tag{2.19}$$

is a global 1-form on the total space of the bundle $\pi : P \to X$. We call θ the connection 1-form of the $U(1)$ bundle P. Furthermore, by (2.17), its exterior derivative is

$$d\theta = -i F_A.$$

The connection 1-form θ splits the tangent space TP of P into vertical and horizontal directions. For the vertical direction, we note that by (2.18), the expression $\frac{\partial}{\partial \psi}$ transforms as a global vector field on $\pi : P \to X$. We define the vertical subbundle V by

$$V = \ker \pi_* = \operatorname{span} \left\{ \frac{\partial}{\partial \psi} \right\}.$$

The horizontal space is given by $H = \ker \theta$. The tangent bundle of P splits as

$$TP = V \oplus H,$$

and $\pi_*|_H : H \to TX$ is isomorphism.

2.1.4.2 Non-Kähler Manifolds Constructed from Principal Bundles

Connections on $U(1)$ principal bundles can be used to construct non-Kähler complex manifolds. This idea was first used by Calabi–Eckmann [CE53], and later generalized by Goldstein–Prokushkin [GO04]. In this section, we will construct the Calabi–Eckmann manifolds.

Our first example will use $\mathbf{P}^1$ as the base manifold. We cover $\mathbf{P}^1$ by the open sets

$$U_0 = \{[Z_0, Z_1] : Z_0 \neq 0\}, \quad U_1 = \{[Z_0, Z_1] : Z_1 \neq 0\},$$

and define coordinates $\zeta = \frac{Z_1}{Z_0}$ on U_0 and $\xi = \frac{Z_0}{Z_1}$ on U_1. The line bundle $L = \mathcal{O}(-1) \to \mathbf{P}^1$ equips the covering $\{U_0, U_1\}$ with the transition function

$$t_{01} : U_0 \cap U_1 \to \mathbf{C}^*, \quad t_{01} = \frac{Z_0}{Z_1}.$$

This data defines a $U(1)$ principal bundle $\pi : P \to \mathbf{P}^1$ by the same covering $\mathbf{P}^1 = U_0 \cup U_1$ and transition function

$$\frac{Z_0}{Z_1} \frac{|Z_1|}{|Z_0|} : U_0 \cap U_1 \to S^1.$$

In the trivialisation $U_0 \times S^1$, we use coordinates $(\zeta, e^{i\psi_0})$, and in the trivialisation $U_1 \times S^1$, we use coordinates $(\xi, e^{i\psi_1})$. On the overlap,

$$e^{i\psi_0} = \frac{\xi}{|\xi|} e^{i\psi_1}.$$

In fact, the space P is diffeomorphic to the sphere S^3. If we write

$$S^3 = \{(z_0, z_1) \in \mathbf{C}^2 : |z_0|^2 + |z_1|^2 = 1\},$$

then a diffeomorphism is given by $F : S^3 \to P$, where

$$F(z_0, z_1) = \left([z_0, z_1], \frac{z_0}{|z_0|} \right) \in U_0 \times S^1, \quad z_0 \neq 0,$$

$$F(0, z_1) = ([0, 1], z_1) \in U_1 \times S^1.$$

The inverse of F is given by

$$F^{-1}(\zeta, e^{i\psi_0}) = \frac{1}{\sqrt{1 + |\zeta|^2}} (e^{i\psi_0}, \zeta e^{i\psi_0}), \quad (\zeta, e^{i\psi_0}) \in U_0 \times S^1,$$

$$F^{-1}([0, 1], e^{i\psi_1}) = (0, e^{i\psi_1}), \quad ([0, 1], e^{i\psi_1}) \in U_1 \times S^1.$$

Next, we define a connection on P.

A metric on $L = \mathcal{O}(-1)$ is defined by two positive functions $h_0 : U_0 \to (0, \infty)$ and $h_1 : U_1 \to (0, \infty)$ satisfying $h_0 = \frac{h_1}{|t_{01}|^2}$. We will take

$$h_0 = 1 + |\zeta|^2, \quad h_1 = 1 + |\xi|^2.$$

The Chern connection of (L, h) is $\nabla = d + A$ with $A = \partial \log h$. As explained in (2.19), a connection on L defines a connection 1-form θ on P given by

$$\theta = d\psi - i A',$$

which satisfies

$$d\theta = -i d A' = -i \bar\partial \partial \log h := \omega_{FS}. \tag{2.20}$$

Next, we add a trivial fiber $S^1 = \{e^{i\phi}\}$ to our space, and consider the manifold

$$M_{1,0} = P \times S^1 \simeq S^3 \times S^1.$$

Using the connection θ, we split the tangent bundle

$$TM_{1,0} = H \oplus \left\langle \frac{\partial}{\partial \psi} \right\rangle \oplus \left\langle \frac{\partial}{\partial \phi} \right\rangle.$$

We can define an almost complex structure J on $M_{1,0}$ by identifying H with $T\mathbf{P}^1$ and using the standard complex structure on ∂_ψ and ∂_ϕ. To be precise, if j is the complex structure on $\mathbf{P}^1$, then

$$J = (\pi^*|_H j) \oplus I, \quad I \frac{\partial}{\partial \psi} = \frac{\partial}{\partial \phi}, \quad I \frac{\partial}{\partial \phi} = -\frac{\partial}{\partial \psi}.$$

The space $T^{1,0}M_{1,0}$ is spanned by pullbacks of $T^{1,0}\mathbf{P}^1$ and

$$\frac{\partial}{\partial \psi} - i \frac{\partial}{\partial \phi}.$$

To show J is integrable, we can apply the Newlander–Nirenberg theorem. If z denotes a local holomorphic coordinate on $\mathbf{P}^1$, then $(1, 0)$-forms on $M_{1,0}$ are locally generated by

$$\{\pi^*dz, \theta + id\phi\}.$$

We note that $\theta + id\phi$ is a $(1, 0)$ form since it sends $\partial_\psi + i\partial_\phi$ to zero and $H = \ker \theta$. For local functions f_1, f_2, then by (2.20) we compute

$$d[f_1 dz + f_2(\theta + id\phi)] = df_1 \wedge dz + df_2 \wedge (\theta + id\phi) + f_2 \omega_{FS}. \qquad (2.21)$$

It follows that for any $\eta \in \Omega^{1,0}(M_{1,0})$, then $(d\eta)^{2,0} = 0$. By the Newlander–Nirenberg theorem, we conclude that $M_{1,0}$ is a complex manifold.

The complex surface $M_{1,0}$ is known as the Hopf surface. Since it is topologically $S^3 \times S^1$, we see that the second Betti number of $M_{1,0}$ is zero. Therefore $M_{1,0}$ is a non-Kähler complex surface.

This same construction can be applied to the manifold $M_{1,1} = P \times P$, which is a product of two copies of the $U(1)$ principal bundle P over $\mathbf{P}^1$. Then $M_{1,1}$ is a complex manifold of complex dimension 3, which is a fibration over $\mathbf{P}^1 \times \mathbf{P}^1$.

$$\pi : M_{1,1} \to \mathbf{P}^1 \times \mathbf{P}^1.$$

Since $M_{1,1} \simeq S^3 \times S^3$, this construction defines a non-Kähler complex structure on $S^3 \times S^3$.

In fact, the threefold $M_{1,1}$ does not even admit a balanced metric [MI82]. Suppose ω is a positive $(1,1)$ form on $M_{1,1}$ such that $d\omega^2 = 0$. Let D be a divisor on the base $\mathbf{P}^1 \times \mathbf{P}^1$. Since

$$\int_{\pi^*(D)} \omega^2 > 0,$$

it follows that the class $[\omega^2] \in H^4(M_{1,1}, \mathbf{R})$ is non-trivial. This is a contradiction, since $H^4(S^3 \times S^3, \mathbf{R}) = 0$.

The construction described above readily generalizes to $M_{p,q} = S^{2p+1} \times S^{2q+1}$, giving fibrations

$$\pi : M_{p,q} \to \mathbf{P}^p \times \mathbf{P}^q.$$

These non-Kähler complex manifolds were discovered in [CE53] and are now named Calabi–Eckmann manifolds. A variant of this construction will be revisited in Sect. 2.2.3.4 to produce T^2 fibrations over Calabi–Yau surfaces [GO04], and these manifolds will play a role as a class of solutions to the Hull–Strominger system [FY08, FY07].

2.2 Calabi–Yau Manifolds with Torsion

Let X be a compact complex manifold of complex dimension n. We assume now and henceforth in these notes that $n \geq 3$. Suppose X admits a nowhere vanishing holomorphic $(n, 0)$ form Ω. Given a Hermitian metric $\omega = i g_{\bar{k}j} dz^j \wedge d\bar{z}^k$, the norm of Ω is defined by

$$\|\Omega\|_\omega^2 \frac{\omega^n}{n!} = i^{n^2} \Omega \wedge \bar{\Omega}. \tag{2.22}$$

Using a local coordinate representation $\Omega = \Omega(z)\, dz^1 \wedge \cdots \wedge dz^n$, this norm is

$$\|\Omega\|_\omega^2 = \Omega(z)\overline{\Omega}(z)(\det g_{\bar{k}j})^{-1}.$$

A Hermitian metric ω on (X, Ω) is said to be conformally balanced if it satisfies

$$d(\|\Omega\|_\omega \omega^{n-1}) = 0. \tag{2.23}$$

We see that the Hermitian metric $\chi = \|\Omega\|_\omega^{1/(n-1)} \omega$ is balanced in the sense of Michelsohn [MI82]. We will call (X, Ω, ω) a Calabi–Yau manifold with torsion.

Though Kähler manifolds provide a class of examples, Calabi–Yau manifolds with torsion need not admit a Kähler metric. We shall see that Calabi–Yau manifolds with torsion, though non-Kähler, still retain interesting structure. The geometry of Hermitian manifolds satisfying condition (2.23) belongs somewhere between Kähler geometry and the general theory of non-Kähler complex manifold described in Sect. 2.1. We note that there are other proposed generalizations of non-Kähler Calabi–Yau manifolds in the literature; see e.g. [GGP08, LE11, TO15].

It was shown by Li–Yau [LY05] that condition (2.23) is equivalent to certain $SU(n)$ structures arising in heterotic string theory [HU186, HU286, ST86, DS14, IP01, GMPW04]. In this section, we will explore the geometric implications of this condition.

2.2.1 Curvature and Holonomy

2.2.1.1 Holonomy

From the point of view of differential geometry, Calabi–Yau manifolds with torsion can be understood by imposing a holonomy constraint. While Kähler Calabi–Yau manifolds are characterized by the Levi-Civita connection having holonomy contained in $SU(n)$, here we consider the holonomy of the Strominger–Bismut connection ∇^+ instead.

Lemma 2.4 ([MI82]) *Let (X, ω) be a Hermitian manifold equipped with a nowhere vanishing holomorphic $(n, 0)$ form Ω. Define $\chi = \|\Omega\|_\omega^{1/(n-1)} \omega$. Then*

$$d_\chi^\dagger \chi = i(\partial \log \|\Omega\|_\omega - \tau) - i(\bar{\partial} \log \|\Omega\|_\omega - \bar{\tau}).$$

Here τ is the torsion 1-form of ω, and $d_\chi^\dagger$ is the L^2 adjoint with respect to χ.

Proof The torsion 1-form of χ is given by

$$T_j^\chi = \|\Omega\|_\omega^{-1/(n-1)} g^{i\bar{k}} \left[\partial_i (\|\Omega\|_\omega^{1/(n-1)} g_{\bar{k}j}) - \partial_j (\|\Omega\|_\omega^{1/(n-1)} g_{\bar{k}i}) \right].$$

Simplifying this expression give

$$T_j^\chi = T_j - \partial_j \log \|\Omega\|_\omega,$$

where T_j is the torsion of ω. We apply the identity (2.13) for the adjoint $\partial_\chi^\dagger$ of χ. $\square$

Next, we interpret the conformally balanced condition in terms of a torsion constraint. This relationship between T and $\log \|\Omega\|_\omega$ will have a recurring role as the key identity in the subsequent computations.

Proposition 2.2 ([MI82]) *Let (X, ω) be a Hermitian manifold equipped with a nowhere vanishing holomorphic $(n, 0)$ form Ω. The conformally balanced condition (2.23) is equivalent to the torsion constraint*

$$T_j = \partial_j \log \|\Omega\|_\omega, \quad \bar{T}_{\bar{j}} = \partial_{\bar{j}} \log \|\Omega\|_\omega.$$

Proof Expanding the conformally balanced condition gives

$$0 = \partial \log \|\Omega\|_\omega \wedge \omega^{n-1} + (n-1)\partial\omega \wedge \omega^{n-2}.$$

A computation shows the following identity

$$(n-1)\partial\omega \wedge \omega^{n-2} = -\tau \wedge \omega^{n-1}.$$

Therefore

$$\partial \log \|\Omega\|_\omega \wedge \omega^{n-1} = \tau \wedge \omega^{n-1}.$$

It follows that $\tau = \partial \log \|\Omega\|_\omega$. $\qquad\square$

Our first application of the torsion constraint will be to construct parallel sections of the canonical bundle.

Lemma 2.5 ([GA16]) *Let (X, ω) be a Hermitian manifold with a nowhere vanishing holomorphic $(n, 0)$ form Ω. Suppose (X, ω, Ω) satisfies the conformally balanced condition (2.23). Then $\psi = \|\Omega\|_\omega^{-1}\Omega$ satisfies*

$$\nabla^+\psi = 0.$$

Thus $\psi \in \Gamma(X, K_X)$ is nowhere vanishing and parallel with respect to the Strominger–Bismut connection ∇^+.

Proof By (2.8) and (2.6), the induced connection ∇^+ on ψ is given by

$$\nabla_i^+\psi = \partial_i\psi - (\Gamma_{i\alpha}^\alpha - T^\alpha{}_{i\alpha})\psi, \quad \nabla_{\bar{i}}^+\psi = \partial_{\bar{i}}\psi - g^{k\bar{m}}\bar{T}_{k\bar{i}\bar{m}}\psi. \qquad (2.24)$$

The unbarred derivative is

$$\nabla_i^+\psi = -\partial_i \log \|\Omega\|_\omega\psi + \|\Omega\|_\omega^{-1}\partial_i\Omega - \Gamma_{i\alpha}^\alpha\psi - T_i\psi.$$

We note that

$$2\partial_i \log \|\Omega\|_\omega = \frac{\partial_i\Omega}{\Omega} - g^{p\bar{q}}\partial_i g_{\bar{q}p} = \frac{\partial_i\Omega}{\Omega} - \Gamma_{i\alpha}^\alpha.$$

and therefore

$$\nabla_i^+ \psi = (\partial_i \log \|\Omega\|_\omega - T_i)\psi.$$

By (2.24), we also have

$$\nabla_{\bar{i}}^+ \psi = (-\partial_{\bar{i}} \log \|\Omega\|_\omega + \bar{T}_{\bar{i}})\psi.$$

If (X, ω, Ω) is conformally balanced, we may use Proposition 2.2 and substitute the torsion constraint $T_i = \partial_i \log \|\Omega\|_\omega$ to conclude $\nabla^+ \psi = 0$. $\quad\square$

Theorem 2.1 ([GA16]) *Let (X, ω) be a compact Hermitian manifold with nowhere vanishing holomorphic $(n, 0)$ form Ω. Then (X, ω, Ω) satisfies the conformally balanced condition (2.23) if and only if there exists $\psi \in \Gamma(X, K_X)$ which is nowhere vanishing and parallel with respect to the Strominger–Bismut connection ∇^+.*

Proof The previous lemma constructs a nowhere vanishing parallel section if (X, ω, Ω) is conformally balanced. On the other hand, suppose there exists a nowhere vanishing section $\psi \in \Gamma(X, K_X)$ such that

$$\nabla^+ \psi = 0.$$

We will follow the proof given in lecture notes of Garcia-Fernandez [GA16]. Write

$$\psi = e^{-f}\Omega,$$

for a complex function f. Since $\nabla^+ g_{\bar{k}j} = 0$, the norm of ψ is constant. Let us assume that $\|\psi\|_\omega = 1$. Then

$$1 = e^{-f-\bar{f}}\|\Omega\|_\omega^2,$$

and

$$f + \bar{f} = 2\log\|\Omega\|_\omega.$$

By the formula (2.24), we obtain

$$0 = \nabla_i^+ \psi = (-\partial_i f - T_i + 2\partial_i \log \|\Omega\|_\omega)\psi,$$
$$0 = \nabla_{\bar{i}}^+ \psi = (-\partial_{\bar{i}} f + \bar{T}_{\bar{i}} f)\psi. \tag{2.25}$$

We know that the real part $\mathrm{Re}\, f$ is $\log\|\Omega\|_\omega$, and we will now show that the imaginary part $\mathrm{Im}\, f$ is constant. For this, we use (2.25) to compute

$$\partial_i(f - \bar{f}) = 2(\partial_i \log \|\Omega\|_\omega - T_i),$$
$$\partial_{\bar{i}}(f - \bar{f}) = -2(\partial_{\bar{i}} \log \|\Omega\|_\omega - \bar{T}_{\bar{i}}).$$

By Lemma 2.4,

$$id(f - \bar{f}) = 2d^+_\chi \chi,$$

for $\chi = \|\Omega\|_\omega^{1/(n-1)} \chi$. Therefore

$$d^+_\chi d(f - \bar{f}) = 0,$$

hence $\langle d(f - \bar{f}), d(f - \bar{f}) \rangle_\chi = 0$ and Im f is constant. Since Re $f = \log \|\Omega\|_\omega$, it follows that

$$df = d \log \|\Omega\|_\omega$$

and (2.25) implies the torsion constraint

$$\partial \log \|\Omega\|_\omega = \tau.$$

By Proposition 2.2, (X, ω, Ω) is conformally balanced. □

As a consequence of the existence of parallel sections, we obtain the following interpretation of the conformally balanced condition in terms of a holonomy constraint.

Corollary 2.1 ([ST86, LY05]) *A compact Hermitian manifold with trivial canonical bundle (X, ω, Ω) satisfies the conformally balanced condition (2.23) if and only if*

$$\mathrm{Hol}(\nabla^+) \subseteq SU(n).$$

2.2.1.2 Curvature

Next, we study the structure of the curvature tensor of Calabi–Yau manifolds with torsion. We start with the curvature of the Bismut connection. By the definition (2.8), we can write $\nabla^+ = d + A$ with

$$A_j{}^P{}_q = \Gamma^P_{iq} - T^P{}_{jq}, \quad A_{\bar{j}}{}^P{}_q = g^{p\bar{k}} \bar{T}_{q\bar{j}\bar{k}}.$$

From this expression, we may compute $Rm^+ = dA + A \wedge A$. The components $(\mathrm{Tr}\, Rm^+)_{\alpha\beta} = R^+_{\alpha\beta}{}^\gamma{}_\gamma$ are

$$(\mathrm{Tr}\, Rm^+)_{kj} = \partial_j T_k - \partial_k T_j, \quad (\mathrm{Tr}\, Rm^+)_{\bar{k}\bar{j}} = -(\partial_{\bar{j}} \bar{T}_{\bar{k}} - \partial_{\bar{k}} \bar{T}_{\bar{j}}), \tag{2.26}$$

$$(\mathrm{Tr}\, Rm^+)_{\bar{k}j} = -\partial_{\bar{k}} T_j - \partial_j \bar{T}_{\bar{k}} + \partial_j \partial_{\bar{k}} \log \|\Omega\|_\omega^2. \tag{2.27}$$

The following characterization is due to Fino and Grantcharov, which indicates that conformally balanced metrics can be viewed as non-Kähler analogs of Kähler Ricci-flat metrics.

Theorem 2.2 ([FG04]) *Let (X, ω) be a compact Hermitian manifold with nowhere vanishing holomorphic $(n, 0)$ form Ω. Then (X, ω, Ω) is conformally balanced if and only if*

$$\operatorname{Tr} Rm^+ = 0.$$

Proof From (2.26) and (2.27), we see that manifolds satisfying the torsion constraint in Proposition 2.2 satisfy $\operatorname{Tr} Rm^+ = 0$. For the other direction, we note that by Lemma 2.4, we can write

$$\operatorname{Tr} Rm^+ = i d d_\chi^\dagger \chi,$$

for $\chi = \|\Omega\|_\omega^{1/(n-1)} \omega$. It follows that if $\operatorname{Tr} Rm^+ = 0$, then $\langle d_\chi^\dagger \chi, d_\chi^\dagger \chi \rangle_\chi = 0$ and hence $d_\chi^\dagger \chi = 0$. By Lemma 2.4, we conclude $\partial \log \|\Omega\|_\omega = \tau$ and hence (X, ω, Ω) is conformally balanced. □

For most subsequent computations, we will be using the Chern connection ∇, so we now turn to curvature of the Chern connection. This tensor satisfies certain useful identities on Calabi–Yau manifolds with torsion that we will now describe.

Proposition 2.3 *The Chern–Ricci curvature of a conformally balanced metric (X, ω, Ω) satisfies*

$$R_{\bar{k}j} = 2\nabla_{\bar{k}} T_j.$$

Proof The Chern–Ricci curvature is given by

$$R_{\bar{k}j} = \partial_j \partial_{\bar{k}} \log \|\Omega\|_\omega^2.$$

Applying the torsion constraint (Proposition 2.2) gives the result. □

As a consequence, we obtain the following identities between Ricci curvatures of the Chern connection.

Proposition 2.4 ([PPZ318]) *A conformally balanced metric (X, ω, Ω) satisfies*

$$R'_{\bar{k}j} = R''_{\bar{k}j} = \frac{1}{2} R_{\bar{k}j},$$

$$R' = \frac{1}{2} R, \quad R = g^{j\bar{k}} \partial_j \partial_{\bar{k}} \log \|\Omega\|_\omega^2.$$

Proof By the Bianchi identity (Lemma 2.3),

$$R'_{\bar{k}j} = g^{p\bar{q}} R_{\bar{k}p\bar{q}j} = g^{p\bar{q}} (R_{\bar{k}j\bar{q}p} + \nabla_{\bar{k}} T_{\bar{q}jp}) = R_{\bar{k}j} - \nabla_{\bar{k}} T_j.$$

Applying the previous proposition gives $R'_{\bar{k}j} = \frac{1}{2} R_{\bar{k}j}$. The identity for $R''_{\bar{k}j}$ is derived similarly. Taking the trace gives the relation between the scalar curvatures R and R'.   □

From the divergence theorem (2.10), we note in passing that the total scalar curvature of the Chern connection of a Calabi–Yau manifold with torsion is positive. In fact,

$$\int_X R\, \omega^n = \int_X (2|\tau|^2)\, \omega^n.$$

We conclude this section with the remark that in Strominger's work [ST86], the condition $d(\|\Omega\|_\omega \omega^{n-1}) = 0$ appeared in another form. The reformulation of this condition in terms of balanced metrics is due to Li and Yau [LY05].

Theorem 2.3 ([LY05]) *Let* (X, ω) *be a Hermitian manifold with nowhere vanishing holomorphic* $(n, 0)$ *form* Ω. *The conformally balanced condition* $d(\|\Omega\|_\omega \omega^{n-1}) = 0$ *is equivalent to the equation*

$$d^\dagger \omega = i(\bar{\partial} - \partial) \log \|\Omega\|_\omega.$$

Proof This follows from combining $d^\dagger \omega = i(\bar{\tau} - \tau)$ (2.13) with $\partial \log \|\Omega\|_\omega = \tau$ (Proposition 2.2).   □

2.2.2 Rigidity Theorems

We note in this section some conditions under which a Calabi–Yau manifold with torsion is actually Kähler. We start with a result of Ivanov–Papadopoulos [IP01]. The proof given here follows the computation of [PPZ318].

Theorem 2.4 ([IP01]) *Let* (X, ω, Ω) *be a compact Calabi–Yau manifold with torsion, so that* $d(\|\Omega\|_\omega \omega^{n-1}) = 0$. *Suppose*

$$i\partial\bar{\partial}\omega = 0.$$

Then ω *is a Kähler metric.*

Proof We start by computing $i\partial\bar{\partial}\omega$. Its components are

$$i\partial\bar{\partial}\omega = \frac{1}{4} (i\partial\bar{\partial}\omega)_{\bar{i}\bar{j}k\ell}\, dz^\ell \wedge dz^k \wedge d\bar{z}^j \wedge d\bar{z}^i,$$

given explicitly by

$$(i\partial\bar\partial\omega)_{\bar i jk\bar\ell} = \partial_\ell\partial_{\bar j}g_{\bar i k} - \partial_\ell\partial_{\bar i}g_{\bar j k} + \partial_k\partial_{\bar i}g_{\bar j\ell} - \partial_k\partial_{\bar j}g_{\bar i\ell}.$$

Using the definition of the curvature tensor (2.15) and the torsion (2.4), we find

$$(i\partial\bar\partial\omega)_{\bar i jk\bar\ell} = -R_{\bar i k j\bar\ell} + R_{\bar j k i\bar\ell} - R_{\bar j\ell i k} + R_{\bar i\ell jk} - g^{s\bar r}T_{\bar r\ell k}\bar T_{\bar s\bar i j}. \tag{2.28}$$

Setting this expression to zero and contracting the indices, we see that pluriclosed metrics satisfy

$$0 = g^{\ell\bar j}g^{k\bar i}(i\partial\bar\partial\omega)_{\bar i jk\bar\ell} = 2R' - 2R + |T|^2.$$

Applying Proposition 2.4, we see that if we further assume that ω is conformally balanced, then

$$g^{j\bar k}\partial_j\partial_{\bar k}\log\|\Omega\|_\omega^2 = |T|^2 \geq 0.$$

The maximum principle for elliptic equations (Proposition 2.1) implies that $\log\|\Omega\|_\omega^2$ must be constant, and hence $|T|^2 = 0$.   □

Next, we state the result of Fino–Tomassini [FT11], which builds on work of Matsuo–Takahashi [MT01]. We follow here the computation given in [PPZ19].

Theorem 2.5 ([FT11, MT01]) *Let (X, Ω, ω) be a compact Calabi–Yau manifold with torsion of dimension $n \geq 3$, so that $d(\|\Omega\|_\omega\omega^{n-1}) = 0$. Suppose*

$$i\partial\bar\partial\omega^{n-2} = 0.$$

Then ω is a Kähler metric.

Proof We assume that $n \geq 4$, since the statement follows from the previous theorem when $n = 3$. Expanding derivatives,

$$i\partial\bar\partial\omega^{n-2} = (n-2)i\partial\bar\partial\omega \wedge \omega^{n-3} + i(n-2)(n-3)T \wedge \bar T \wedge \omega^{n-4}.$$

We will wedge this expression with ω to obtain an equation on top forms. For this, we use the general identities

$$\Phi \wedge \omega^{n-2} = \frac{1}{2n(n-1)}\left\{g^{i\bar j}g^{k\bar\ell}\Phi_{\bar\ell j k\bar i}\right\}\omega^n, \tag{2.29}$$

and

$$\Psi \wedge \omega^{n-3} = -\frac{i}{6n(n-1)(n-2)}\left\{g^{i\bar j}g^{k\bar\ell}g^{m\bar n}\Psi_{\bar n\bar\ell jmki}\right\}\omega^n, \tag{2.30}$$

for any $\Phi \in \Omega^{2,2}(X, \mathbf{R})$ and $\Psi \in \Omega^{3,3}(X, \mathbf{R})$, where we use the component convention (2.1). Applying these identities gives

$$\omega \wedge i\partial\bar{\partial}\omega^{n-2}$$
$$= \left[\frac{(n-2)}{2n(n-1)} g^{i\bar{j}} g^{k\bar{\ell}} (i\partial\bar{\partial}\omega)_{\bar{\ell}\bar{j}ki} + \frac{(n-3)}{6n(n-1)} g^{i\bar{j}} g^{k\bar{\ell}} g^{m\bar{n}} (T \wedge \bar{T})_{\bar{n}\bar{\ell}\bar{j}mki} \right] \omega^n.$$
(2.31)

Symmetrizing the components of the torsion tensor T, we see that

$$(T \wedge \bar{T})_{\bar{n}\bar{\ell}\bar{j}mki} = T_{\bar{j}mi} \bar{T}_{k\bar{n}\bar{\ell}} + T_{\bar{\ell}mi} \bar{T}_{kj\bar{n}} + T_{\bar{n}mi} \bar{T}_{k\bar{\ell}\bar{j}} + T_{\bar{j}km} \bar{T}_{i\bar{n}\bar{\ell}} + T_{\bar{\ell}km} \bar{T}_{ij\bar{n}}$$

$$+ T_{\bar{n}km} \bar{T}_{i\bar{\ell}\bar{j}} + T_{\bar{j}ik} \bar{T}_{m\bar{n}\bar{\ell}} + T_{\bar{\ell}ik} \bar{T}_{mj\bar{n}} + T_{\bar{n}ik} \bar{T}_{m\bar{\ell}\bar{j}}.$$
(2.32)

Setting (2.31) to zero and substituting the expression (2.28) for $i\partial\bar{\partial}\omega$ and (2.32) for $T \wedge \bar{T}$, we obtain the following identity

$$0 = \frac{(n-2)}{2n(n-1)} (2R' - 2R + |T|^2) + \frac{(n-3)}{6n(n-1)} (6|\tau|^2 - 3|T|^2),$$

satisfied by any astheno-Kähler metric ω. We now use the conformally balanced condition by applying Proposition 2.4, which gives $2R' - 2R = -g^{j\bar{k}} \partial_j \partial_{\bar{k}} \log \|\Omega\|_\omega$. Simplifying, we obtain

$$(n-2) g^{j\bar{k}} \partial_j \partial_{\bar{k}} \log \|\Omega\|_\omega = |T|^2 + 2(n-3)|\tau|^2 \geq 0.$$

By the maximum principle for elliptic equations (Proposition 2.1) we must have $|T|^2 + 2(n-3)|\tau|^2 = 0$. Hence $|T|^2 = 0$ and ω is Kähler. □

There are more theorems of this nature; for other conditions on balanced metrics which imply that it is Kähler, see [FIUV09, LY12, LY17].

A folklore conjecture in the field (e.g. [FV16]) speculates that if a Calabi–Yau with torsion (X, Ω, ω) admits another metric ω_2 which is pluriclosed, then X must be a Kähler. If ω_2 is instead assumed to be astheno-Kähler, then X need not be Kähler [FGV, LU17].

2.2.3 Examples

2.2.3.1 Kähler Calabi–Yau

We have already seen that conformally balanced metrics generalize Kähler Ricci-flat metrics, since they are characterized by vanishing of the Ricci curvature of ∇^+, and ∇^+ coincides with the Levi-Civita connection for Kähler metrics. We note here a simple direct proof that Kähler Ricci-flat metrics are conformally balanced.

Let (X, Ω) be a Kähler Calabi–Yau manifold. By Yau's theorem [YA78], there exists a Kähler metric ω with zero Ricci curvature. In this case, $\|\Omega\|_\omega$ is constant, since

$$i\partial\bar\partial \log \|\Omega\|_\omega^2 = i\partial\bar\partial \log \Omega(z)\overline{\Omega(z)} - i\partial\bar\partial \log \det g_{\bar k j} = 0,$$

and hence $g^{j\bar k}\partial_j\partial_{\bar k} \log \|\Omega\|_\omega^2 = 0$. By the maximum principle, $\|\Omega\|_\omega$ is constant. Since ω is Kähler, we have $d\omega^{n-1} = 0$, and hence $d(\|\Omega\|_\omega\omega^{n-1}) = 0$.

2.2.3.2   Complex Lie Groups

Next, we study invariant metrics on complex Lie groups, which provide a class of natural non-Kähler metrics. Let G be a complex Lie group. Choose a positive definite inner product on the Lie algebra $\mathbf{g}$, and let $e_1, \ldots, e_n \in \mathbf{g}$ be an orthonormal frame of left-invariant holomorphic vector fields on G. The structure constants of the Lie algebra $\mathbf{g}$ in this basis will be denoted

$$[e_a, e_b] = c^d{}_{ab}e_d.$$

Taking the dual frame $e^1, \ldots, e^n$, we may define a left-invariant metric ω by

$$\omega = i \sum_a e^a \wedge \bar e^a.$$

We note that this metric cannot be Kähler unless G is trivial. Indeed, taking the exterior derivative gives

$$\partial e^a = \frac{1}{2}c^a{}_{bd}e^d \wedge e^b. \tag{2.33}$$

Therefore

$$i\bar\partial\omega = \frac{1}{2}\overline{c^a{}_{bd}}\, e^a \wedge \bar e^d \wedge \bar e^b,$$

$$i\partial\bar\partial\omega = \frac{1}{4}\overline{c^a{}_{bd}}c^a{}_{rs}\, e^s \wedge e^r \wedge \bar e^d \wedge \bar e^b, \tag{2.34}$$

so this invariant metric is not Kähler or pluriclosed in general. We take the Calabi–Yau form to be

$$\Omega = e^1 \wedge \cdots \wedge e^n.$$

which is a nowhere vanishing holomorphic $(n, 0)$ form. Using (2.22), we see that

$$\|\Omega\|_\omega = 1.$$

Checking whether ω is conformally balanced reduces to verifying that $d\omega^{n-1} = 0$. This implies a condition of the structure constants, which does not hold for arbitrary Lie groups, but still admits plenty of examples. We say that G is unimodular if its structure constants satisfy

$$\sum_p c^p{}_{pa} = 0.$$

This condition is well-defined on G and does not depend on the choice of frame. It was noted by Abbena and Grassi [AG86] that $d\omega^{n-1} = 0$ if and only if G is unimodular. Indeed, from (2.33) we see that $T^a{}_{bd} = c^a{}_{bd}$. Hence G is unimodular if and only if $T_j = 0$, which holds if and only if ω is conformally balanced by Proposition 2.2.

Thus unimodular complex Lie groups admit left invariant conformally balanced metrics. An explicit example is given by $SL(2, \mathbf{C})$. To obtain a compact threefold, we may quotient out by a discrete group and let $X = SL(2, \mathbf{C})/\Lambda$.

We claim that X does not admit a Kähler metric. For this, we use the fact that $SL(2, \mathbf{C})$ admits a basis e^a such that $c^a{}_{bd} = \epsilon_{abd}$ the Levi-Civita symbol. Let $\omega = i\delta_{ba} e^a \wedge \bar{e}^b$, and compute

$$(\omega^2)_{\bar{b}\bar{d}rs} = 2(\delta_{\bar{d}s}\delta_{\bar{b}r} - \delta_{\bar{d}r}\delta_{\bar{b}s}).$$

In dimension 3, we have the contracted epsilon identity

$$\epsilon_{ars}\epsilon_{abd} = \delta_{rb}\delta_{sd} - \delta_{rd}\delta_{bs}. \tag{2.35}$$

Therefore, by (2.34),

$$(i\partial\bar{\partial}\omega)_{\bar{b}\bar{d}rs} = \delta_{ds}\delta_{br} - \delta_{dr}\delta_{bs}.$$

We see that ω^2 and $i\partial\bar{\partial}\omega$ are proportional to each other.

$$i\partial\bar{\partial}\omega = \frac{1}{2}\omega^2. \tag{2.36}$$

This in particular illustrates another difference with Kähler geometry, where ω^2 always represents a non-zero cohomology class. Now suppose X admits a Kähler metric χ. Then

$$0 = \int_X i\partial\bar{\partial}\omega \wedge \chi = \frac{1}{2}\int_X \omega^2 \wedge \chi \tag{2.37}$$

which is a contradiction since $\omega^2 \wedge \chi > 0$.

For more examples of complex Lie groups, Fei–Yau [FY15, Proposition 3.7] classify complex unimodular Lie algebras of dimension 3 and study the Hull–Strominger system in each case. A theorem of Wang [WA54] states that the only compact parallelizable manifolds admitting Kähler metrics are the complex tori.

2.2.3.3   Iwasawa Manifold

We consider the action of $a, b, c \in \mathbf{Z}[i]$ on $\mathbf{C}^3$ given by

$$(x, y, z) \mapsto (x + a, y + c, z + \bar{a}y + b). \tag{2.38}$$

Let X be the quotient of $\mathbf{C}^3$ under this action. The manifold X is an example of an Iwasawa manifold. We have a projection

$$\pi : X \to T^4 = \mathbf{C}/\Lambda \times \mathbf{C}/\Lambda, \quad \pi(x, y, z) = (x, y).$$

Here Λ is the lattice generated by $1, i$. The fibers are isomorphic to tori $\pi^{-1}(x, y) = T^2$. Hence M is a torus fibration over T^4. The form

$$\Omega = dz \wedge dx \wedge dy,$$

is defined on X, and is holomorphic nowhere vanishing. We define

$$\theta = dz - \bar{x}dy.$$

This form on $\mathbf{C}^3$ is invariant under the action (2.38), and is thus well-defined on X. Consider the family of metrics

$$\omega_u = e^u \hat{\omega} + i\theta \wedge \bar{\theta}, \quad \hat{\omega} = idx \wedge d\bar{x} + idy \wedge d\bar{y},$$

where $u : T^4 \to \mathbf{R}$ is an arbitrary function on the base T^4. A computation shows that

$$\|\Omega\|_{\omega_u} = e^{-u},$$

and

$$d(\|\Omega\|_{\omega_u} \omega_u^2) = 0.$$

Thus (X, ω_u, Ω) is conformally balanced. However, X does not admit a Kähler metric. Let ω_0 be metric with $u = 0$. Direct computation gives

$$i\partial\bar{\partial}\omega_0 = \frac{\hat{\omega}^2}{2}.$$

We can rule out the existence of a Kähler metric χ by considering $\int_X i \partial \bar{\partial} \omega_0 \wedge \chi$ as in the previous section, see (2.37).

2.2.3.4 Goldstein–Prokushkin Fibrations

In this section, we describe a construction of Goldstein–Prokushkin [GO04] which utilizes $U(1)$ principal bundles to generalize the previous example. Let $(S, \hat{\omega}, \Omega)$ be a Kähler Calabi–Yau surface equipped with two $(1, 1)$ form $\omega_1, \omega_2 \in 2\pi H^2(S, \mathbf{Z})$, which are anti-self-dual with respect to $\hat{\omega}$.

$$\star \omega_1 = -\omega_1, \quad \star \omega_2 = -\omega_2.$$

There exists line bundles L_1, L_2 over S with connections A_1, A_2 whose curvature $i F_{A_1}, i F_{A_2}$ is equal to ω_1, ω_2. As detailed in Sect. 2.1.4, the line bundles L_1, L_2 can be compactified to form S^1 principal bundles $P_1 \to S$, $P_2 \to S$ equipped with connection 1-forms θ_1, θ_2 satisfying

$$d\theta_i = -\omega_i.$$

Let X denote the total space of the $S^1 \times S^1$ principal bundle $\pi : X \to S$ whose fibers are the product of the fibers of P_1, P_2. Locally, points of X are given by $(z, e^{i\psi_1}, e^{i\psi_2})$. As we discussed in Sect. 2.1.4, we have the global vector fields

$$\frac{\partial}{\partial \psi_1}, \frac{\partial}{\partial \psi_2},$$

which span the vertical space $V = \ker \pi_*$, and satisfy

$$\theta_1 \left(\frac{\partial}{\partial \psi_1} \right) = 1, \quad \theta_2 \left(\frac{\partial}{\partial \psi_2} \right) = 1.$$

The horizontal space is given by

$$H = \ker \theta_1 \cap \ker \theta_2,$$

and the tangent space admits the decomposition

$$TX = H \oplus V.$$

Furthermore

$$\pi_\star|_H : H \to TS$$

is an isomorphism. It follows that the complex structure j_S on S induces an almost complex structure on H. We define on X the almost complex structure

$$J = (\pi^*|_H j_S) \oplus I, \quad I\frac{\partial}{\partial \psi_1} = \frac{\partial}{\partial \psi_2}, \quad I\frac{\partial}{\partial \psi_2} = -\frac{\partial}{\partial \psi_1}.$$

We define the 1-form

$$\theta = -\theta_1 - i\theta_2.$$

Since $\theta|_H = 0$ and $\theta(\partial_{\psi_1} + i\partial_{\psi_2}) = 0$, we see that $\theta(V) = 0$ for any $V \in T^{0,1}X$. Thus θ is a $(1, 0)$ form. Furthermore,

$$d\theta = \pi^*(\omega_1 + i\omega_2).$$

Similarly to our discussion of Eq. (2.21) in Sect. 2.1.4.2, we can use that $(1, 0)$ forms are locally generated by $\{\pi^*dz^1, \pi^*dz^2, \theta\}$ to apply the Newlander–Nirenberg theorem and establish that J integrable. Thus X is a compact complex manifold of dimension 3.

In fact, X is a Calabi–Yau manifold with torsion. Let

$$\Omega = \theta \wedge \pi^*\Omega_S,$$

which is a nowhere vanishing $(3, 0)$ form. The form Ω is holomorphic since $d\Omega = 0$.

For $u \in C^\infty(S, \mathbf{R})$, we consider the family of metrics

$$\omega_u = \pi^*(e^u\hat{\omega}) + i\theta \wedge \bar{\theta}.$$

These metrics will be revisited, as they form the Fu–Yau ansatz of solutions to the Hull–Strominger system [FY08]. We compute

$$i\Omega \wedge \bar{\Omega} = i\theta \wedge \bar{\theta} \wedge \pi^*(\Omega_S \wedge \overline{\Omega_S}) = i\theta \wedge \bar{\theta} \wedge \pi^*\left(\|\Omega_S\|_{\hat{\omega}}^2 \frac{\hat{\omega}^2}{2}\right),$$

$$\omega_u^2 = \pi^*(e^{2u}\hat{\omega}^2) + 2\pi^*(e^u\hat{\omega}) \wedge i\theta \wedge \bar{\theta}, \quad \omega_u^3 = 3\pi^*(e^{2u}\hat{\omega}^2) \wedge i\theta \wedge \bar{\theta}.$$

Since $(S, \hat{\omega})$ is Kähler Ricci-flat, then $\|\Omega_S\|_{\hat{\omega}}$ is constant, which we may normalize such that

$$\|\Omega\|_{\omega_u} = e^{-u}. \tag{2.39}$$

We can now compute

$$d(\|\Omega\|_{\omega_u}\omega_u^2) = d(\pi^*(e^u\hat{\omega}^2) + 2\pi^*\hat{\omega} \wedge i\theta \wedge \bar{\theta})$$
$$= 2\pi^*\hat{\omega} \wedge i\pi^*(\omega_1 + i\omega_2) \wedge \bar{\theta} - 2\pi^*\hat{\omega} \wedge i\theta \wedge \pi^*(\omega_1 - i\omega_2)$$
$$= 0,$$

since

$$\hat{\omega} \wedge \omega_1 = \hat{\omega} \wedge \omega_2 = 0,$$

as ω_1, ω_2 are anti-self-dual. Thus (X, ω_u, Ω) is Calabi–Yau with torsion. In fact, X is non-Kähler unless $\omega_1 = \omega_2 = 0$. To see this, we compute

$$i\partial\bar{\partial}\omega_0 = -\bar{\partial}\theta \wedge \partial\bar{\theta} = -(\pi^*\omega_1 + i\pi^*\omega_2)(\pi^*\omega_1 - i\pi^*\omega_2) = -\pi^*(\omega_1^2 + \omega_2^2).$$

Since ω_1, ω_2 are anti-self-dual,

$$i\partial\bar{\partial}\omega_0 = \pi^*(\omega_1 \wedge \star\omega_1 + \omega_2 \wedge \star\omega_2).$$

If X admits a Kähler metric χ, then

$$0 = \int_X i\partial\bar{\partial}\omega_0 \wedge \chi = \int_X \pi^*(\omega_1 \wedge \star\omega_1 + \omega_2 \wedge \star\omega_2) \wedge \chi,$$

which is strictly positive unless $\|\omega_1\|_{\hat{\omega}}^2 = \|\omega_2\|_{\hat{\omega}}^2 = 0$.

2.2.3.5 Fei Twistor Space

As our last example, we outline a construction of Fei [FE16, FE15] which generalizes earlier constructions of Calabi [CA58] and Gray [GR69]. The example will be a T^4 fibration over a Riemann surface.

We first describe the base of the fibration. Let (Σ, φ) be a Riemann surface equipped with a nonconstant holomorphic map $\varphi : \Sigma \to \mathbf{P}^1$ satisfying $\varphi^*\mathcal{O}(2) = K_\Sigma$. This condition is known to imply that the genus of Σ must be at least three. As a concrete example, we may take Σ to be a minimal surface in T^3 with φ being the Gauss map [FHP17]. By the work of Meeks [ME90] and Traizet [TR08], there exists minimal surfaces of genus $g \geq 3$ in T^3.

Using stereographic coordinates, we may write $\varphi = (\alpha, \beta, \gamma)$ with $(\alpha, \beta, \gamma) \in S^2 \subseteq \mathbf{R}^3$. Fixing the Fubini-Study metric ω_{FS} on $\mathbf{P}^1$, we pullback via φ an orthonormal basis of sections of $\mathcal{O}(2)$ to obtain 1-forms μ_1, μ_2, μ_3. We then equip Σ with the metric

$$\hat{\omega} = i\mu_1 \wedge \bar{\mu}_1 + i\mu_2 \wedge \bar{\mu}_2 + i\mu_3 \wedge \bar{\mu}_3.$$

This metric has Gauss curvature κ given by

$$\kappa \hat{\omega} = -\varphi^* \omega_{FS},$$

hence $\kappa \leq 0$ and κ vanishes at branch points of φ.

We now describe the fibers. Let (T^4, g) be the 4-torus with flat metric, which we will view as a hyperkähler manifold with complex structures I, J, K satisfying $IJ = K$ and $I^2 = J^2 = K^2 = -1$, and corresponding Kähler metrics ω_I, ω_J, ω_K. At each $z \in \Sigma$, we use the map $\varphi = (\alpha, \beta, \gamma)$ to equip T^4 with the complex structure

$$\alpha I + \beta J + \gamma K.$$

If j_Σ denotes the complex structure on Σ, we may form the product $X = \Sigma \times T^4$ and equip it with the complex structure

$$J_0 = j_\Sigma \oplus (\alpha I + \beta J + \gamma K).$$

This complex structure is integrable, thus X is a compact complex manifold of dimension 3. In fact, X has trivial canonical bundle, and we can give an explicit expression for a nowhere vanishing holomorphic $(3, 0)$ form

$$\Omega = \mu_1 \wedge \omega_I + \mu_2 \wedge \omega_J + \mu_3 \wedge \omega_K.$$

Let

$$\omega' = \alpha \omega_I + \beta \omega_J + \gamma \omega_K$$

be the Kähler metric corresponding to the complex structure $\alpha I + \beta J + \gamma K$ on T^4. The Fei ansatz ω_f on X is the following family of conformally balanced metrics.

Proposition 2.5 ([FE16, FE15]) *Given any $f \in C^\infty(\Sigma, \mathbf{R})$, the Hermitian metric given by*

$$\omega_f = e^{2f} \hat{\omega} + e^f \omega',$$

is conformally balanced. Furthermore, $\|\Omega\|_{\omega_f} = e^{-2f}$.

Thus X is Calabi–Yau with torsion, and in fact, it is non-Kähler.

2.2.3.6 Other Examples

We have now discussed many examples of Calabi–Yau manifolds with balanced metrics, many of which were already listed in the pioneering work of Michelsohn [MI82]. There are also example which will not be studied in these notes. For example, there is the construction of Fu et al. [FLY12] on connected sums of $S^3 \times S^3$. There are parallelizable examples on nilmanifolds and solvmanifolds [UG07, OUV17, FIUV09, FG04, UV14, UV15]. Non-compact examples are constructed in [FY09, FE17, FIUV14]. There are also examples from the physics literature, e.g. [BD02, BBDG03, DRS99, HIS16, MS11].

2.3 Anomaly Flow with Zero Slope

In this section, we will discuss a geometric flow which preserves the geometry described in Sect. 2.2. The material in this section can be found in joint work with Phong and Zhang [PPZ218, PPZ318, PPZ19].

A central problem in complex geometry is to detect when a given complex manifold admits a Kähler metric. We would like to study this question on Calabi–Yau manifolds with torsion. Motivated by Sect. 2.2.2, we will deform conformally balanced metrics towards astheno-Kähler ($i\partial\bar{\partial}\omega^{n-2} = 0$).

Together with Phong and Zhang [PPZ19], we introduce the flow

$$\frac{d}{dt}(\|\Omega\|_\omega \omega^{n-1}) = i\partial\bar{\partial}\omega^{n-2},$$

$$d(\|\Omega\|_{\omega(0)}\omega(0)^{n-1}) = 0. \tag{2.40}$$

We call this evolution equation the Anomaly flow with zero slope. The name comes from an extension of the flow which adds higher order correction terms proportional to a parameter α', which is used to study the Hull–Strominger system and the cancellation of anomalies in theoretical physics. We will discuss the Anomaly flow when α' terms are included in Sect. 2.4.

The first thing to note is that the conformally balanced property is preserved by the flow

$$d(\|\Omega\|_{\omega(t)}\omega(t)^{n-1}) = 0,$$

which follows from taking the exterior derivative of (2.40). In fact, the balanced class of the initial metric

$$[\|\Omega\|_{\omega(0)}\omega(0)^{n-1}] \in H^{n-1,n-1}_{BC}(X, \mathbf{R})$$

is also preserved, since

$$\frac{d}{dt}[\|\Omega\|_\omega \omega^{n-1}] = [i\partial\bar{\partial}\omega^{n-2}] = 0. \tag{2.41}$$

Here $H_{BC}^{n-1,n-1}(X)$ is the Bott–Chern cohomology of X, given by

$$H_{BC}^{n-1,n-1}(X) = \frac{\{\alpha \in \Omega^{n-1,n-1}(X) : d\alpha = 0\}}{\{i\partial\bar{\partial}\beta : \beta \in \Omega^{n-2,n-2}(X)\}}.$$

Stationary points ω_∞ of the flow satisfy both

$$d(\|\Omega\|_{\omega_\infty}\omega_\infty^{n-1}) = 0, \quad i\partial\bar{\partial}\omega_\infty^{n-2} = 0,$$

hence by Theorem 2.5, they are Kähler. The Anomaly flow with zero slope thus deforms balanced metrics to a Kähler metric in a given balanced class.

2.3.1 Evolution of the Metric

The first question to ask about the flow (2.40) is whether it exists for a short-time, and if so, we would like an explicit expression for the evolution equation of the metric $\omega = ig_{\bar{k}j}dz^j \wedge d\bar{z}^k$.

We begin by deriving the evolution of the determinant of the metric.

Lemma 2.6 *Suppose $\omega(t) = ig_{\bar{k}j}dz^j \wedge d\bar{z}^k$ satisfies the evolution equation*

$$\frac{d}{dt}(\|\Omega\|_\omega\omega^{n-1}) = \Psi(t), \tag{2.42}$$

for $\Psi(t) \in \Omega^{n-1,n-1}(X, \mathbf{R})$. Then the norm of Ω evolves by

$$\frac{d}{dt}\|\Omega\|_\omega = -\frac{n}{(n-2)}\frac{\Psi \wedge \omega}{\omega^n},$$

which follows from the identity

$$\mathrm{Tr}\,\dot{\omega} = \frac{2n}{(n-2)\|\Omega\|_\omega}\frac{\Psi \wedge \omega}{\omega^n}.$$

From now on, traces will always be taken with respect to the evolving metric ω. Explicitly,

$$\mathrm{Tr}\,\alpha = i^{-1}g^{j\bar{k}}\alpha_{\bar{k}j},$$

for a $(1, 1)$ form $\alpha = \alpha_{\bar{k}j}dz^j \wedge d\bar{z}^k$.

Proof Using the well-known formula

$$\delta \det g_{\bar{k}j} = (\det g_{\bar{k}j})g^{j\bar{k}}(\delta g)_{\bar{k}j},$$

we differentiate

$$\frac{d}{dt}\|\Omega\|_\omega = \frac{d}{dt}(\Omega\bar{\Omega})^{1/2}(\det g)^{-1/2} = -\frac{1}{2}\|\Omega\|_\omega \mathrm{Tr}\,\dot{\omega}.$$

Expanding (2.42), we obtain

$$\left(\frac{d}{dt}\|\Omega\|_\omega\right)\omega^{n-1} + (n-1)\|\Omega\|_\omega\dot{\omega}\wedge\omega^{n-2} = \Psi.$$

Substituting the variation of $\|\Omega\|_\omega$ gives

$$-\frac{1}{2}\|\Omega\|_\omega(\mathrm{Tr}\,\dot{\omega})\omega^{n-1} + (n-1)\|\Omega\|_\omega\dot{\omega}\wedge\omega^{n-2} = \Psi. \tag{2.43}$$

Next, we wedge this equation with ω to obtain the following equation of top forms.

$$-\frac{1}{2}\|\Omega\|_\omega(\mathrm{Tr}\,\dot{\omega})\omega^n + (n-1)\|\Omega\|_\omega\frac{(\mathrm{Tr}\,\dot{\omega})}{n}\omega^n = \Psi\wedge\omega.$$

From this equation we can solve for $\mathrm{Tr}\,\dot{\omega}$. □

Lemma 2.7 *Suppose* $\omega(t)$ *satisfies*

$$\frac{d}{dt}(\|\Omega\|_\omega\omega^{n-1}) = \Psi(t),$$

for $\Psi(t) \in \Omega^{n-1,n-1}(X, \mathbf{R})$. *Then the metric evolves by*

$$\partial_t\omega = \left[\frac{n}{(n-2)\|\Omega\|_\omega}\frac{\Psi\wedge\omega}{\omega^n}\right]\omega - \frac{1}{(n-1)!\|\Omega\|_\omega}\star\Psi.$$

Proof To extract $\partial_t\omega$, we will apply the Hodge star operator $\star$ with respect to ω to the expanded equation (2.43).

$$-\frac{(n-1)!}{2}\|\Omega\|_\omega(\mathrm{Tr}\,\dot{\omega})\omega + (n-1)!\|\Omega\|_\omega(-\partial_t\omega + (\mathrm{Tr}\,\dot{\omega})\omega) = \star\Psi$$

Here we used the identities $\star\omega^{n-1} = (n-1)!\omega$ and

$$[\star(\alpha\wedge\omega^{n-2})]_{\bar{q}p} = -(n-2)!\alpha_{\bar{q}p} + i(n-2)!(\mathrm{Tr}\,\alpha)g_{\bar{q}p}, \tag{2.44}$$

for any $\alpha \in \Omega^{1,1}(X)$. This last identity can be found in e.g. [HU305, PPZ318]. Therefore

$$\partial_t \omega = \frac{1}{2}(\mathrm{Tr}\,\dot{\omega})\omega - \frac{1}{(n-1)!\|\Omega\|_\omega} \star \Psi.$$

Substituting the previous lemma gives the desired expression.   □

For the Anomaly flow with zero slope, the form Ψ is given by

$$\Psi = i\partial\bar\partial\omega^{n-2} = (n-2)i\partial\bar\partial\omega \wedge \omega^{n-3} + i(n-2)(n-3)T \wedge \bar{T} \wedge \omega^{n-4}. \quad (2.45)$$

To obtain an explicit expression for the evolution of the metric, we must expand the torsion terms.

Theorem 2.6 ([PPZ19]) *Suppose* $\omega(t)$ *solves the Anomaly flow*

$$\frac{d}{dt}(\|\Omega\|_\omega \omega^{n-1}) = i\partial\bar\partial(\omega^{n-2}), \quad d(\|\Omega\|_{\omega(0)}\omega(0)^{n-1}) = 0.$$

If $n = 3$, then the metric evolves via

$$\partial_t g_{\bar k j} = \frac{1}{2\|\Omega\|_\omega}\left[-\tilde{R}_{\bar k j} + g^{m\bar\ell}g^{s\bar r}T_{\bar r m j}\bar{T}_{s\bar\ell\bar k} \right],$$

and if $n \geq 4$, then

$$\partial_t g_{\bar k j} = \frac{1}{(n-1)\|\Omega\|_\omega}\left[-\tilde{R}_{\bar k j} + \frac{1}{2(n-2)}(|T|^2 - 2|\tau|^2)\,g_{\bar k j} \right.$$
$$\left. -\frac{1}{2}g^{q\bar p}g^{s\bar r}T_{\bar k q s}\bar{T}_{j\bar p\bar r} + g^{s\bar r}(T_{\bar k j s}\bar{T}_{\bar r} + T_s\bar{T}_{j\bar k\bar r}) + T_j\bar{T}_{\bar k} \right]. \quad (2.46)$$

The metric evolution can be compared with other flows in Hermitian geometry, e.g. [ST10, ST11, TW15, US16, ZH16]. The expression when $n = 3$ is similar to the metric evolution in the Streets–Tian pluriclosed flow [ST10], though they differ by the presence of the determinant of the metric $\|\Omega\|_\omega$. We note that the Anomaly flow is a flow of balanced metrics while the pluriclosed flow is a flow of pluriclosed metrics, so these flows exist in different realms of Hermitian geometry. Such torsion-type terms appearing in (2.46) also appear in other Ricci flows preserving other types of geometry, such as for example the metric evolution in the G2 Laplacian flow [KA09, BR05].

Proof We will derive the expression assuming that $n \geq 4$, as the case $n = 3$ is easier and follows a similar argument. We use the notation

$$\mathrm{Tr}\,\Phi = i^{-2}g^{p\bar q}g^{j\bar k}\Phi_{\bar k j\bar q p}, \quad \mathrm{Tr}\,\Psi = i^{-3}g^{j\bar k}g^{p\bar q}g^{s\bar r}\Psi_{\bar r s\bar q p\bar k j},$$

for $\Phi \in \Omega^{2,2}(X)$ and $\Psi \in \Omega^{3,3}(X)$. We begin by computing

$$(\star i \partial \bar{\partial} \omega^{n-2})_{\bar{q}p}$$

$$= (n-2)[\star (i \partial \bar{\partial} \omega \wedge \omega^{n-3})]_{\bar{q}p} + i(n-2)(n-3)[\star (T \wedge \bar{T} \wedge \omega^{n-4})]_{\bar{q}p}$$

$$= i(n-2)! g^{s\bar{r}} (i \partial \bar{\partial} \omega)_{\bar{r}s\bar{q}p} + i \frac{(n-2)!}{2} (\mathrm{Tr}\, i \partial \bar{\partial} \omega) g_{\bar{q}p}$$

$$+ i \frac{(n-2)!}{2} g^{i\bar{j}} g^{s\bar{r}} (T \wedge \bar{T})_{\bar{r}s\bar{j}i\bar{q}p} - \frac{(n-2)!}{6} (\mathrm{Tr}\, T \wedge \bar{T}) g_{\bar{q}p}. \qquad (2.47)$$

This follows from (2.45) and the following identities for the Hodge star operator

$$[\star (\Phi \wedge \omega^{n-3})]_{\bar{q}p} = i(n-3)! g^{s\bar{r}} \Phi_{\bar{r}s\bar{q}p} + i \frac{(n-3)!}{2} (\mathrm{Tr}\, \Phi) g_{\bar{q}p},$$

$$[\star (\Psi \wedge \omega^{n-4})]_{\bar{q}p} = \frac{(n-4)!}{2} g^{i\bar{j}} g^{s\bar{r}} \Psi_{\bar{r}s\bar{j}i\bar{q}p} + i \frac{(n-4)!}{6} (\mathrm{Tr}\, \Psi) g_{\bar{q}p}, \quad (2.48)$$

which hold for any $\Phi \in \Omega^{2,2}(X, \mathbf{R})$ and $\Psi \in \Omega^{3,3}(X, \mathbf{R})$. For a proof of these Hodge star identities, see [PPZ19].

Next, we compute using (2.29) and (2.30),

$$\frac{i \partial \bar{\partial} \omega^{n-2} \wedge \omega}{\omega^n} = (n-2) \frac{i \partial \bar{\partial} \omega \wedge \omega^{n-2}}{\omega^n} + i(n-2)(n-3) \frac{T \wedge \bar{T} \wedge \omega^{n-3}}{\omega^n}$$

$$= \frac{(n-2)}{2n(n-1)} \mathrm{Tr}\,(i \partial \bar{\partial} \omega) + \frac{i(n-3)}{6n(n-1)} \mathrm{Tr}\,(T \wedge \bar{T}). \qquad (2.49)$$

We now substitute (2.47) and (2.49) into Lemma 2.7. The $\mathrm{Tr}\,(i \partial \bar{\partial} \omega)$ terms cancel exactly, and we are left with

$$\partial_t g_{\bar{q}p} = -\frac{1}{(n-1)\|\Omega\|_\omega} g^{s\bar{r}} (i \partial \bar{\partial} \omega)_{\bar{r}s\bar{q}p} - \frac{1}{2(n-1)\|\Omega\|_\omega} g^{i\bar{j}} g^{s\bar{r}} (T \wedge \bar{T})_{\bar{r}s\bar{j}i\bar{q}p}$$

$$- \frac{i}{6(n-1)(n-2)\|\Omega\|_\omega} \mathrm{Tr}\,(T \wedge \bar{T})\, g_{\bar{q}p}. \qquad (2.50)$$

By identity (2.28), we have

$$g^{s\bar{r}} (i \partial \bar{\partial} \omega)_{\bar{r}\bar{q}sp} = -\tilde{R}_{\bar{q}p} + R'_{\bar{q}p} - R_{\bar{q}p} + R''_{p\bar{q}} - g^{s\bar{r}} g^{n\bar{m}} T_{\bar{m}ps} \bar{T}_{n\bar{r}\bar{q}}.$$

We now use that the evolving metrics are conformally balanced. In this case, by Proposition 2.4, we have

$$g^{s\bar{r}} (i \partial \bar{\partial} \omega)_{\bar{r}s\bar{q}p} = \tilde{R}_{\bar{q}p} - g^{s\bar{r}} g^{n\bar{m}} T_{\bar{m}sp} \bar{T}_{n\bar{r}\bar{q}}. \qquad (2.51)$$

Substituting (2.51) and (2.32) into (2.50) and expanding the torsion terms gives the explicit expression for $\partial_t g_{\bar{q}p}$. □

As a consequence of Theorem 2.6, the Anomaly flow with zero slope exists for a short-time from any initial metric. Indeed, from (2.15) we have

$$\tilde{R}_{\bar{m}\ell} = -g^{j\bar{k}}\partial_j\partial_{\bar{k}}g_{\bar{m}\ell} + g^{j\bar{k}}g^{s\bar{r}}\partial_{\bar{k}}g_{\bar{m}s}\partial_j g_{\bar{r}\ell}, \tag{2.52}$$

and so $\tilde{R}_{\bar{m}\ell}(g)$ is an elliptic operator in g. There is a slight subtlety, which is that the proof of Theorem 2.6 only shows that the Anomaly flow with zero slope is parabolic when restricted to variations in the space of conformally balanced metrics. One way to resolve this issue is by using the Hamilton–Nash–Moser [HA82] implicit function theorem, and we refer to [PPZ116, PPZ19] for details.

Corollary 2.2 ([PPZ19]) *Let ω_0 be a conformally balanced Hermitian metric. There exists an $\epsilon > 0$ such that Anomaly flow with zero slope admits a unique solution on $[0, \epsilon)$ with $\omega(0) = \omega_0$.*

2.3.2 Non-Kähler Examples

We outline here some simple examples to illustrate possible behaviors of the flow.

2.3.2.1 Iwasawa Manifold

Let $\pi : X \to T^4$ be the Iwasawa manifold considered in Sect. 2.2.3.3 with ansatz $\omega_u = e^u\hat{\omega} + i\theta \wedge \bar{\theta}$, where

$$\hat{\omega} = idx \wedge d\bar{x} + idy \wedge d\bar{y}, \quad \theta = dz - \bar{x}dy,$$

and $u(x, y)$ is a smooth function $u : T^4 \to \mathbf{R}$. We will show that this ansatz is preserved by the Anomaly flow. We previously computed that $\|\Omega\|_{\omega_u} = e^{-u}$, and so

$$\|\Omega\|_{\omega_u}\omega_u^2 = e^u\hat{\omega}^2 + 2i\hat{\omega} \wedge \theta \wedge \bar{\theta}.$$

Furthermore,

$$i\partial\bar{\partial}\omega_u = i\partial\bar{\partial}e^u \wedge \hat{\omega} + \frac{\hat{\omega}^2}{2}.$$

The Anomaly flow with zero slope $\partial_t(\|\Omega\|_\omega \omega^2) = i\partial\bar\partial\omega$ reduces to

$$\partial_t e^u = \frac{1}{2}(\Delta_{\hat\omega} e^u + 1). \tag{2.53}$$

The flow exists for all time by linear parabolic theory. The functional defined by

$$M(\omega(t)) = \int_X \|\Omega\|_{\omega(t)}\, \omega(t)^3,$$

satisfies in this case

$$\frac{d}{dt}M(t) = \frac{d}{dt}\int_X 3e^u \hat\omega^2 \wedge i\theta \wedge \bar\theta$$

$$= 3\int_X i\partial\bar\partial(e^u\hat\omega \wedge i\theta \wedge \bar\theta) + \frac{3}{2}\int_X \hat\omega^2 \wedge i\theta \wedge \bar\theta$$

$$= \frac{1}{2}\int_X (\hat\omega + i\theta \wedge \bar\theta)^3 > 0.$$

It follows that $M(t) \to \infty$ linearly as $t \to \infty$. The functional $M(\omega)$ is sometimes called the dilaton functional, and was introduced in [GRST18] to develop a variational formulation of the Hull–Strominger system.

Since (2.53) is a linear parabolic equation and $\int e^u \to \infty$ as $t \to \infty$, we also have that $e^u \to \infty$ everywhere on T^4 as $t \to \infty$. The geometric statement is that $\|\Omega\|_{\omega_u} \to 0$ everywhere on the base T^4. The flow cannot converge in this case since the Iwasawa manifold does not admit a Kähler metric.

2.3.2.2 Compact Quotients of $SL(2, \mathbf{C})$

Next, we study quotients of $SL(2, \mathbf{C})$ by a lattice Λ as described in Sect. 2.2.3.2. Let $\{e_a\}$ be a left-invariant basis of holomorphic vector fields with $[e_a, e_b] = \epsilon_{abd}e_d$. We will study the ansatz

$$\omega = \rho\hat\omega, \quad \hat\omega = ie^a \wedge \bar e^a,$$

where $\rho > 0$ is a constant. This ansatz was used by Fei–Yau to solve the Hull–Strominger system on complex Lie groups [FY15].

As computed in (2.36),

$$i\partial\bar\partial\omega = \rho\frac{\hat\omega^2}{2}.$$

Next, we compute using the definition of the norm (2.22) and obtain

$$\|\Omega\|_\omega = \rho^{-3/2}.$$

Thus

$$\|\Omega\|_\omega \omega^2 = (\rho^{-3/2}\rho^2)\hat\omega^2.$$

Using the ansatz $\omega = \rho\hat\omega$ on $X = SL(2, \mathbf{C})/\Lambda$, the Anomaly flow with zero slope becomes the ODE

$$\frac{d}{dt}(\rho^{1/2}) = \frac{1}{2}\rho,$$

whose solution is given by

$$\rho(t) = \frac{1}{(\rho(0)^{-1/2} - \frac{t}{2})^2}.$$

We see that the flow develops a singularity as $\rho \to \infty$ in finite time. In particular, there exists $T < \infty$ such that $\|\Omega\|_\omega \to 0$ as $t \to T$. The flow cannot converge since X does not admit a Kähler metric.

2.3.3 Kähler Manifolds

The previous two examples illustrate how the Anomaly flow can develop singularities on non-Kähler manifolds. If the manifold is already known to admit a Kähler metric, the flow should detect it. Since there are many different Kähler metrics on a given Kähler manifold, the flow must select a single one in the limit. We will explain this mechanism in this section and explain how the flow may provide insight in studying the relation between the Kähler cone and the balanced cone.

Let X be a compact complex manifold with Kähler metric $\hat\chi = i\hat\chi_{\bar k j}dz^j \wedge d\bar z^k$ and nowhere vanishing holomorphic $(n, 0)$ form Ω. We will start the Anomaly flow with zero slope with the initial data

$$\|\Omega\|_{\omega(0)}\omega(0)^{n-1} = \hat\chi^{n-1}. \tag{2.54}$$

This equation determines the initial metric $\omega(0)$, which is manifestly conformally balanced and is explicitly given by the following lemma.

Lemma 2.8 Let $\chi \in \Omega^{1,1}(X, \mathbf{R})$ be a Hermitian metric and $\Omega \in \Omega^{n,0}(X)$ be nowhere vanishing. The equation

$$\|\Omega\|_\omega \omega^{n-1} = \chi^{n-1} \tag{2.55}$$

admits a unique Hermitian metric solution ω given by

$$\omega = \|\Omega\|_\chi^{-2/(n-2)}\chi.$$

Proof We let

$$\omega = \|\Omega\|_\omega^{-1/(n-1)}\chi, \tag{2.56}$$

and so we only need to solve for the determinant. Taking the determinant of both sides of (2.55) and raising to the power of $\frac{-1}{(n-1)}$ gives

$$\|\Omega\|_\omega^{-n/(n-1)}(\det\omega)^{-1} = (\det\chi)^{-1}.$$

Recall that $\|\Omega\|_\omega^2 = \Omega\bar\Omega(\det\omega)^{-1}$. Multiplying both sides by $\Omega\bar\Omega$, we obtain

$$\|\Omega\|_\omega^2\|\Omega\|_\omega^{-n/(n-1)} = \|\Omega\|_\chi^2.$$

Therefore

$$\|\Omega\|_\omega^{1/(n-1)} = \|\Omega\|_\chi^{2/(n-2)}, \tag{2.57}$$

and the existence result follows from (2.56). For uniqueness, suppose ω and $\tilde\omega$ solve (2.55). Then (2.57) determines $\|\Omega\|_\omega = \|\Omega\|_{\tilde\omega}$ and so $\tilde\omega^{n-1} = \omega^{n-1}$, from which it follows [MI82] that $\omega = \tilde\omega$. □

We claim that the solution to the Anomaly flow with zero slope and initial data (2.54) is given by

$$\|\Omega\|_{\omega(t)}\omega(t)^{n-1} = \chi(t)^n, \tag{2.58}$$

where

$$\chi = \hat\chi + i\partial\bar\partial\varphi > 0,$$

and the scalar potential φ satisfies

$$\dot\varphi = e^{-f}\frac{(\hat\chi + i\partial\bar\partial\varphi)^n}{\hat\chi^n}, \quad \varphi(x,0) = 0,$$

(we use the notation $\dot\varphi = \partial_t\varphi$), with

$$e^{-f} = \frac{1}{(n-1)\|\Omega\|_{\hat\chi}^2}.$$

Indeed, the ansatz (2.58) solves the equation of the flow. To see this, we compute

$$\frac{d}{dt}\|\Omega\|_\omega \omega^{n-1} = (n-1)\dot\chi \wedge \chi^{n-2}$$

$$= (n-1)i\partial\bar\partial\dot\varphi \wedge \chi^{n-2}.$$

The equation for $\dot\varphi$ can be rearranged as

$$\dot\varphi = \frac{1}{(n-1)\|\Omega\|_\chi^2}.$$

Therefore

$$\frac{d}{dt}\|\Omega\|_\omega \omega^{n-1} = i\partial\bar\partial(\|\Omega\|_\chi^{-2}) \wedge \chi^{n-2}.$$

On the other hand, by Lemma 2.8, we have

$$i\partial\bar\partial\omega^{n-2} = i\partial\bar\partial(\|\Omega\|_\chi^{-2}\chi^{n-2})$$

$$= i\partial\bar\partial(\|\Omega\|_\chi^{-2}) \wedge \chi^{n-2}.$$

It follows that the ansatz (2.58) satisfies

$$\frac{d}{dt}\|\Omega\|_\omega \omega^{n-1} = i\partial\bar\partial\omega^{n-2}.$$

By uniqueness of solutions, the ansatz (2.58) is preserved by the Anomaly flow with zero slope. To summarize our discussion, we state the following result.

Theorem 2.7 ([PPZ19]) *Let X be a compact complex manifold of dimension n with a nowhere vanishing holomorphic $(n,0)$ form Ω. Suppose X admits a Kähler metric $\hat\chi$. Then the Anomaly flow $\frac{d}{dt}\|\Omega\|_\omega\omega^{n-1} = i\partial\bar\partial\omega^{n-2}$ with initial metric satisfying*

$$\|\Omega\|_{\omega(0)}\omega(0)^{n-1} = \hat\chi^{n-1} \tag{2.59}$$

reduces to the following scalar flow of potentials

$$\dot\varphi = e^{-f}\frac{\det(\hat\chi_{\bar k j} + \varphi_{\bar k j})}{\det\hat\chi_{\bar k j}}, \quad \varphi(x,0) = 0, \tag{2.60}$$

with the positivity condition $\hat\chi + i\partial\bar\partial\varphi > 0$, where $e^f = (n-1)\|\Omega\|_{\hat\chi}^2$. The evolving metric in the Anomaly flow is given by

$$\omega(t) = \|\Omega\|_{\chi(t)}^{-2/(n-2)}\chi(t), \quad \chi(t) = \hat\chi + i\partial\bar\partial\varphi. \tag{2.61}$$

The Monge–Ampère flow (2.60) arising here shares similarities with the Kähler–Ricci flow and the MA^{-1} flow. The Kähler–Ricci flow was introduced by Cao [CA852] and has since been an area of active research in Kähler geometry (e.g. [CSW18, DL17, GZ17, PS06, PT17, ST07, SW13, TZ06, TZ16]). The MA^{-1} flow was recently introduced by Collins–Hisamoto–Takahashi [CHT18], and is expected to produce optimal degenerations on Fano manifolds which do not admit Kähler-Einstein metrics.

Unlike the Kähler–Ricci flow, the logarithm does not appear in the speed of evolution $\dot{\varphi}$, and unlike the MA^{-1} flow, the determinant of χ appears in the numerator instead of the denominator. For general parabolic equations, changes in speed can have major implications in the analysis, see [FGP18] for a recent example of this phenomenon in Kähler geometry. Though the analysis of (2.60) does differ from the Kähler–Ricci flow and MA^{-1} flow, in [PPZ19] we show that a smooth solution to the flow exists for all time t.

In contrast to the previous examples in section Sect. 2.3.2, in this case we can easily show that $\|\Omega\|_\omega$ stays bounded above and below along the flow. Differentiating (2.60),

$$\partial_t \dot{\varphi} = e^{-f} \left\{ \frac{\det \chi_{\bar{k}j}}{\det \hat{\chi}_{\bar{k}j}} \right\} \chi^{j\bar{k}} \partial_j \partial_{\bar{k}} \dot{\varphi}.$$

This is a linear parabolic equation for $\dot{\varphi}$. It follows from the maximum principle for parabolic equations (e.g. Proposition 1.7 in [SW13]) that

$$\inf_X \dot{\varphi}(x, 0) \leq \dot{\varphi}(x, t) \leq \sup_X \dot{\varphi}(x, 0).$$

Since $\varphi(x, 0) = 0$, we have

$$\inf_X e^{-f} \leq \dot{\varphi}(x, t) \leq \sup_X e^{-f}.$$

By (2.60), we have

$$e^f \inf_X e^{-f} \leq \frac{\det \chi_{\bar{k}j}}{\det \hat{\chi}_{\bar{k}j}} \leq e^f \sup_X e^{-f}.$$

By (2.57),

$$\|\Omega\|_{\omega(t)} = \|\Omega\|_\chi^{2(n-1)/(n-2)} = \|\Omega\|_{\hat{\chi}}^{2(n-1)/(n-2)} \left(\frac{\det \hat{\chi}}{\det \chi} \right)^{(n-1)/(n-2)}.$$

Therefore

$$C^{-1} \leq \|\Omega\|_{\omega(t)} \leq C,$$

along the flow, where $C > 0$ only depends on $\|\Omega\|_{\hat{\chi}}$ and n. The degeneration of $\|\Omega\|_\omega$ exhibited for non-Kähler examples in Sect. 2.3.2 does not occur in this case.

Estimating $\|\Omega\|_{\omega(t)}$ is only the first step in the study of the flow. From here, we can use a priori estimates and techniques from fully nonlinear PDE to establish long-time existence and convergence. We refer to [PPZ19] for full details. The result is

Theorem 2.8 ([PPZ19]) *Let X be a compact complex manifold of dimension n with a nowhere vanishing holomorphic $(n, 0)$ form Ω. Suppose X admits a Kähler metric $\hat{\chi}$. Then the Anomaly flow $\frac{d}{dt}\|\Omega\|_\omega \omega^{n-1} = i\partial\bar{\partial}\omega^{n-2}$ with initial metric satisfying*

$$\|\Omega\|_{\omega(0)}\omega(0)^{n-1} = \hat{\chi}^{n-1}$$

exists for all time, and smoothly converges to a Kähler metric ω_∞.

In fact, ω_∞ is given explicitly by

$$\omega_\infty = \|\Omega\|_{\chi_\infty}^{-2/(n-2)}\chi_\infty,$$

where χ_∞ is the unique Kähler Ricci-flat metric in the cohomology class $[\hat{\chi}]$, and

$$\|\Omega\|_{\chi_\infty}^2 = \frac{n!}{[\hat{\chi}]^n}\int_X i^{n^2}\Omega\wedge\bar{\Omega}.$$

To conclude this section, we note that we cannot expect the Anomaly flow on Kähler manifolds to converge starting from an arbitrary metric. This is due to the relationship between the Kähler cone and the balanced cone. Indeed, an initial conformally balanced metric determines a balanced class

$$[\|\Omega\|_{\omega(0)}\omega(0)^{n-1}] \in H_{BC}^{n-1,n-1}(X),$$

and the evolving metric $\omega(t)$ remains in this class (2.41). Stationary points of the flow are Kähler metrics, so convergence of the flow would produce a Kähler metric in the balanced class of the initial metric. However, there exists Kähler manifolds with balanced classes which do not admit any Kähler metric [FX14, TO09]. Understanding which balanced classes come from Kähler classes is an interesting problem in Hermitian geometry [FX14], and we hope that future work studying the Anomaly flow and its singularities will provide insight.

2.4   Anomaly Flow with α' Corrections

We will now restrict our attention to Calabi–Yau threefolds. In this section, we modify the Anomaly flow (2.40) by adding α' correction terms. The parameter $\alpha \in \mathbf{R}$ will be referred to as the slope parameter.

Let X be a compact complex manifold of dimension $n = 3$. Suppose X admits a nowhere vanishing holomorphic $(3, 0)$ form Ω. We first study the case of threefolds with vanishing second Chern class, so we assume that $c_1(X) = c_2(X) = 0$. Consider the flow

$$\frac{d}{dt}(\|\Omega\|_\omega \omega^2) = i\partial\bar\partial\omega - \frac{\alpha'}{4} \operatorname{Tr} Rm \wedge Rm, \qquad (2.62)$$

$$d(\|\Omega\|_{\omega(0)}\omega(0)^{n-1}) = 0.$$

Recall that we use the notation Rm for the endomorphism-valued $(1, 1)$ form which is the curvature of the Chern connection of ω. When $\alpha' = 0$ and $n = 3$, this flow becomes (2.40) from Sect. 2.3. Stationary points ω_∞ satisfy

$$\frac{\alpha'}{4} \operatorname{Tr} Rm \wedge Rm = i\partial\bar\partial\omega_\infty, \quad d(\|\Omega\|_{\omega_\infty}\omega_\infty^2) = 0,$$

which can be viewed as a sort of non-Kähler analog of the Kähler–Einstein equation

$$\operatorname{Tr} Rm = \lambda\omega, \quad d\omega = 0.$$

More generally, if $c_2(X) \neq 0$, we can add a cancellation term $\Phi \in \Omega^{2,2}(X, \mathbf{R})$ with $[\Phi] = c_2(X)$, and consider the flow

$$\frac{d}{dt}(\|\Omega\|_\omega \omega^2) = i\partial\bar\partial\omega - \frac{\alpha'}{4}(\operatorname{Tr} Rm \wedge Rm - \Phi(t)), \qquad (2.63)$$

$$d(\|\Omega\|_{\omega(0)}\omega(0)^2) = 0.$$

Flows of type (2.63) are called Anomaly flows, as introduced in joint work with Phong and Zhang [PPZ218, PPZ318]. The motivation for studying this evolution equation comes from theoretical physics, which we describe next.

2.4.1 Hull–Strominger System

Our motivation for adding the α' correction terms comes from heterotic string theory. The Hull–Strominger system [HU186, ST86] is the following system of equations on a Calabi–Yau threefold

$$F \wedge \omega^2 = 0, \quad F^{0,2} = F^{2,0} = 0, \qquad (2.64)$$

$$i\partial\bar\partial\omega - \frac{\alpha'}{4}(\operatorname{Tr} Rm \wedge Rm - \operatorname{Tr} F \wedge F) = 0, \qquad (2.65)$$

$$d(\|\Omega\|_\omega \omega^2) = 0. \qquad (2.66)$$

The system is a coupled equation for a Hermitian meric ω on X and a metric h on a given holomorphic vector bundle $E \to X$. Here Rm, F are the curvature forms of unitary connections of ω, h, viewed as endomorphism valued 2-forms.

Equation (2.64) is the Hermitian-Yang-Mills equation, which admits solutions as long as E is stable of degree zero with respect to ω by the Donaldson-Uhlenbeck-Yau theorem [DO85, UY85] (see [LY86, BU88] for its extension to the Hermitian setting). Equation (2.65) is the Green-Schwarz anomaly cancellation equation from theoretical physics [GS87]. All together, the system was introduced by Hull and Strominger as a model for the heterotic string admitting non-zero torsion, generalizing the equation proposed by Candelas–Horowitz–Strominger–Witten [CA851] where the threefold is required to be Kähler with Ricci-flat metric.

For example, Kähler Calabi–Yau threefolds provide solutions to the Hull–Strominger system. In this case, we take the gauge bundle E to be the tangent bundle $E = T^{1,0}X$, and $h = \omega$ to be Kähler Ricci-flat. Then (2.64) and (2.65) hold automatically, and by the argument in Sect. 2.2.3.1, we see that ω is conformally balanced.

Going beyond Kähler geometry, there are many diverse examples of solutions using various gauge bundles E. The first solutions in the mathematics literature were obtained by Li and Yau [LY05] by perturbing the Kähler solutions, and the first solutions on non-Kähler manifolds were obtained by Fu and Yau [FY08]. Since then, there have been constructions of parallelizable examples [FIUV14, FIUV14, FY15, OUV17, GR11], solutions on Kähler manifolds for arbitrary admissible gauge bundles [AG121, AG122], solutions on fibrations over a Riemann surface [FHP17], and non-compact examples [FY09, FE17, HIS16].

The Hull–Strominger system is interesting from the point of view of canonical metrics on non-Kähler Calabi–Yau threefolds, as it is a curvature constraint (2.65) combined with a closedness condition (2.66). There are also other proposed optimal metrics in non-Kähler complex geometry: e.g. constant Chern scalar curvature [ACS17], vanishing Chern–Ricci curvature [TW10, TW17, STW17], Chern–Ricci flat balanced [FE17], just to name a few.

As a system of partial differential equations, the Hull–Strominger system is fully nonlinear. It can be viewed as an analog of the σ_2 equation, but as a full system for the metric tensor $g_{\bar{k}j}$. There has been much progress in the study of scalar σ_k-type equations in complex geometry e.g. [BL05, CJY15, DDT17, DL15, DK17, DPZ18, HMW10, PPZ116], but very little is known about PDE systems which are nonlinear in second derivatives.

To study the Hull–Strominger system, it was proposed in [PPZ218] to use the Anomaly flow with $\Phi = \mathrm{Tr}\, F \wedge F$ coupled to the Donaldson heat flow [DO85].

$$h^{-1}\partial_t h = -\Lambda_\omega F,$$

$$\frac{d}{dt}(\|\Omega\|_\omega \omega^2) = i\partial\bar\partial\omega - \frac{\alpha'}{4}(\mathrm{Tr}\, Rm \wedge Rm - \mathrm{Tr}\, F \wedge F),$$

$$d(\|\Omega\|_{\omega(0)}\omega(0)^2) = 0.$$

Stationary points solve the Hull–Strominger system. The Anomaly flow, when restricted to certain ansatzes, provides new nonlinear equations arising naturally from geometry and physics [PPZ217, PPZ317]. We will describe some of these new equations in the following sections.

2.4.2 Evolution of the Metric

We now derive the evolution of the metric tensor $\omega = i g_{\bar{k}j} dz^j \wedge d\bar{z}^k$ under the Anomaly flow (2.63). The argument given here is similar to the one from Sect. 2.3.1. We write

$$\frac{d}{dt}(\|\Omega\|_\omega \omega^2) = \Psi,$$

with

$$\Psi = \left[i \partial \bar{\partial} \omega - \frac{\alpha'}{4} (\mathrm{Tr}\, Rm \wedge Rm - \Phi) \right].$$

By Lemma 2.6, we already know that the trace of the evolution of the metric is given by

$$\mathrm{Tr}\,\dot{\omega} = \frac{6}{\|\Omega\|_\omega} \frac{\Psi \wedge \omega}{\omega^3},$$

which combined with identity (2.29) is

$$\mathrm{Tr}\,\dot{\omega} = \frac{1}{2\|\Omega\|_\omega} \mathrm{Tr}\,\Psi. \tag{2.67}$$

As in (2.43), we expand the flow to the following expression

$$-\frac{1}{2}(\mathrm{Tr}\,\dot{\omega})\omega^2 + 2\dot{\omega} \wedge \omega - \frac{1}{\|\Omega\|_\omega}\Psi = 0. \tag{2.68}$$

We apply the Hodge star operator $\star$ with respect to ω to both sides of the equation. By identities (2.44), (2.48), and $\star \omega^2 = 2\omega$, the components of the resulting $(1, 1)$ form are given by

$$0 = \star \left[-\frac{1}{2}(\mathrm{Tr}\,\dot{\omega})\omega^2 + 2\dot{\omega} \wedge \omega - \frac{1}{\|\Omega\|_\omega}\Psi \right]_{\bar{k}j}$$

$$= -2i \partial_t g_{\bar{k}j} + i(\mathrm{Tr}\,\dot{\omega})g_{\bar{k}j} - \frac{1}{\|\Omega\|_\omega}\left[-i g^{s\bar{r}}\Psi_{\bar{r}ks j} + \frac{i}{2}(\mathrm{Tr}\,\Psi)g_{\bar{k}j} \right]. \tag{2.69}$$

Substituting the expression for $\text{Tr}\,\dot{\omega}$ (2.67) into (2.69), we see that the $\text{Tr}\,\Psi$ terms cancel and the evolution of the metric is

$$\frac{d}{dt}g_{\bar{k}j} = \frac{1}{2\|\Omega\|_\omega} g^{s\bar{r}}\Psi_{\bar{r}ksj}.$$

From here, we can derive an explicit expression for the evolution of the metric.

Theorem 2.9 ([PPZ318]) *Suppose $\omega(t)$ solves the Anomaly flow (2.63). Then the metric evolves by*

$$\frac{d}{dt}g_{\bar{k}j} = \frac{1}{2\|\Omega\|_\omega}\left[-\tilde{R}_{\bar{k}j} + g^{s\bar{r}}g^{n\bar{m}}T_{\bar{m}sj}\bar{T}_{n\bar{r}\bar{k}} - \frac{\alpha'}{4}g^{s\bar{r}}(R_{[\bar{k}s}{}^\alpha{}_\beta R_{\bar{r}j]}{}^\beta{}_\alpha - \Phi_{\bar{r}\bar{k}sj})\right], \tag{2.70}$$

where $[,]$ denotes antisymmetrization in both barred and unbarred indices.

Proof We have already established

$$\frac{d}{dt}g_{\bar{k}j} = \frac{1}{2\|\Omega\|_\omega}\left[-g^{s\bar{r}}(i\partial\bar{\partial}\omega)_{\bar{r}s\bar{k}j} - \frac{\alpha'}{4}g^{s\bar{r}}(\text{Tr}\,Rm \wedge Rm - \Phi)_{\bar{r}ksj}\right].$$

By (2.51), we have an expression for $g^{s\bar{r}}(i\partial\bar{\partial}\omega)_{\bar{r}ksj}$ in terms of Ricci curvature and torsion. This gives the desired expression. □

We note that (2.70) is a fully nonlinear system, as it is quadratic in the curvature. For other geometric flows which are quadratic in the curvature, see e.g. [FR85, GGI13, OL09]. Since the flow is fully nonlinear, we cannot expect short-time existence for arbitrary initial data. However, from (2.70), we see that the right-hand side is parabolic if the α' correction terms are small. The full details are provided in [PPZ218].

Theorem 2.10 ([PPZ218]) *Let ω_0 be a conformally balanced Hermitian metric on X satisfying $|\alpha'Rm| < \frac{1}{2}$. Then there exists $T > 0$ such that the Anomaly flow (2.63) admits a unique solution $\omega(t)$ on $[0, T)$ with $\omega(0) = \omega_0$.*

Given any metric $g_{\bar{k}j}$, we can find $\lambda \gg 1$ so that $\lambda g_{\bar{k}j}$ satisfies $|\alpha'Rm| \ll 1$. This is simply because $Rm(\lambda g) = Rm(g)$ (with Rm defined as in (2.14)). Thus to guarantee short-time existence starting from a given metric, we can rescale the size of the manifold, or choose a small value for α'. For several examples [FHP17, PPZ418], the condition $|\alpha'Rm| \ll 1$ is preserved along the flow, which suggests that it is a natural condition.

2.4.3 Anomaly Flow with Fu–Yau Ansatz

2.4.3.1 Scalar Reduction

In this section, we return to the construction of Goldstein–Prokushkin described in Sect. 2.2.3.4. We first recall the setup.

The base of the fibration $(S, \hat{\omega}, \Omega_S)$ is a Calabi–Yau surface with Kähler Ricci-flat metric $\hat{\omega}$ and nowhere vanishing holomorphic $(2, 0)$ form Ω_S. Let $\omega_1, \omega_2 \in 2\pi H^2(S, \mathbf{Z})$ be anti-self-dual $(1, 1)$ forms. Using this data, Goldstein and Prokushkin [GO04] constructed a T^2 fibration $\pi : X \to S$ which is non-Kähler but admits conformally balanced metrics. Their construction builds on earlier ideas of Calabi and Eckmann [CE53], which we discussed in detail in Sect. 2.1.4.2.

We recall that the connections of the $U(1)$ principal bundles forming the S^1 fibers of X define $\theta \in \Omega^{1,0}(X)$ satisfying

$$\partial\theta = 0, \ \ \bar{\partial}\theta = \omega_1 + i\omega_2.$$

Furthermore,

$$\Omega = \Omega_S \wedge \theta$$

is a nowhere vanishing holomorphic $(3, 0)$ form on X, and the family of metrics

$$\omega_u = e^u \hat{\omega} + i\theta \wedge \bar{\theta}, \tag{2.71}$$

is conformally balanced for any $u : S \to \mathbf{R}$. These metrics were used by Fu and Yau [FY08, FY07] to solve the Hull–Strominger system on the threefold X.

In this section, we will start the Anomaly flow with a metric of this form, and check whether the ansatz is preserved. For this, we compute (see (2.39))

$$\|\Omega\|_{\omega_u} = e^{-u}, \ \ \ \|\Omega\|_{\omega_u} \omega_u^2 = e^u \hat{\omega}^2 + 2\hat{\omega} \wedge i\theta \wedge \bar{\theta}, \tag{2.72}$$

and

$$i\partial\bar{\partial}\omega_u = i\partial\bar{\partial}e^u \wedge \hat{\omega} - \bar{\partial}\theta \wedge \partial\bar{\theta} = i\partial\bar{\partial}e^u \wedge \hat{\omega} - (\omega_1^2 + \omega_2^2). \tag{2.73}$$

Next, we must compute the curvature terms. This calculation was done by Fu and Yau in [FY08].

Theorem 2.11 ([FY08]) *The curvature of the Chern connection of ω_u satisfies*

$$\mathrm{Tr}\, Rm(\omega_u) \wedge Rm(\omega_u) = \mathrm{Tr}\, Rm(\hat{\omega}) \wedge Rm(\hat{\omega}) + 2\partial\bar{\partial}u \wedge \partial\bar{\partial}u + 4i\partial\bar{\partial}(e^{-u}\rho),$$

where $\rho \in \Omega^{1,1}(S, \mathbf{R})$ is given by $\rho = \rho_{\bar{k}j}\, dz^j \wedge d\bar{z}^k$ with

$$\rho_{\bar{k}j} = \frac{i}{2}\hat{g}^{p\bar{q}}(\omega_1 - i\omega_2)_{\bar{q}j}(\omega_1 + i\omega_2)_{\bar{k}p}. \tag{2.74}$$

Proof We work in a local coordinate chart. Since $\bar{\partial}(\omega_1 + i\omega_2) = 0$, there are local functions φ_1, φ_2 such that

$$\bar{\partial}(\varphi_i dz^i) = \omega_1 + i\omega_2, \tag{2.75}$$

where z^1, z^2 are local holomorphic coordinates on the base S. Define

$$\theta_0 = \theta - \varphi_1 dz^1 - \varphi_2 dz^2.$$

Then $\{dz^1, dz^2, \theta_0\}$ is a local holomorphic frame of $\Omega^{1,0}(X)$. The metric can be written as

$$\omega_u = (e^u \hat{g}_{\bar{k}j} + \overline{\varphi_k}\varphi_j)i dz^j \wedge d\bar{z}^k$$
$$+ \overline{\varphi_k} i\theta_0 \wedge d\bar{z}^k + \varphi_k\, i dz^k \wedge \overline{\theta_0} + i\theta_0 \wedge \overline{\theta_0}.$$

Let $B = (\varphi_1, \varphi_2)$. Then the metric in this local frame is given by

$$g = \begin{bmatrix} (e^u \hat{g} + B^* B) & B^* \\ B & 1 \end{bmatrix}.$$

Its inverse is

$$g^{-1} = \begin{bmatrix} e^{-u}\hat{g}^{-1} & -e^{-u}\hat{g}^{-1}B^* \\ -e^{-u}B\hat{g}^{-1} & 1 + e^{-u}B\hat{g}^{-1}B^* \end{bmatrix}.$$

The curvature in this frame is $Rm = \bar{\partial}g^{-1}\partial g$. Computing at a point $p \in X$, we may assume that $p = 0$ and $B(0) = 0$. The curvature at p is then

$$Rm = \begin{bmatrix} R_{\bar{1}1} & R_{\bar{1}2} \\ R_{\bar{2}1} & R_{\bar{2}2} \end{bmatrix},$$

with

$$R_{\bar{1}1} = \bar{\partial}\partial u \cdot I + \hat{Rm} - e^{-u}\hat{g}^{-1}\partial B^* \wedge \bar{\partial}B$$
$$R_{\bar{2}1} = -\bar{\partial}B \wedge \partial u - \bar{\partial}B\hat{g}^{-1}\partial\hat{g} + \bar{\partial}\partial B$$
$$R_{\bar{1}2} = \bar{\partial}(e^{-u}\hat{g}^{-1}\partial B^*)$$
$$R_{\bar{2}2} = -e^{-u}\bar{\partial}B\hat{g}^{-1}\partial B^*.$$

We must compute

$$\mathrm{Tr}\, Rm \wedge Rm = \mathrm{Tr}\, R_{\bar{1}1} R_{\bar{1}1} + \mathrm{Tr}\, R_{\bar{1}2} R_{\bar{2}1} + \mathrm{Tr}\, R_{\bar{2}1} R_{\bar{1}2} + \mathrm{Tr}\, R_{\bar{2}2} R_{\bar{2}2}.$$

Expanding this out, we obtain the following expression.

$$\begin{aligned}
\mathrm{Tr}&Rm \wedge Rm \\
&= 2(\bar{\partial}\partial u)^2 + \mathrm{Tr}\,\hat{Rm}^2 + e^{-2u}\mathrm{Tr}\,(\hat{g}^{-1}\partial B^*\bar{\partial}B\hat{g}^{-1}\partial B^*\bar{\partial}B) \\
&\quad + 2\partial\bar{\partial}u\mathrm{Tr}\,\hat{Rm} - 2e^{-u}\bar{\partial}\partial u\mathrm{Tr}\,\hat{g}^{-1}\partial B^*\bar{\partial}B - 2e^{-u}\mathrm{Tr}(\hat{Rm}\hat{g}^{-1}\partial B^*\bar{\partial}B) \\
&\quad - 2\mathrm{Tr}(\bar{\partial}(e^{-u}\hat{g}^{-1}\partial B^*)\bar{\partial}B\partial u) - 2\mathrm{Tr}(\bar{\partial}(e^{-u}\hat{g}^{-1}\partial B^*)\bar{\partial}B\hat{g}^{-1}\partial\hat{g}) \\
&\quad + 2\mathrm{Tr}(\bar{\partial}(e^{-u}\hat{g}^{-1}\partial B^*)\bar{\partial}\partial B) + e^{-2u}\bar{\partial}B\hat{g}^{-1}\partial B^*\bar{\partial}Bg^{-1}\partial B^*.
\end{aligned}$$

Using the identities

$$\begin{aligned}
-2\mathrm{Tr}\,\bar{\partial}(e^{-u}\hat{g}^{-1}\partial B^*)\bar{\partial}B\hat{g}^{-1}\partial\hat{g} &= -2\bar{\partial}\mathrm{Tr}(e^{-u}\hat{g}^{-1}\partial B^*\bar{\partial}B\hat{g}^{-1}\partial\hat{g}) \\
&\quad + 2\mathrm{Tr}(e^{-u}\hat{g}^{-1}\partial B^*\bar{\partial}B\,\hat{Rm}),
\end{aligned}$$

and

$$\begin{aligned}
-2e^{-u}\bar{\partial}\partial u\,\mathrm{Tr}(\hat{g}^{-1}\partial B^*\bar{\partial}B) &= -2\bar{\partial}\mathrm{Tr}(e^{-u}\hat{g}^{-1}\partial B^*\bar{\partial}B\partial u) \\
&\quad + 2\mathrm{Tr}\,\bar{\partial}(e^{-u}\hat{g}^{-1}\partial B^*)(\bar{\partial}B\partial u),
\end{aligned}$$

as well as $\mathrm{Tr}\,\hat{Rm} = 0$, we cancel a few terms and are left with

$$\begin{aligned}
\mathrm{Tr}\, Rm \wedge Rm &= 2(\bar{\partial}\partial u)^2 + \mathrm{Tr}\hat{Rm}^2 - 2\bar{\partial}\mathrm{Tr}(e^{-u}\hat{g}^{-1}\partial B^*\bar{\partial}B\hat{g}^{-1}\partial\hat{g}) \\
&\quad - 2\bar{\partial}\mathrm{Tr}(e^{-u}\hat{g}^{-1}\partial B^*\bar{\partial}B\partial u) + 2\bar{\partial}\mathrm{Tr}(e^{-u}\hat{g}^{-1}\partial B^*\bar{\partial}\partial B).
\end{aligned}$$

Using $\partial\hat{g}^{-1} = -\hat{g}^{-1}\partial\hat{g}\,\hat{g}^{-1}$, this expression simplifies to

$$\mathrm{Tr}\, Rm \wedge Rm = 2(\bar{\partial}\partial u)^2 + \mathrm{Tr}\,\hat{Rm} \wedge \hat{Rm} + 2\bar{\partial}\partial\mathrm{Tr}(e^{-u}\hat{g}^{-1}\partial B^* \wedge \bar{\partial}B).$$

We have by definition

$$\partial B^* \wedge \bar{\partial}B = \begin{pmatrix} \partial_i\overline{\varphi_1}\partial_{\bar{k}}\varphi_1 & \partial_i\overline{\varphi_1}\partial_{\bar{k}}\varphi_2 \\ \partial_i\overline{\varphi_2}\partial_{\bar{k}}\varphi_1 & \partial_i\overline{\varphi_2}\partial_{\bar{k}}\varphi_1 \end{pmatrix} dz^i \wedge d\bar{z}^k.$$

Using (2.75), we obtain (2.74). □

We now add a gauge bundle to the system. Let E_S be a stable vector bundle of degree zero over the base Kähler surface $(S, \hat{\omega})$. By the Donaldson-Uhlenbeck-Yau theorem [DO85, UY85], we may equip E_S with a metric H_S satisfying

$$F(H_S) \wedge \hat{\omega} = 0.$$

On the threefold, we consider the bundle $E = \pi^* E_S \to X$ with metric $H = \pi^* H_S$. This metric is Hermitian–Yang–Mills with respect to the Fu–Yau ansatz ω_u, since

$$F(H) \wedge \omega_u^2 = 0$$

for any $u \in C^\infty(S, \mathbf{R})$.

Putting together everything computed so far, we have

$$i\partial\bar{\partial}\omega_u - \frac{\alpha'}{4}(\mathrm{Tr}Rm(\omega_u) \wedge Rm(\omega_u) - \mathrm{Tr}\,F(H) \wedge F(H))$$

$$= i\partial\bar{\partial}(e^u\hat{\omega} - \alpha'e^{-u}\rho) - \frac{\alpha'}{2}(\partial\bar{\partial}u) \wedge (\partial\bar{\partial}u) + \mu, \qquad (2.76)$$

where $\mu \in \Omega^{2,2}(S, \mathbf{R})$ is given by

$$\mu = \frac{\alpha'}{4}(\mathrm{Tr}\,F(H_S) \wedge F(H_S) - \mathrm{Tr}Rm(\hat{\omega}) \wedge Rm(\hat{\omega})) - (\omega_1^2 + \omega_2^2).$$

Combining (2.72) and (2.76), we see that the Anomaly flow reduces to the following scalar fully nonlinear PDE on the base manifold S.

$$\frac{d}{dt}e^u\,\hat{\omega}^2 = i\partial\bar{\partial}(e^u\hat{\omega} - \alpha'e^{-u}\rho) + \frac{\alpha'}{2}(i\partial\bar{\partial}u)^2 + \mu. \qquad (2.77)$$

This evolution equation can also be written as

$$\frac{d}{dt}e^u = \frac{1}{2}\left[\Delta_{\hat{\omega}}e^u - \alpha'\frac{i\partial\bar{\partial}(e^{-u}\rho)}{\hat{\omega}^2/2!} + \alpha'\hat{\sigma}_2(i\partial\bar{\partial}u) + \frac{\mu}{\hat{\omega}^2/2!}\right].$$

Here $\hat{\sigma}_2(i\partial\bar{\partial}u) = (i\partial\bar{\partial}u)^2\hat{\omega}^{-2}$ is the determinant of the complex Hessian of u with respect to $\hat{\omega}$.

By standard parabolic theory, this equation admits a short-time solution as long as

$$\omega' = e^u\hat{\omega} + \alpha'e^{-u}\rho + \frac{\alpha'}{2}i\partial\bar{\partial}u > 0.$$

2.4.3.2 Stationary Points

For stationary points of (2.77) to exist, integrating both sides shows that we require

$$\int_S \mu = 0,$$

which is the cohomological constraint

$$\frac{\alpha'}{4}\int_S \left[\mathrm{Tr}\,Rm(\hat{\omega}) \wedge Rm(\hat{\omega}) - \mathrm{Tr}\,F(H_S) \wedge F(H_S)\right] = \int_S [|\omega_1|^2 + |\omega_2|^2]\frac{\hat{\omega}^2}{2!}.$$

It is possible to construct data $(S, E_S, \omega_1, \omega_2, \alpha')$ satisfying this condition. Indeed, since we assume $c_1(S) = c_1(E_S) = 0$, the constraint is

$$\frac{\alpha'}{4}[c_2(S) - c_2(E_S)] = \int_S \left[\left|\frac{\omega_1}{2\pi}\right|_{\hat{\omega}}^2 + \left|\frac{\omega_2}{2\pi}\right|_{\hat{\omega}}^2\right]\frac{\hat{\omega}^2}{2}.$$

Note that when seeking solutions to the Hull–Strominger system, after rescaling $\omega_u \mapsto \lambda\omega_u$ in (2.65) we can assume that $\frac{\alpha'}{4} \in \mathbf{Z}$. Explicit examples are exhibited in [FY08, FY07]; when $\alpha' > 0$, we may take S to be a $K3$ surface and use the theory of stable bundles over $K3$ surfaces to construct E_S, and when $\alpha' < 0$ we may take S to be either a torus T^4 or a $K3$ surface.

The main theorem of Fu–Yau guarantees the existence of smooth solutions to the Hull–Strominger system when the cohomological condition $\int_S \mu = 0$ is satisfied.

Theorem 2.12 ([FY08, FY07]) *Let $(S, \hat{\omega})$ be a Kähler surface, $\alpha' \in \mathbf{R}$, $\rho \in \Omega^{1,1}(S, \mathbf{R})$, and $\mu \in \Omega^{2,2}(S, \mathbf{R})$. Suppose μ satisfies the condition $\int_S \mu = 0$. Then there exists a smooth function $u : S \to \mathbf{R}$ solving*

$$0 = i\partial\bar{\partial}(e^u\hat{\omega} - \alpha'e^{-u}\rho) + \frac{\alpha'}{2}(i\partial\bar{\partial}u)^2 + \mu,$$

such that $\omega' = e^u\hat{\omega} + \alpha'e^{-u}\rho + \frac{\alpha'}{2}i\partial\bar{\partial}u > 0$.

For further work relating to the Fu-Yau solutions, we refer to [CHZ118, CHZ218, GA40, LE11, PPZ117, PPZ116, PPZ216, PPZ118].

2.4.3.3   Long-Time Existence

The first observation in the Anomaly flow with Fu-Yau ansatz is the following conserved quantity.

Lemma 2.9 *Let $\omega(t) = e^{u(t)}\hat{\omega} + i\theta \wedge \bar{\theta}$ be a solution to the Anomaly flow with the cohomology condition $\int_S \mu = 0$ satisfied. Then the conservation law*

$$\frac{d}{dt}\int_X \|\Omega\|_\omega \omega^3 = 0,$$

holds along the flow.

Proof In the case of the Fu-Yau ansatz $\omega = e^u \hat{\omega} + i\theta \wedge \bar{\theta}$, by (2.72) we have

$$\int_X \|\Omega\|_\omega \omega^3 = \int_X 3 e^u \hat{\omega}^2 \wedge i\theta \wedge \bar{\theta}.$$

Using $\int_S \mu = 0$, from (2.77) we see that

$$\frac{d}{dt} \int_S e^u \hat{\omega}^2 = 0$$

is a conserved quantity. $\square$

Together with D.H. Phong and X.-W. Zhang, we prove the following result.

Theorem 2.13 ([PPZ418]) *There exists $L_0 \gg 1$ depending only on $(S, \hat{\omega})$, μ, ρ, α' with the following property. Suppose $\int_S \mu = 0$. Start the Anomaly flow on the fibration $\pi : X \to S$ with initial data*

$$\omega(0) = L\hat{\omega} + i\theta \wedge \bar{\theta},$$

for any constant $L \geq L_0$. Then the flow exists for all time, and converges to a solution to the Hull–Strominger system.

For initial data with small L, we suspect that the flow will develop singularities. We will discuss in Sect. 2.4.4.1 an example of the Anomaly flow over Riemann surfaces where this behavior is observed.

Different choices of L correspond to different balanced classes of the stationary point. We know that the balanced class $[\|\Omega\|_\omega \omega^2] \in H^4(X, \mathbf{R})$ is preserved by the Anomaly flow, and in this case

$$[\|\Omega\|_\omega \omega^2] = [e^u \hat{\omega}^2] + 2[\hat{\omega} \wedge i\theta \wedge \bar{\theta}].$$

The class $[e^u \hat{\omega}^2] \in H^4(S, \mathbf{R})$ is a top cohomology class on the Kähler surface S, and is therefore parametrized by the integrals

$$\int_S e^u \hat{\omega}^2 \in \mathbf{R}.$$

Therefore the choice of $\int_S e^u \hat{\omega}^2$ in the initial data is related to the choice of balanced class of the evolving metric.

As an aside, we note that in general, the conservation of the balanced class $[\|\Omega\|_\omega \omega^2] \in H^{2,2}_{BC}(X)$ along the Anomaly flow should lead to conserved quantities, which may also be useful when studying the flow beyond the Fu–Yau ansatz. The Bott–Chern cohomology of complex manifolds differs in general from the de Rham cohomology, and we refer to [AT13, AN13, ADT16] for recent progress on computing Bott–Chern cohomology.

2.4.4   Nonlinear Blow-Up

In this section, we briefly describe a few more examples and illustrate some of the nonlinear phenomena which can occur.

2.4.4.1   Fibrations over Riemann Surfaces

We return to the construction of fibrations $p : X \to \Sigma$ over a Riemann surface $(\Sigma, \hat{\omega})$ of genus $g \geq 3$ described in Sect. 2.2.3.5. We recall that these were non-Kähler threefolds, and the Fei ansatz metrics

$$\omega_f = e^{2f} \hat{\omega} + e^f \omega',$$

are conformally balanced for any smooth function $f : \Sigma \to \mathbf{R}$.

It is not immediately clear that this family of metrics will be preserved by the Anomaly flow. It turns out that this is indeed the case, and the flow reduces to a single scalar parabolic PDE for f on the base Σ of the fibration. The key computation in [FE15, FHP17] gives the identity

$$i\partial\bar{\partial}\omega_f - \frac{\alpha'}{4}\mathrm{Tr}\, Rm(\omega_f) \wedge Rm(\omega_f) = (i\partial\bar{\partial}u - \kappa u\hat{\omega}) \wedge \omega',$$

where

$$u = e^f + \frac{\alpha'}{2}\kappa e^{-f}.$$

and $\kappa \leq 0$ is the Gauss curvature of the background metric $\hat{\omega}$. Since

$$\|\Omega\|_{\omega_f}\omega_f^2 = 2\mathrm{vol}_{T^4} + 2e^f \hat{\omega} \wedge \omega',$$

we can factor out ω' in the formulation of the Anomaly flow as $(2, 2)$ forms, and the flow reduces to

$$\partial_t e^f = \frac{1}{2}\left[\hat{g}^{z\bar{z}}\partial_z\partial_{\bar{z}}\left(e^f + \frac{\alpha'}{2}\kappa e^{-f}\right) - \kappa\left(e^f + \frac{\alpha'}{2}\kappa e^{-f}\right)\right], \qquad (2.78)$$

on the Riemann surface $(\Sigma, \hat{\omega})$. The flow admits a short-time solution as long as

$$e^f - \frac{\alpha'}{2}\kappa e^{-f} > 0,$$

which is automatic if $\alpha' > 0$. In [FHP17], together with T. Fei and Z. Huang, we study the asymptotics of the flow.

Theorem 2.14 ([FHP17]) *There exists $L_0 \gg 1$ depending on $(\Sigma, \hat{\omega})$ and α' with the following property. Start Anomaly flow with initial data*

$$\omega(0) = L^2 \hat{\omega} + L\omega',$$

for any constant $L \geq L_0$. Then the flow exists for all time and

$$\frac{\omega_f}{\frac{1}{3!} \int_X \|\Omega\|_{\omega_f} \omega_f^3} \to p^* \omega_\Sigma,$$

where $\omega_\Sigma = q_1^2 \hat{\omega}$ is a smooth metric on Σ, and $q_1 > 0$ is the first eigenfunction of the operator $-\Delta_{\hat{\omega}} + 2\kappa$.

In the above theorem, we have long-time existence, but unlike Theorem 2.13, $\|\Omega\|_{\omega_f} \to 0$ as $t \to \infty$. This can be understood by the fact that there are no stationary points in the large radius regime $e^f \gg 1$. We note that the result in [FHP17] is more general than the one stated above; the asymptotic behavior holds if the initial data satisfies $u(x, 0) \geq 0$.

For initial data with small L, finite-time blow-up can occur. Indeed, following [FHP17], we consider the case when $\alpha' > 0$. If

$$L^2 < \frac{8\alpha' \pi^2 (g - 1)^2}{\|\kappa\|_{L^\infty(\Sigma)} \text{Vol}(\Sigma, \hat{\omega})^2}, \tag{2.79}$$

then the flow encounters a singularity in finite time. To see this, we compute using the evolution equation (2.78), and use that $\kappa \leq 0$ and that the Laplacian integrates to zero.

$$\frac{d}{dt} \int_\Sigma e^f \hat{\omega} = \frac{1}{2} \int_\Sigma |\kappa| e^f \hat{\omega} - \frac{\alpha'}{4} \int_\Sigma \kappa^2 e^{-f} \hat{\omega}.$$

By the Cauchy–Schwarz inequality and the Gauss–Bonnet theorem,

$$(4\pi(g - 1))^2 = \left(\int_\Sigma |\kappa| \hat{\omega} \right)^2 \leq \left(\int_\Sigma e^f \hat{\omega} \right) \left(\int_\Sigma \kappa^2 e^{-f} \hat{\omega} \right).$$

Therefore

$$\frac{d}{dt} \left[\int_\Sigma e^f \hat{\omega} \right] \leq \frac{\|\kappa\|_{L^\infty(\Sigma)}}{2} \left[\int_\Sigma e^f \hat{\omega} \right] - \frac{\alpha'}{4} (4\pi(g - 1))^2 \left[\int_\Sigma e^f \hat{\omega} \right]^{-1}.$$

The ODE for $A(t) = \int e^f$ is then

$$\frac{d}{dt} A^2 \leq \|\kappa\|_{L^\infty} A^2 - 8\alpha' \pi^2 (g - 1)^2,$$

which can be rearranged as

$$\frac{d}{dt}\left((\|\kappa\|_{L^\infty} A^2 - 8\alpha'\pi^2(g-1)^2)e^{-\|\kappa\|_\infty t} \right) \leq 0.$$

Therefore

$$|\kappa\|_{L^\infty} A(t)^2$$

$$\leq 8\alpha'\pi^2(g-1)^2 - \left[8\alpha'\pi^2(g-1)^2 - \|\kappa\|_{L^\infty} \text{Vol}(\Sigma)^2 L^2 \right] \exp(\|\kappa\|_{L^\infty} t),$$

and we see that the flow must terminate in finite time if (2.79) holds. In fact, $\|\Omega\|_{\omega_f} \to \infty$ in finite time.

2.4.4.2 Lie Groups

For our final example, we will study the Anomaly flow using unitary connections beyond the Chern connection. Let X be a complex Lie group of dimension $n = 3$, and let $\{e_1, e_2, e_3\}$ be a frame of holomorphic vector fields. Let $\{e^1, e^2, e^3\}$ be the dual frame of holomorphic $(1, 0)$ forms. Denote the structure constants by

$$[e_a, e_b] = c^d{}_{ab} e_d.$$

Consider the Hermitian metric

$$\hat{\omega} = i \sum_a e^a \wedge \bar{e}^a.$$

A section of $T^{1,0}X$ can be expressed as $V = V^a e_a$. By definition (2.8), Strominger–Bismut connection ∇^+ of $\hat{\omega}$ acts in the frame $\{e_a\}$ by

$$\nabla^+_b V^a = \nabla^C_b V^a - T^a{}_{bc} V^c, \quad \nabla^+_{\bar{b}} V^a = \nabla^C_{\bar{b}} V^a + \bar{T}_{c\bar{b}a} V^c,$$

where we now denote the Chern connection by ∇^C for clarity. Since $g_{\bar{a}b} = \delta_{ab}$ in this frame, $\nabla^C = d$. Furthermore,

$$T = i\partial\omega = -\frac{1}{2} c^a{}_{bd} e^d \wedge e^b \wedge \bar{e}^a.$$

Therefore

$$\nabla^+_b V^a = \partial_b V^a + c^a{}_{bd} V^d, \quad \nabla^+_{\bar{b}} V^a = \partial_{\bar{b}} V^a - \overline{c^d{}_{ba}} V^d.$$

Along the Gauduchon line $\nabla^{(\kappa)} = (1 - \kappa)\nabla^C + \kappa\nabla^+$, we have

$$\nabla_b^{(\kappa)} V^a = \partial_b V^a + A^{(\kappa)}{}_b{}^a{}_c V^c, \quad \nabla_{\bar{b}}^{(\kappa)} V^a = \partial_{\bar{b}} V^a + A^{(\kappa)}{}_{\bar{b}}{}^a{}_c V^c,$$

with

$$A^{(\kappa)}{}_b{}^a{}_d = \kappa\, c^a{}_{bd}, \quad A^{(\kappa)}{}_{\bar{b}}{}^a{}_d = -\kappa\, \overline{c^d{}_{ba}}.$$

The curvature form is defined by $Rm = dA + A \wedge A$. More specifically,

$$Rm = \frac{1}{2} R_{kj}{}^a{}_b\, e^j \wedge e^k + \frac{1}{2} R_{\bar{k}\bar{j}}{}^a{}_b\, \bar{e}^j \wedge \bar{e}^k + R_{\bar{k}j}{}^a{}_b\, e^j \wedge \bar{e}^k,$$

where the components are

$$R_{kj}{}^a{}_b = \partial_{e_j} A_k{}^a{}_b - \partial_{e_k} A_j{}^a{}_b - c^r{}_{jk} A_r{}^a{}_b + A_j{}^a{}_c A_k{}^c{}_b - A_k{}^a{}_c A_j{}^c{}_b,$$

$$R_{\bar{k}\bar{j}}{}^a{}_b = \partial_{\bar{e}_j} A_{\bar{k}}{}^a{}_b - \partial_{\bar{e}_k} A_{\bar{j}}{}^a{}_b - \overline{c^r{}_{jk}} A_{\bar{r}}{}^a{}_b + A_{\bar{j}}{}^a{}_c A_{\bar{k}}{}^c{}_b - A_{\bar{k}}{}^a{}_c A_{\bar{j}}{}^c{}_b,$$

$$R_{\bar{k}j}{}^a{}_b = \partial_{e_j} A_{\bar{k}}{}^a{}_b - \partial_{\bar{e}_k} A_j{}^a{}_b + A_j{}^a{}_c A_{\bar{k}}{}^c{}_b - A_{\bar{k}}{}^a{}_c A_j{}^c{}_b.$$

Using the expression for the connection $A^{(\kappa)}$ on the Gauduchon line, the components are explicitly

$$R_{kj}{}^p{}_q = -\kappa c^r{}_{jk} c^p{}_{rq} + \kappa^2 c^p{}_{jr} c^r{}_{kq} - \kappa^2 c^p{}_{kr} c^r{}_{jq},$$

$$R_{\bar{k}\bar{j}}{}^p{}_q = \kappa \overline{c^r{}_{jk} c^q{}_{rp}} + \kappa^2 \overline{c^r{}_{jp} c^q{}_{kr}} - \kappa^2 \overline{c^r{}_{kp} c^q{}_{jr}},$$

$$R_{\bar{k}j}{}^p{}_q = \kappa^2 (-c^p{}_{jr} \overline{c^q{}_{kr}} + \overline{c^r{}_{kp}} c^r{}_{jq}).$$

The surprising computation of Fei–Yau [FY15] shows that $\mathrm{Tr}\, Rm \wedge Rm$ is actually a $(2, 2)$ form, and its $(2, 2)$ part is given by

$$(\mathrm{Tr}\, Rm \wedge Rm)_{\bar{k}\bar{\ell}ij} = 2\kappa^2 (2\kappa - 1)\overline{c^r{}_{k\ell} c^s{}_{rp}} c^q{}_{ij} c^s{}_{qp}.$$

We refer to [FY15] for the full calculation.

We now specialize to the Lie group $SL(2, \mathbf{C})$ with structure constants $c^i{}_{jk} = \epsilon_{ijk}$ the Levi-Civita symbol. Let $\Omega = e^1 \wedge e^2 \wedge e^3$. We also fix $\kappa = 1$ for simplicity, so that we only consider the Strominger–Bismut connection ∇^+. In this case, by two applications of the contracted epsilon identity (2.35), we derive

$$(\mathrm{Tr}\, Rm^+ \wedge Rm^+)_{\bar{k}\bar{\ell}ij} = 2\overline{c^r{}_{k\ell}} c^q{}_{ij}\, [\,\overline{c^s{}_{rp}} c^s{}_{qp}\,]$$

$$= 2\overline{c^r{}_{k\ell}} c^q{}_{ij}\, [2\delta_{rq}]$$

$$= 4(\delta_{ki}\delta_{\ell j} - \delta_{kj}\delta_{\ell i}).$$

Since $\hat{\omega} = i\delta_{ik}e^k \wedge \bar{e}^i$, we have

$$(\text{Tr } Rm^+ \wedge Rm^+)_{\bar{k}\bar{\ell}ij} = 2(\hat{\omega}^2)_{\bar{k}\bar{\ell}ij}.$$

By (2.36), we know $i\partial\bar{\partial}\hat{\omega}$ is also proportional to $\hat{\omega}^2$.

$$i\partial\bar{\partial}\hat{\omega} = \frac{1}{2}\hat{\omega}^2.$$

By scaling the metric $\hat{\omega}$, we see that the diagonal ansatz

$$\omega(t) = \lambda^2(t)\hat{\omega},$$

is preserved by the Anomaly flow

$$\frac{d}{dt}(\|\Omega\|_\omega\omega^2) = i\partial\bar{\partial}\omega - \frac{\alpha'}{4}\text{Tr } Rm^+ \wedge Rm^+,$$

and becomes the ODE

$$\frac{d}{dt}\lambda = \frac{1}{2}(\lambda^2 - \alpha').$$

In the large radius regime, if we start with

$$\omega(0) = L\hat{\omega}$$

where $L \gg 1$, then $\|\Omega\|_{\omega(t)} \to 0$ in finite-time. Outside of this region, the behavior is sensitive to initial data and sign of α'. For example, if $\alpha' > 0$, then for small initial λ, we may have that $\|\Omega\|_{\omega(t)} \to \infty$ in finite-time.

Acknowledgements I would like to first thank D.H. Phong, my former Ph.D. advisor, for guiding me through this material over the course of many years, and whose style shaped the presentation of this course. I thank Xiangwen Zhang and Teng Fei, whose joint work is discussed here, for countless inspiring discussions on the content of these notes. I also thank Daniele Angella, Giovanni Bazzoni, Slawomir Dinew, Kevin Smith, Freid Tong, and Yuri Ustinovskiy for valuable comments and corrections. These lecture notes were prepared for a course given at the CIME Summer School on complex non-Kähler geometry in 2018, and I would like to thank D. Angella, L. Arosio and E. Di Nezza for the invitation and for organizing a wonderful conference.

References

[AG86] E. Abbena, A. Grassi, Hermitian left invariant metrics on complex Lie groups and cosymplectic Hermitian manifolds. Bollettino della Unione Matematica Italiana-A **5**(6), 371–379 (1986)

[AG121] B. Andreas, M. Garcia-Fernandez, Solutions of the Strominger system via stable bundles on Calabi-Yau threefolds. Commun. Math. Phys. **315**, 153–168 (2012)

[AG122] B. Andreas, M. Garcia-Fernandez, Heterotic non-Kähler geometries via polystable bundles on Calabi-Yau threefolds. J. Geom. Phys. **62**(2), 183–188 (2012)

[AN13] D. Angella, The cohomologies of the Iwasawa manifold and of its small deformations. J. Geom. Anal. **23**(3), 1355–1378 (2013)

[AT13] D. Angella, A. Tomassini, On the $\partial\bar{\partial}$-Lemma and Bott-Chern cohomology. Invent. Math. **192**(1), 71–81 (2013)

[ADT16] D. Angella, G. Dloussky, A. Tomassini, On Bott-Chern cohomology of compact complex surfaces. Annali di Matematica Pura ed Applicata **195**(1), 199–217 (2016)

[ACS17] D. Angella, S. Calamai, C. Spotti, On the Chern-Yamabe problem. Math. Res. Lett. **24**(3), 645–677 (2017)

[BD02] K. Becker, K. Dasgupta, Heterotic strings with torsion. J. High Energy Phys. **11**, 006 (2002)

[BBDG03] K. Becker, M. Becker, K. Dasgupta, P. Green, Compactifications of heterotic theory on non-Kahler complex manifolds. I. J. High Energy Phys. **4**(04), 1–59 (2003)

[BI89] J.M. Bismut, A local index theorem for non Kahler manifolds. Math. Ann. **284**(4), 681–699 (1989)

[BL05] Z. Blocki, Weak solutions to the complex Hessian equation. Ann. Inst. Fourier **55**(5), 1735–1756 (2005)

[BR05] R. Bryant, Some remarks on G2-structures, in *Proceedings of Gokova Geometry-Topology Conference* (2005), pp. 75–109

[BU88] N. Buchdahl, Hermitian-Einstein connections and stable vector bundles over compact complex surfaces. Math. Ann. **280**, 625–684 (1988)

[CA58] E. Calabi, Construction and properties of some 6-dimensional almost complex manifolds. Trans. Am. Math. Soc. **87**(2), 407–438 (1958)

[CE53] E. Calabi, B. Eckmann, A class of compact complex manifolds which are not algebraic. Ann. Math. **58**, 494–500 (1953)

[CA851] P. Candelas, G. Horowitz, A. Strominger, E. Witten, Vacuum configurations for superstrings. Nucl. Phys. B **258**, 46–74 (1985)

[CA852] H.-D. Cao, Deformation of Kahler matrics to Kahler-Einstein metrics on compact Kahler manifolds. Invent. Math. **81**(2), 359–372 (1985)

[CSW18] X.X. Chen, S. Sun, B. Wang, Kahler–Ricci flow, Kahler–Einstein metric, and K-stability. Geom. Topol. **22**(6), 3145–3173 (2018)

[CHZ118] J. Chu, L. Huang, X. Zhu, The Fu-Yau equation in higher dimensions (2018). arXiv:1801.09351

[CHZ218] J. Chu, L. Huang, X. Zhu, The Fu-Yau equation on compact astheno-Kahler manifolds (2018). arXiv:1803.01475

[CHT18] T. Collins, T. Hisamoto, R. Takahashi, The inverse Monge-Ampere flow and applications to Kahler-Einstein metrics (2018). arXiv:1712.01685

[CJY15] T. Collins, A. Jacob, S.-T. Yau, (1,1) forms with specified Lagrangian phase: a priori estimates and algebraic obstructions (2015). arXiv:1508.01934

[DRS99] K. Dasgupta, G. Rajesh, S. Sethi, M theory, orientifolds and g-flux. J. High Energy Phys. **8**, 023 (1999)

[DS14] X. de la Ossa, E. Svanes, Holomorphic bundles and the moduli space of N=1 supersymmetric heterotic compactifications. J. High Energy Phys. **2014**, 123 (2014)

[DL17] E. Di Nezza, C. Lu, Uniqueness and short time regularity of the weak Kahler-Ricci flow. Adv. Math. **305**, 953–993 (2017)

[DK17] S. Dinew, S. Kolodziej, Liouville and Calabi-Yau type theorems for complex Hessian equations. Am. J. Math. **139**(2), 403–415 (2017)

[DL15] S. Dinew, C.H. Lu, Mixed Hessian inequalities and uniqueness in the class $\mathscr{E}(X, \omega, m)$. Math. Z. **279**(3–4), 753–766 (2015)

[DDT17] S. Dinew, H.S. Do, T.D. To, A viscosity approach to the Dirichlet problem for degenerate complex Hessian type equations (2017). arXiv:1712.08572

[DPZ18] S. Dinew, S. Plis, X. Zhang, Regularity of degenerate Hessian equation (2018). arXiv:1805.05761

[DO85] S.K. Donaldson, Anti self-dual Yang-Mills connections over complex algebraic surfaces and stable vector bundles. Proc. Lond. Math. Soc. **50**(1), 1–26 (1985)

[FE16] T. Fei, A construction of non-Kähler Calabi-Yau manifolds and new solutions to the Strominger system. Adv. Math. **302**(22), 529–550 (2016)

[FE15] T. Fei, Stable forms, vector cross products and their applications in geometry (2015). arXiv:1504.02807

[FE17] T. Fei, Some torsional local models of heterotic strings. Commun. Anal. Geom. **25**(5), 941–968 (2017)

[FY15] T. Fei, S.T. Yau, Invariant solutions to the Strominger system on complex Lie groups and their quotients. Commun. Math. Phys. **338**(3), 1–13 (2015)

[FGP18] T. Fei, B. Guo, D.H. Phong, On convergence criteria for the coupled flow of Li-Yuan-Zhang (2018). arXiv:1808.06968

[FHP17] T. Fei, Z. Huang, S. Picard, A construction of infinitely many solutions to the Strominger system (2017). arXiv:1703.10067 (preprint)

[FHP17] T. Fei, Z. Huang, S. Picard, The anomaly flow over Riemann surfaces (2017). arXiv:1711.08186

[FIUV09] M. Fernandez, S. Ivanov, L. Ugarte, R. Villacampa, Non-Kahler heterotic string compactifications with non-zero fluxes and constant dilaton. Commun. Math. Phys. **288**, 677–697 (2009)

[FIUV14] M. Fernandez, S. Ivanov, L. Ugarte, R. Villacampa, Non-Kahler heterotic string solutions with non-zero fluxes and non-constant dilaton. J. High Energy Phys. **2014**(6), 73 (2014)

[FG04] A. Fino, G. Grantcharov, Properties of manifolds with skew-symmetric torsion and special holonomy. Adv. Math. **189**(2), 439–450 (2004)

[FT11] A. Fino, A. Tomassini, On astheno-Kahler metrics. J. Lond. Math. Soc. **83**(2), 290–308 (2011)

[FV16] A. Fino, L. Vezzoni, On the existence of balanced and SKT metrics on nilmanifolds. Proc. Am. Math. Soc. **144**(6), 2455–2459 (2016)

[FGV] A. Fino, G. Grantcharov, L. Vezzoni, Astheno-Kähler and balanced structures on fibrations. Int. Math. Res. Not. arXiv:1608.06743

[FR85] D. Friedan, Nonlinear models in 2+ epsilon dimensions. Ann. Phys. **163**(2), 318–419 (1985)

[FX14] J.X. Fu, J. Xiao, Relations between the Kahler cone and the balanced cone of a Kahler manifold. Adv. Math. **263**, 230–252 (2014)

[FY07] J.X. Fu, S.T. Yau, A Monge-Ampère type equation motivated by string theory. Commun. Anal. Geom. **15**(1), 29–76 (2007)

[FY08] J.X. Fu, S.T. Yau, The theory of superstring with flux on non-Kähler manifolds and the complex Monge-Ampère equation. J. Differ. Geom. **78**(3), 369–428 (2008)

[FY09] J.X. Fu, L.S. Tseng, S.T. Yau, Local heterotic torsional models. Commun. Math. Phys. **289**, 1151–1169 (2009)

[FLY12] J.-X. Fu, J. Li, S.-T. Yau, Balanced metrics on non-Kähler Calabi-Yau threefolds. J. Differ. Geom. **90**(1), 81–129 (2012)

[FWW13] J.X. Fu, Z.-Z. Wang, D.-M. Wu, Semilinear equations, the γ_k function, and generalized Gauduchon metrics. J. Eur. Math. Soc. **15**, 659–680 (2013)

[GA16] M. Garcia-Fernandez, Lectures on the Strominger system. Travaux Math. **XXIV**, 7–61 (2016). Special Issue: School GEOQUANT at the ICMAT

[GA40] M. Garcia-Fernandez, T-dual solutions of the Hull-Strominger system on non-Kahler threefolds (2018). arXiv:1810.04740

[GRST18] M. Garcia-Fernandez, R. Rubio, C. Shahbazi, C. Tipler, Canonical metrics on holomorphic Courant algebroids (2018). arXiv:1803.01873

[GA75] P. Gauduchon, La constante fondamentale d'un fibre en droites au-dessus d'une variete hermitienne compacte. C.R. Acad. Sci. Paris Ser. T. **281**, 393–396 (1975)

[GA77] P. Gauduchon, Le theoreme de l'excentricite nulle. C. R. Acad. Sci. Paris Ser. A-B **285**(5), 387–390 (1977)

[GA97] P. Gauduchon, Hermitian connections and Dirac operators. Bollettino della Unione Matematica Italiana-B **11**(2), 257–288 (1997)

[GMPW04] J. Gauntlett, D. Martelli, S. Pakis, D. Waldram, G-structures and wrapped NS5-branes. Commun. Math. Phys. **247**(2), 421–445 (2004)

[GGI13] K. Gimre, C. Guenther, J. Isenberg, A geometric introduction to the two-loop renormalization group flow. J. Fixed Point Theory Appl. **14**(1), 3–20 (2013)

[GO04] E. Goldstein, S. Prokushkin, Geometric model for complex non-Kähler manifolds with $SU(3)$ structure. Commun. Math. Phys. **251**(1), 65–78 (2004)

[GR11] G. Grantcharov, Geometry of compact complex homogeneous spaces with vanishing first Chern class. Adv. Math. **226**, 3136–3159 (2011)

[GGP08] D. Grantcharov, G. Grantcharov, Y.S. Poon, Calabi-Yau connections with torsion on toric bundles. J. Differ. Geom. **78**(1), 13–32 (2008)

[GR69] A. Gray, Vector cross products on manifolds. Trans. AMS **141**, 465–504 (1969)

[GS87] M. Green, J. Schwarz, Anomaly cancellations in supersymmetric $D = 10$ gauge theory and superstring theory. Phys. Lett. B. **149**, 117–122 (1987)

[GZ17] V. Guedj, A. Zeriahi, Regularizing properties of the twisted Kahler-Ricci flow. J. Reine Angew. Math. **729**, 275–304 (2017)

[HIS16] N. Halmagyi, D. Israel, E. Svanes, The Abelian heterotic conifold. J. High Energy Phys. **7**, 29 (2016)

[HA82] R. Hamilton, Three-manifolds with positive Ricci curvature. J. Differ. Geom. **17**, 255–306 (1982)

[HL11] Q. Han, F. Lin, *Elliptic Partial Differential Equations*, vol. 1 (American Mathematical Society, Providence, 2011)

[HMW10] Z. Hou, X.N. Ma, D. Wu, A second order estimate for complex Hessian equations on a compact Kähler manifold. Math. Res. Lett. **17**, 547–561 (2010)

[HU186] C. Hull, Superstring compactifications with torsion and space-time supersymmetry, in *Proceedings of the First Torino Meeting on Superunification and Extra Dimensions*, ed. by R. D' Auria, P. Fre (World Scientific, Singapore, 1986)

[HU286] C. Hull, Compactifications of the heterotic superstring. Phys. Lett. B **178**, 357–364 (1986)

[HU305] D. Huybrechts, *Complex Geometry: An Introduction* Universitext (Springer, Berlin, 2005)

[IP01] S. Ivanov, G. Papadopoulos, Vanishing theorems and string backgrounds. Classical Quant. Gravity **18**, 1089–1110 (2001)

[JY93] J. Jost, S.-T. Yau, A nonlinear elliptic system for maps from Hermitian to Riemannian manifolds and rigidity theorems in Hermitian geometry. Acta Math. **170**(2), 221–254 (1993)

[KA09] S. Karigiannis, Flows of G2-structures, I. Quar. J. Math. **60**(4), 487–522 (2009)

[LU17] A. Latorre, L. Ugarte, On non-Kähler compact complex manifolds with balanced and astheno-Kähler metrics. C. R. Acad. Sci. Paris, Ser. I **355**, 90–93 (2017)

[LE11] H. Lee, Strominger's system on non-Kahler Hermitian manifolds. Ph.D. Dissertation. University of Oxford (2011)

[LY86] J. Li, S.T. Yau, Hermitian-Yang-Mills connections on non-Kahler manifolds, in *Mathematical Aspects of String Theory*. Advanced Series in Mathematical Physics (World Scientific Publishing, Singapore, 1986), pp. 560–573

[LY05] J. Li, S.T. Yau, The existence of supersymmetric string theory with torsion. J. Differ. Geom. **70**(1), 143–181 (2005)

[LY12] K.F. Liu, X.K. Yang, Geometry of Hermitian manifolds. Int. J. Math. **23**(06), 1250055 (2012)

[LY17] K.F. Liu, X.K. Yang, Ricci curvatures on Hermitian manifolds. Trans. Am. Math. Soc. **369**(7), 5157–5196 (2017)

[MS11] D. Martelli, J. Sparks, Non-Kahler heterotic rotations. Adv. Theor. Math. Phys. **15**(1), 131–174 (2011)

[MT01] K. Matsuo, T. Takahashi, On compact astheno-Kähler manifolds. Colloq. Math. **2**(89), 213–221 (2001)

[ME90] W.H. Meeks III, The theory of triply periodic minimal surfaces. Indiana Univ. Math. J. **39**(3), 877–936 (1990)

[MI82] M.L. Michelsohn, On the existence of special metrics in complex geometry. Acta Math. **149**, 261–295 (1982)

[OL09] T.A. Oliynyk, The second-order renormalization group flow for nonlinear sigma models in two dimensions. Classical Quant. Gravity **26**(10), 105020 (2009)

[OUV17] A. Otal, L. Ugarte, R. Villacampa, Invariant solutions to the Strominger system and the heterotic equations of motion on solvmanifolds. Nucl. Phys. B **920**, 442–474 (2017)

[PS06] D.H. Phong, J. Sturm, On stability and the convergence of the Kahler-Ricci flow. J. Differ. Geom. **72**(1), 149–168 (2006)

[PT17] D.H. Phong, T.D. To, Fully non-linear parabolic equations on compact Hermitian manifolds (2017). arXiv:1711.10697

[PPZ117] D.H. Phong, S. Picard, X.W. Zhang, On estimates for the Fu-Yau generalization of a Strominger system (2017). J. Reine Angew. Math. arXiv:1507.08193

[PPZ217] D.H. Phong, S. Picard, X.W. Zhang, The Anomaly flow on unimodular Lie groups (2017). arXiv:1705.09763

[PPZ118] D.H. Phong, S. Picard, X.W. Zhang, Fu-Yau Hessian equations (2018). arXiv:1801.09842

[PPZ19] D.H. Phong, S. Picard, X.W. Zhang, A flow of conformally balanced metrics with Kähler fixed points (2019). arXiv:1805.01029

[PPZ116] D.H. Phong, S. Picard, X.W. Zhang, A second order estimate for general complex Hessian equations. Anal. PDE **9**(7), 1693–1709 (2016)

[PPZ216] D.H. Phong, S. Picard, X.W. Zhang, The Fu-Yau equation with negative slope parameter. Invent. Math. **209**(2), 541–576 (2016)

[PPZ317] D.H. Phong, S. Picard, X.W. Zhang, New curvature flows in complex geometry. Surveys Differ. Geom. **22**(1), 331–364 (2017)

[PPZ218] D.H. Phong, S. Picard, X.W. Zhang, Geometric flows and Strominger systems. Math. Z. **288**, 101–113 (2018)

[PPZ318] D.H. Phong, S. Picard, X.W. Zhang, Anomaly flows. Commun. Anal. Geom. **26**(4), 955–1008 (2018)

[PPZ418] D.H. Phong, S. Picard, X.W. Zhang, The Anomaly flow and the Fu-Yau equation. Ann. PDE **4**(2), 13 (2018)

[ST07] J. Song, G. Tian, The Kahler-Ricci flow on surfaces of positive Kodaira dimension. Invent. Math. **170**(3), 609–653 (2007)

[SW13] J. Song, B. Weinkove, An introduction to the Kähler–Ricci flow, in *An Introduction to the Kähler-Ricci Flow* (Springer, Cham, 2013), pp. 89–188

[ST10] J. Streets, G. Tian, A parabolic flow of pluriclosed metrics. Int. Math. Res. Not. **16**, 3101–3133 (2010)

[ST11] J. Streets, G. Tian, Hermitian curvature flow. J. Eur. Math. Soc. **13**(3), 601–634 (2011)

[ST86] A. Strominger, Superstrings with torsion. Nuclear Phys. B **274**(2), 253–284 (1986)

[STW17] G. Szekelyhidi, V. Tosatti, B. Weinkove, Gauduchon metrics with prescribed volume form. Acta Math. **219**(1), 181–211 (2017)

[TZ06] G. Tian, Z. Zhang, On the Kahler-Ricci flow on projective manifolds of general type. Chin. Ann. Math. Ser. B **27**(2), 179–192 (2006)

[TZ16] G. Tian, Z. Zhang, Regularity of Kahler-Ricci flows on Fano manifolds. Acta Math. **216**(1), 127–176 (2016)

[TO09] V. Tosatti, Limits of Calabi-Yau metrics when the Kahler class degenerates. J. Eur. Math. Soc. **11**, 755–776 (2009)

[TO15] V. Tosatti, Non-Kahler Calabi-Yau manifolds. Contemp. Math. **644**, 261–277 (2015)

[TW10] V. Tosatti, B. Weinkove, The complex Monge-Ampere equation on compact Hermitian manifolds. J. Am. Math. Soc. **23**(4), 1187–1195 (2010)

[TW15] V. Tosatti, B. Weinkove, On the evolution of a Hermitian metric by its Chern-Ricci form. J. Differ. Geom. **99**(1), 125–163 (2015)

[TW17] V. Tosatti, B. Weinkove, The Monge-Ampère equation for (n-1)-plurisubharmonic functions on a compact Kähler manifold. J. Am. Math. Soc. **30**(2), 311–346 (2017)

[TR08] M. Traizet, On the genus of triply periodic minimal surfaces. J. Differ. Geom. **79**(2), 243–275 (2008)

[UG07] L. Ugarte, Hermitian structures on six-dimensional nilmanifolds. Transform. Groups **12**(1), 175–202 (2007)

[UV14] L. Ugarte, R. Villacampa, Non-nilpotent complex geometry of nilmanifolds and heterotic supersymmetry. Asian J. Math. **18**(2), 229–246 (2014)

[UV15] L. Ugarte, R. Villacampa, Balanced Hermitian geometry on 6-dimensional nilmanifolds. Forum Math. **27**(2), 1025–1070 (2015)

[UY85] K. Uhlenbeck, S.T. Yau, On the existence of Hermitian-Yang-Mills connections in stable vector bundles. Commun. Pure Appl. Math. **39**(Suppl.), S257–S293 (1985). Frontiers of the mathematical sciences: 1985 (New York, 1985)

[US16] Y. Ustinovskiy, The Hermitian curvature flow on manifolds with non-negative Griffiths curvature (2016). arXiv:1604.04813

[WA54] H.-C. Wang, Closed manifolds with homogeneous complex structure. Am. J. Math. **76**(1), 1–32 (1954)

[YA78] S.-T. Yau, On the Ricci curvature of a compact Kähler manifold and the complex Monge-Ampère equation, I. Commun. Pure Appl. Math. **31**, 339–411 (1978)

[ZH16] T. Zheng, A parabolic Monge-Ampere type equation of Gauduchon metrics. Int. Math. Res. Not. **2019**(17), 5497–5538 (2019)

Chapter 3
Non-Kählerian Compact Complex Surfaces

Andrei Teleman

Abstract This text follows the lecture series given by the author in the CIME School "Complex non-Kähler geometry" (Cetraro, July 9–13, 2018) and is dedicated to the classification of non-Kählerian surfaces. In the first three sections we present the classical theory:

- The Enriques Kodaira classification for surfaces and the classes of non-Kählerian surfaces,
- Class VII surfaces and their general properties,
- Kato surfaces: construction, classification and moduli.

In Sect. 3.4 we explain the main ideas and techniques used in the proofs of our results on the existence of cycles of curves on class VII surfaces with small b_2. Section 3.5 deals with criteria for the existence of smooth algebraic deformations of the singular surface obtained by contracting a cycle of rational curves in a minimal class VII surface. We included an Appendix in which we introduce several fundamental objects in non-Kählerian complex geometry (the Picard group of a compact complex manifold, the Gauduchon degree, the Kobayashi-Hitchin correspondence for line bundles, unitary flat line bundles), and we prove basic properties of these objects.

A. Teleman (✉)
Institut de Mathématiques de Marseille, CNRS, Centrale Marseille, I2M, UMR 7373,
Aix-Marseille Université, Marseille, France
e-mail: andrei.teleman@univ-amu.fr

© Springer Nature Switzerland AG 2019
D. Angella et al. (eds.), *Complex Non-Kähler Geometry*, Lecture Notes
in Mathematics 2246, https://doi.org/10.1007/978-3-030-25883-2_3

3.1   The Enriques-Kodaira Classification: Classes of Non-Kählerian Surfaces

3.1.1   The Kodaira Dimension and the Algebraic Dimension

Recall that the Kodaira dimension of a connected, compact complex manifold X can be defined as follows:

$$\text{kod}(X) := \begin{cases} -\infty & \text{if } \forall n \in \mathbb{N}^* \; h^0(\mathcal{K}^{\otimes n}) = 0, \\ \min\left\{ k \in \mathbb{N} \middle| \; \left(\frac{h^0(\mathcal{K}^{\otimes n})}{n^k}\right)_n \text{ is bounded} \right\} & \text{if } \exists n \in \mathbb{N}^* \; h^0(\mathcal{K}^{\otimes n}) > 0 \end{cases}.$$

Therefore the Kodaira dimension measures the growth of the plurigenera

$$P_n(X) := h^0(\mathcal{K}_X^{\otimes n})$$

of X as $n \to \infty$. Note that $\text{kod}(X) = 0$ if and only if $0 \leqslant P_n(X) \leqslant 1$ for any n, and there exists n such that $P_n(X) = 1$. For $d > 0$ one has $\text{kod}(X) = d$ if and only if the sequence $(P_n(X))_n$ has polynomial growth of degree d for $n \to \infty$. Recall also that the algebraic dimension $a(X)$ of a compact complex manifold X is the transcendence degree of its field of meromorphic functions $\mathcal{M}(X)$. One has the following general inequality which compares the three dimensions associated with complex manifolds:

$$\text{kod}(X) \leqslant a(X) \leqslant \dim(X).$$

The equality $a(X) = \dim(X)$ holds if and only if X is Moishezon, i.e. it has a modification which is a projective algebraic manifold. The plurigenera, the Kodaira dimension and the algebraic dimension are bimeromorphic invariants of complex manifolds.

A compact complex manifold is called Kählerian if it admits a Kähler metric. Kählerianity implies strong topological properties. For instance, using Hodge theory, it follows that the Betti numbers of any of any Kählerian compact complex n-dimensional manifold X satisfy the conditions: $b_{2k}(X) > 0$ for $0 \leqslant k \leqslant n$, and $b_{2k+1}(X) \in 2\mathbb{N}$ for any k.

Throughout this article by a complex surface we mean a compact, connected 2-dimensional complex manifold. For surfaces we have a simple Kählerianity criterion:

Theorem 3.1.1 *A complex surface is Kählerian (it admits a Kähler metric) if and only if $b_1(X)$ is even.*

Theorem 3.1.1 has been first proved indirectly, using the Enriques-Kodaira classification of complex surfaces, classical Kählerianity criteria for elliptic surfaces, and Siu's theorem stating that any K3 surface is Kählerian [Siu]. A different proof, which gives directly and uniformly the existence of a Kähler metric on any surface with even first Betti number, is due to Buchdahl [Bu2], [Bu3].

For surfaces we also have a simple algebraicity condition: a complex surface is projective algebraic if only if $a(X) = 2$.

3.1.2 Elliptic Surfaces

A surface X is called elliptic if it admits a surjective map $f : X \to Y$ to a Riemann surface Y whose generic fibre is an elliptic curve. Such a map is called an elliptic fibration.

Definition 3.1.2 (See [BHPV, section V.5], [Pl]) An elliptic fibration $f : X \to Y$ is called

(1) relatively minimal, if X contains no vertical (-1)-curve.
(2) elliptic bundle, if it is a locally trivial fibre bundle on Y with an elliptic curve as standard fibre.
(3) principal elliptic bundle, if it is an elliptic bundle satisfying one of the following two equivalent conditions:

   (a) the structure group of the bundle reduces to the group of translations of the standard fibre.
   (b) $R^1 f_*(\mathcal{O}_X) \simeq \mathcal{O}_Y$.

(4) elliptic quasi-bundle, if one of the following equivalent two conditions is satisfied:

   (a) any fibre of f is smooth elliptic, or a multiple of a smooth elliptic curve.
   (b) the holomorphic type of a general fibre of f is constant.

(5) principal elliptic quasi-bundle, if it is an elliptic quasi-bundle and

$$R^1 f_*(\mathcal{O}_X) \simeq \mathcal{O}_Y.$$

Principal elliptic quasi-bundles have a simple structure [Pl, Proposition 1.8]:

Proposition 3.1.3 *Any principal elliptic quasi-bundle $f : X \to Y$ can be obtained by applying a finite sequence of logarithmic transformations to a topologically trivial principal elliptic bundle over Y.*

This result can be used to describe effectively moduli spaces of surfaces which are total spaces of principal elliptic quasi-bundles, so it can be regarded as a classification theorem for this class of surfaces. On the other hand, by Plantiko [Pl, Lemma 1.1], we know that *any* relatively minimal elliptic fibration with *non-Kählerian* total space X is a principal elliptic quasi-bundle:

Proposition 3.1.4 *Let* $f : X \to Y$ *be a relatively minimal elliptic fibration with non-Kählerian total space* X. *Then* f *is a principal elliptic quasi-bundle, so it can be obtained by applying a finite sequence of logarithmic transformations to a topologically trivial principal elliptic bundle over* Y.

Therefore, by Proposition 3.1.4, the classification of non-Kählerian minimal elliptic surfaces reduces to the classification of pairs $(f_0 : X_0 \to Y, \Sigma)$, where f_0 is a topologically trivial principal elliptic bundle, and Σ a finite sequence of logarithmic transformations chosen such that the first Betti number of the resulting surface is odd. This condition can be written down explicitly [Pl, Lemma 1.10]. Therefore, in conclusion, the classification of non-Kählerian minimal elliptic surfaces is well understood.

3.1.3   The Enriques-Kodaira Classification

The Enriques-Kodaira list gives a coarse classification of minimal complex surfaces taking into account their Kodaira dimension. We explain briefly this list pointing out and describing the classes of non-Kählerian surfaces.

3.1.3.1   Surfaces with $\mathrm{kod}(X) = -\infty$

Any minimal surface X with $\mathrm{kod}(X) = -\infty$ belongs to one of the following classes:

1. Minimal rational surfaces: surfaces biholomorphic to $\mathbb{P}^2$ or Hirzebruch surfaces.
2. Ruled surfaces of genus $g \geqslant 1$. A surface in this class is biholomorphic to $\mathbb{P}(E)$, where E is a holomorphic rank 2-bundle on a Riemann surface of genus $g \geqslant 1$.
3. Minimal class VII surfaces. A class VII surface is a complex surface X with $b_1(X) = 1$ and $\mathrm{kod}(X) = -\infty$.

The surfaces in the first two classes are algebraic. Class VII surfaces are non-Kählerian, and are not classified yet. This important gap makes the Enriques-Kodaira classification incomplete. On the other hand the above list shows that, in the Kählerian case, the condition $\mathrm{kod}(X) = -\infty$ is very restrictive: there are only few families of Kählerian surfaces with this property, and these families can be described explicitly. We will see that a similar result is expected in the non-Kählerian case: if the standard conjecture on class VII surfaces is true, this class will be the union of well understood subclasses, which can be described and classified explicitly. In other words, the classification of class VII surfaces is a very challenging, longstanding, still unsolved problem, but, if the expected conjecture is proved, we will have a clear and explicit classification of these surfaces.

3.1.3.2   Surfaces with kod$(X) = 0$

Any minimal surface X with kod$(X) = 0$ belongs to one of the following classes:

1. Bidimensional tori.
2. K3 surfaces.
3. Bielliptic surfaces.
4. Enriques surfaces.
5. Primary Kodaira surfaces.
6. Secondary Kodaira surfaces.

The surfaces in the first four classes are Kählerian. The set of possible algebraic dimensions of bidimensional tori is $\{0, 1, 2\}$, so this class contains both algebraic and non-algebraic surfaces. The same holds for K3 surfaces.

A bielliptic surface is the quotient of a product of elliptic curves by a finite group, and is an elliptic quasi-bundle over an elliptic curve. An Enriques surface is the quotient of an elliptic, algebraic K3 surface by an involution, and is an elliptic fibration over a rational curve. Bielliptic surfaces and Enriques surfaces are all projective algebraic.

Primary and secondary Kodaira surfaces are all non-Kählerian. Primary Kodaira surfaces are topologically non-trivial elliptic principal bundles over elliptic curves. A secondary Kodaira surface is the quotient of a primary Kodaira surface by a finite group, and is a principal elliptic quasi-bundle over a rational curve. The classification of Kodaira surfaces is well understood.

3.1.3.3   Surfaces with kod$(X) = 1$

Any surface X with kod$(X) = 1$ is elliptic. Note that there are many families of elliptic surfaces X with kod$(X) \neq 1$. For instance some tori, some K3 surfaces, all bielliptic surfaces, all Enriques surfaces, all primary and secondary Kodaira surfaces, and also some class VII surfaces are elliptic, but the Kodaira dimension of all these surfaces is not 1. Surfaces with kod$(X) = 1$ are also called properly elliptic surfaces.

3.1.3.4   Surfaces with kod$(X) = 2$

A surface X with kod$(X) = 2$ is called surface of general type. All these surfaces are projective algebraic. The classification of these surfaces leads to two interesting research topics which have been intensively studied with impressive success since many decades [BHPV, chapter VII]:

- Problems related to the "geography" of the Chern numbers of minimal surfaces of general type: which pairs (a, b) of positive integers can be realized as the Chern numbers $(c_1^2(X), c_2(X))$ of a minimal surface X of general type?

- Moduli problems: for a fixed pair (a, b) of positive integers describe the Gieseker moduli space $\mathcal{M}_{ab}$ of minimal surfaces X of general type with $c_1^2(X) = a$, $c_2(X) = b$.

The moduli space $\mathcal{M}_{ab}$ is quasi-projective. The proof is based on the fundamental properties of the pluricanonical maps of surfaces of general type. Recall that the canonical model of a minimal surface of general type X is the (possibly singular) normal surface X_{can} obtained by blowing down the connected components of the union of the (-2)-curves of X. The 5-canonical map

$$f_5 : X \to \mathbb{P}(H^0(\mathcal{K}_X^{\otimes 5})^\vee)$$

is everywhere defined, and induces an embedding

$$\kappa_5 : X_{\text{can}} \hookrightarrow \mathbb{P}(H^0(\mathcal{K}_X^{\otimes 5})^\vee)$$

of the canonical model X_{can} [BHPV, Theorem VII 5.1]. Therefore X can be identified with the resolution of singularities of a normal projective subvariety X'_{can} of a projective space $\mathbb{P}_X$, the pair $(X'_{\text{can}}, \mathbb{P}_X)$ being canonically associated with X. This shows that, although surfaces of general type are not fully classified, they are explicit algebraic geometric objects, and are much better understood than class VII surfaces.

The conclusion of this section is: taking into account the Kodaira dimension, there are three classes of minimal, non-Kählerian surfaces:

1. minimal class VII surfaces; they have Kodaira dimension $-\infty$.
2. primary and secondary Kodaira surfaces; they have Kodaira dimension 0.
3. non-Käherian properly elliptic surfaces; they have Kodaira dimension 1.

The surfaces in the second and third classes are all principal elliptic quasi-bundles, and can be easily classified using Proposition 3.1.4. Class VII surfaces are not classified yet, and this is a fundamental gap in the Enriques-Kodaira list. The first properties of these surfaces and the standard conjectures on their classification, will be presented in the next section.

3.2   Class VII Surfaces

3.2.1   Topological Properties

Let X be a class VII surface. The coefficients formula relating homology to cohomology gives

$$H^1(X, \mathbb{Z}) = \text{Hom}(H_1(X, \mathbb{Z}), \mathbb{Z}) \simeq \mathbb{Z}, \tag{3.1}$$

so the cohomology group $H^1(X, \mathbb{Z})$ is always an infinite cyclic group, although $H_1(X, \mathbb{Z})$ might have torsion.

Using Corollary 12 proved in Appendix, we see that $b_1(X) = 1$ implies $q(X) = 1$. Combining with $p_g(X) := h^0(\mathcal{K}_X) = 0$ we obtain

$$\chi(\mathcal{O}_X) = 1 - q(X) + p_g(X) = 0.$$

The Noether formula [BHPV] gives now $c_1^2(X) + c_2(X) = 0$. On the other hand we have $c_2(X) = \chi(X)$, and in our case the topological Euler-Poincaré characteristic $\chi(X)$ of X coincides with $b_2(X)$. Therefore we obtain the following simple general formula for the Chern numbers of a class VII surface:

$$- c_1^2(X) = c_2(X) = b_2(X). \tag{3.2}$$

In general, for a non-Kählerian surface X, we have $b_+(X) = 2p_g(X)$ [BHPV], so the vanishing of $p_g(X)$ gives $b_+(X) = 0$, in other words the intersection form

$$I_X : {}^{H^2(X, \mathbb{Z})}\!/_{\text{Tors}} \times {}^{H^2(X, \mathbb{Z})}\!/_{\text{Tors}} \to \mathbb{Z}$$

of the differentiable 4-manifold X is negative definite. By Donaldson first theorem on the intersection forms of differentiable 4-manifolds, it follows that I_X is standard over $\mathbb{Z}$, i.e., putting $b := b_2(X)$, there exists a basis $(e_1, \ldots, e_b)$ of the free $\mathbb{Z}$-module $H^2(X, \mathbb{Z})/\text{Tors}$ such that $I_X(e_i, e_j) = -\delta_{ij}$. Decomposing the class $\bar{c}_1(\mathcal{K}_X) := c_1(\mathcal{K}_X) + \text{Tors}$ with respect to such a basis we obtain

$$\bar{c}_1(\mathcal{K}_X) = \sum_{i=1}^{b} k_i e_i$$

with $k_i \in \mathbb{Z}$. On the other hand the class $c_1(\mathcal{K}_X) = -c_1(X)$ is a lift of the Stiefel-Whitney class $w_2(X)$, so it is a characteristic element for the intersection form I_X, i.e. it satisfies the identity

$$\bar{c}_1(\mathcal{K}_X) \cdot h \equiv h \cdot h \bmod 2, \ \forall h \in H^2(X, \mathbb{Z})/\text{Tors}.$$

Replacing h with e_i, we see that $k_i \in 2\mathbb{Z} + 1$ for $1 \leqslant i \leqslant b$. On the other hand we have

$$b = -c_1^2(\mathcal{K}_X) = \sum_{i=1}^{b} k_i^2,$$

so $k_i \in \{\pm 1\}$ for any i. Therefore, replacing e_i by $-e_i$ for some indices i if necessary, we obtain a basis $(e_1, \ldots, e_b)$ of $H^2(X, \mathbb{Z})/\text{Tors}$ satisfying

$$I_X(e_i, e_j) = -\delta_{ij}, \quad \bar{c}_1(\mathcal{K}_X) = \sum_{i=1}^{b} e_i. \tag{3.3}$$

A basis satisfying these two conditions is unique up to order, and will be called *a standard basis* of $H^2(X, \mathbb{Z})/\text{Tors}$.

Remark 3.2.1 Let X be a class VII surface with $b_2(X) = b$. The set

$$B_X := \{e \in H^2(X, \mathbb{Z})/\text{Tors}| \ e^2 = e \cdot c_1(\mathcal{K}_X) = -1\}$$

has b elements. The following data are equivalent:

- a standard basis of $H^2(X, \mathbb{Z})/\text{Tors}$.
- a bijection $\{1, \ldots, b\} \to B_X$.
- a total order on B_X.

3.2.2 Analytic Properties: The Picard Group and the Gauduchon Degree

Let X be a class VII surface. Since the Frölicher spectral sequence of a complex surface degenerates at the first level [BHPV] we obtain

$$b_1(X) = q(X) + h^0(\Omega_X^1),$$

so, since $b_1(X) = q(X) = 1$, we have $h^0(\Omega_X^1) = 0$. Consider the short exact sequence of sheaves

$$0 \to \mathbb{C} \hookrightarrow \mathcal{O}_X \xrightarrow{d} \Omega_{\text{cl}}^1 \to 0$$

where Ω_{cl}^1 is the sheaf of closed holomorphic 1-forms. We obtain the cohomology exact sequence

$$0 \to H^0(\Omega_{X\text{cl}}^1) \to H^1(X, \mathbb{C}) \to H^1(\mathcal{O}_X). \tag{3.4}$$

Using the sheaf inclusion $\Omega_{X\text{cl}}^1 \hookrightarrow \Omega_X^1$ and the vanishing of $H^0(\Omega_X^1)$ we obtain $H^0(\Omega_{X\text{cl}}^1) = 0$, so (3.4) shows that the canonical map $H^1(X, \mathbb{C}) \to H^1(\mathcal{O}_X)$ is injective. Since both cohomology spaces are 1-dimensional we obtain

Remark 3.2.2 Let X be a class VII surface. The canonical linear map $H^1(X, \mathbb{C}) \to H^1(\mathcal{O}_X)$ is an isomorphism.

The cohomology group $H^1(X, \mathbb{C}^*)$ has an interesting geometric interpretation: it can be identified with the group of isomorphism classes of flat holomorphic connections with structure group $\mathbb{C}^*$. Comparing the cohomology exact sequences associated with the short exact sequences

$$0 \to \mathbb{Z} \xrightarrow{2\pi i \cdot} \mathcal{O}_X \to \mathcal{O}_X^* \to 0, \quad 0 \to \mathbb{Z} \xrightarrow{2\pi i \cdot} \mathbb{C} \to \mathbb{C}_X^* \to 0$$

of sheaves on X, and using the notations introduced in Appendix we obtain

Remark 3.2.3 Let X be a class VII surface. The canonical group morphism

$$H^1(X, \mathbb{C}^*) \to H^1(\mathcal{O}_X^*) = \mathrm{Pic}(X)$$

is a monomorphism which identifies $H^1(X, \mathbb{C}^*)$ with $\mathrm{Pic}^T(X)$. In other words, any holomorphic line bundle with torsion Chern class on X admits a unique flat holomorphic connection. The obtained isomorphism $H^1(X, \mathbb{C}^*) \xrightarrow{\simeq} \mathrm{Pic}^T(X)$ induces an isomorphism

$$H^1(X, \mathbb{C}) \Big/ {}_{2\pi i H^1(X, \mathbb{Z})} \xrightarrow{\simeq} \mathrm{Pic}^0(X)$$

between the connected components of the unit elements of the two Lie groups.

Using (3.1) and choosing an isomorphism $\varepsilon : \mathbb{Z} \xrightarrow{\simeq} H^1(X, \mathbb{Z})$, we obtain induced isomorphisms

$$\varepsilon_{\mathbb{R}} : \mathbb{R} \xrightarrow{\simeq} H^1(X, \mathbb{R}), \quad \varepsilon_{\mathbb{C}} : \mathbb{C} \xrightarrow{\simeq} H^1(X, \mathbb{C}), \quad \varepsilon_0 : \mathbb{C}^* \xrightarrow{\simeq} \mathrm{Pic}^0(X).$$

By Corollary 12 proved in Appendix it follows that, for any Gauduchon metric g on X, we have a formula the form

$$\deg_g(\varepsilon_0(\zeta)) = C(\varepsilon, g) \log |\zeta| \ \forall \zeta \in \mathbb{C}^*,$$

where $C(\varepsilon, g)$ is a non-zero real constant which depends smoothly on g [LT]. Therefore the sign of $C(\varepsilon, g)$ depends only on ε. We choose ε such that $C(\varepsilon, g) > 0$ for any Gauduchon metric g on X. In this way we obtain *canonical* isomorphisms

$$\mathbb{Z} \xrightarrow{\simeq} H^1(X, \mathbb{Z}), \quad \mathbb{R} \xrightarrow{\simeq} H^1(X, \mathbb{R}), \quad \mathbb{C} \xrightarrow{\simeq} H^1(X, \mathbb{C}), \quad \mathbb{C}^* \xrightarrow{\simeq} \mathrm{Pic}^0(X),$$

and, denoting by $[\mathcal{L}_\zeta] \in \mathrm{Pic}^0(X)$ the element which corresponds to ζ via the fourth canonical isomorphism, we have the identity

$$\deg_g(\mathcal{L}_\zeta) = C_g \log |\zeta| \ \forall \zeta \in \mathbb{C}^*, \tag{3.5}$$

where C_g is a positive constant depending smoothly on g.

3.2.3   The Classification of Class VII Surfaces with $b_2 = 0$

Class VII surfaces with $b_2 = 0$ are classified. Before stating the result, recall that a Hopf surface is a quotient of the form $\mathbb{C}^2 \setminus \{0\}/G$, where G acts properly discontinuously on $\mathbb{C}^2 \setminus \{0\}$. Any such surface belongs to the class VII. A Hopf surface $X = \mathbb{C}^2 \setminus \{0\}/G$ is called primary if G is infinite cyclic, i.e. if $\pi_1(X, x_0) \simeq \mathbb{Z}$. A non-primary Hopf surface is called secondary.

An Inoue surface [In] is a quotient of the form $\mathbb{C} \times H/G$, where H is the half-plane $\{z \in \mathbb{C} | \Im(z) > 0\}$, and G is a solvable group of affine transformations of the complex plane leaving invariant and acting properly discontinuously on $\mathbb{C} \times H$. The classification theorem for class VII surfaces with $b_2 = 0$ [Te1] states:

Theorem 3.2.4 *Any class* VII *surface with* $b_2 = 0$ *is biholomorphic to either a Hopf or an Inoue surface.*

Note that Hopf surfaces and Inoue surfaces are fully classified, so this result solves the classification problem for $b_2 = 0$.

3.3   Kato Surfaces

3.3.1   Definition and Construction of Kato Surfaces

A spherical shell in a complex surface X is an open subset $U \subset X$ which is biholomorphic to a standard neighbourhood of S^3 in $\mathbb{C}^2$. A spherical shell $U \subset X$ is called global if $X \setminus U$ is connected.

Definition 3.3.1 A Kato surface is a minimal class VII surface with $b_2 > 0$ which contains a global spherical shell.

Kato surfaces are well understood: they can be all obtained using a simple two-step construction procedure: iterated blow up of the standard ball $B \subset \mathbb{C}^2$ at the origin, followed by a holomorphic surgery. More precisely, let b be a positive integer, and $\pi : \Pi \to B$ be an iterated blow up of order b of $B \subset \mathbb{C}^2$ at 0 obtained in the following way:

- We start by blowing up B at 0.
- If $b > 1$ we continue with $b - 1$ successive blow ups respecting the following rule: at every stage we blow up a point taken on the last exceptional divisor.

Let $\sigma : B \to U$ be a biholomorphism onto a neighbourhood of a point $x \in \Pi$ chosen on the last exceptional divisor, such that the following two conditions are fulfilled:

- $\sigma(0) = x$.
- σ extends to a biholomorphism between neighbourhoods of the compact closures $\bar{B} \subset \mathbb{C}^2, \bar{U} \subset \Pi$.

Fig. 3.1 The construction of
a Kato surface

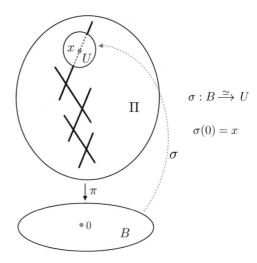

$$\sigma : B \xrightarrow{\simeq} U$$

$$\sigma(0) = x$$

The surface $X(\pi, x, \sigma)$ obtained by identifying the two components of the boundary

$$\partial(\Pi \setminus U) = \pi^{-1}(\partial \bar{B}) \cup \partial(\bar{U})$$

via the obvious extension of $\sigma \circ \pi$ (see Fig. 3.1) is a Kato surface with $b_2 = b$. By results of Kato [Ka1, Ka2, Ka3] and Dloussky [Dl1, Dl2] any Kato surface can be obtained in this way.

The isomorphism type of $X(\pi, x, \sigma)$ depends only of the conjugacy class of the germ $(\pi \circ \sigma)_0$ (Dloussky). This germ is *a contracting germ*. The conjugacy classes of contracting germs of this special type have been classified by Dloussky [Dl2] and Favre [Fa], who gave *normal forms* for these germs. These results show that the conjugacy classes of contracting germs of this type depend on a finite number of parameters, so the moduli space of Kato surfaces with a fixed second Betti number is finite dimensional. This correspondence between Kato surfaces and conjugacy classes of contracting germs allowed Oeljeklaus-Toma to construct moduli spaces of Kato surfaces with a fixed configuration of curves [OT]. In Sect. 3.3.4 we will describe the moduli spaces of Kato surfaces with $b_2 = 1$ and $b_2 = 2$.

Remark 3.3.2 Let (π, x, σ) be a triple defining a Kato surface with $b_2 = b > 0$. The maps

$$\Pi \hookleftarrow \Pi \setminus U \to X(\pi, x, \sigma)$$

induce isomorphisms in 2-homology, so $H_2(X(\pi, x, \sigma), \mathbb{Z})$ comes with a canonical basis induced by the obvious basis of $H_2(\Pi, \mathbb{Z})$ formed by the homology classes created by blow ups, ordered by the "order of creation" [Dl1]. This shows that, using the notation introduced in Remark 3.2.1, the set

$$B_{X(\pi,x,\sigma)} := \{e \in H^2(X, \mathbb{Z}) \mid e^2 = e \cdot c_1(\mathcal{K}_X) = -1\} \subset H^2(X(\pi, x, \sigma), \mathbb{Z})$$

comes with a canonical total order. This order depends effectively on the triple (π, x, σ), but the associated cyclic permutation of B_X is a biholomorphic invariant. Therefore, for any Kato surface X, the set B_X is endowed with a well defined cyclic permutation $c_X : B_X \to B_X$, which is independent of the choices of a triple (π, x, σ) and a biholomorphism $X \simeq X(\pi, x, \sigma)$. This invariance property can be proved using [DlTe1] as follows: Let $B^X := D_X(B_X) \subset H_2(X, \mathbb{Z})$ be image of B_X under Poincaré duality, and $\pi : \tilde{X} \to X$ be the universal cover of X. The set

$$\tilde{B}^X := \{h \in H_2(\tilde{X}, \mathbb{Z}) | \, \pi_*(h) \in B^X\} \subset H_2(\tilde{X}, \mathbb{Z})$$

is naturally endowed with

- a free action of the group $\mathrm{Aut}_X(\tilde{X}) \simeq \mathbb{Z}$,
- a canonical total order defined by

$$h_1 \leqslant h_2 \text{ if } \tilde{E}_{h_1} \supset \tilde{E}_{h_2},$$

where, for a class $h \in \tilde{B}^X$, $\tilde{E}_h$ stands for the canonical effective divisor representing the class h in Borel-Moore homology [DlTe1, Theorem 1].

The canonical cyclic permutation of $c_X : B_X \to B_X$ is induced by the successor map $\tilde{B}^X \to \tilde{B}^X$ with respect to this total order, taking into account the obvious bijections $\tilde{B}^X/\mathbb{Z} \xrightarrow{\simeq} B^X \xrightarrow{\simeq} B_X$.

This remark shows that, for a Kato surface X, fixing a class $e \in B_X$ gives a well defined standard basis of $H^2(X, \mathbb{Z})$ obtained by applying to e the powers of c_X.

3.3.2 Subclasses of Kato Surfaces

Taking into account the structure of curves, Kato surfaces are divided in five classes:

3.3.2.1 Enoki Surfaces

A surface $X \in \mathrm{VII}^{\min}_{b_2>0}$ is called Enoki surface, if it contains a non-empty, homologically trivial, effective divisor. Such a surface contains a homologically trivial cycle C of $b_2(X)$ rational curves. If $b_2 = 1$, C consists of a single homologically trivial rational curve with a simple singularity. For $b_2 \geqslant 2$ the irreducible components of C are smooth rational curves of self-intersection -2. By the main result of [En] such a surface is biholomorphic to a compactification of a holomorphic affine line bundle over an elliptic curve.

An Enoki surface is called *generic* if C is its maximal reduced effective divisor. Equivalently, X is a generic Enoki if it is biholomorphic to a compactification of a *non-linear* holomorphic *affine line bundle* over an elliptic curve. Figure 3.2 shows the configuration of curves on a generic Enoki surface. In this picture, and in the

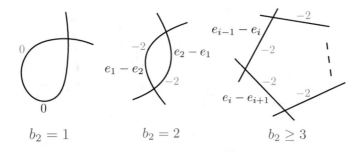

$$b_2 = 1 \qquad\qquad b_2 = 2 \qquad\qquad b_2 \geq 3$$

Fig. 3.2 Curves on generic Enoki surfaces

following pictures showing configuration of curves on Kato surfaces, next to each curve C we indicated its self-intersection number C^2 and the decomposition of the dual class $[C] \in H^2(X, \mathbb{Z})$ with respect to a suitable standard basis $(e_1, \ldots, e_b)$ (see Sect. 3.2.1).

3.3.2.2 Parabolic Inoue (Special Enoki) Surfaces

A surface $X \in \mathrm{VII}^{\min}_{b_2>0}$ is called a parabolic Inoue, or a special Enoki surface, if it is an Enoki surface which also contains an elliptic curve E. Figure 3.3 shows the possible configurations of curves on parabolic Inoue surfaces with $b_2 \leqslant 3$. For such a surface one has $E^2 = -b_2(X)$, $[E] = -\sum_{i=1}^b e_i$, and X is biholomorphic to a compactification of a holomorphic *linear line bundle* over E, such that E corresponds to the zero-section of this line bundle. One has $\mathcal{K}_X = \mathcal{O}(-C - E)$, so $C + E$ is an anti-canonical (AC) divisor on X. Such a surface also admits a non-trivial holomorphic vector field.

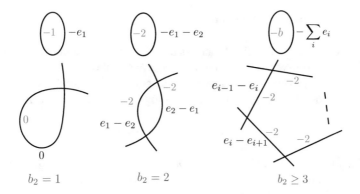

$$b_2 = 1 \qquad\qquad b_2 = 2 \qquad\qquad b_2 \geq 3$$

Fig. 3.3 Curves on parabolic Inoue surfaces with $b_2 \leqslant 3$

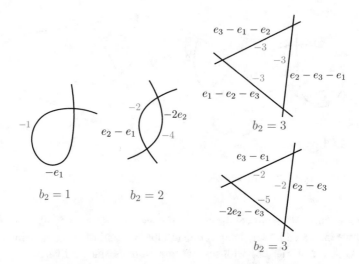

Fig. 3.4 Curves on half Inoue surfaces with $b_2 \leqslant 3$

3.3.2.3 Half Inoue Surfaces

A surface $X \in \mathrm{VII}^{\min}_{b_2>0}$ is called *half Inoue surface*, if it contains a cycle of $b_2(X)$ rational curves C with $C^2 = -b_2(X)$. The maximal reduced divisor of a half Inoue surface is C, and the associated class $[C] \in H^2(X, \mathbb{Z})$ decomposes with respect to a standard basis as $[C] = -\sum_{i=1}^{b} e_i$. One has $\mathcal{K}_X^{\otimes 2} \simeq \mathcal{O}(-2C)$, but $\mathcal{K}_X \not\simeq \mathcal{O}(-C)$. The image of $H_1(C, \mathbb{Z})$ in $H_1(X, \mathbb{Z})$ has index 2. In all other cases, if C is a cycle on a Kato surface X, the canonical morphism $H_1(C, \mathbb{Z}) \to H_1(X, \mathbb{Z})$ is an isomorphism. Figure 3.4 shows the possible configurations of curves on half Inoue surfaces with $b_2 \leqslant 3$.

3.3.2.4 Inoue-Hirzebruch Surfaces

A surface $X \in \mathrm{VII}^{\min}_{b_2>0}$ is called Inoue-Hirzebruch surface, if it contains two cycles C_1, C_2 of rational curves. If this is the case, these two cycles are disjoint, their sum is the maximal reduced effective divisor of X, and contains $b_2(X)$ curves. With respect to a standard basis the corresponding cohomology classes decompose as

$$[C_1] = -e_I, \ [C_2] = -e_J,$$

where (I, J) is a partition of $\{1, \ldots, b\}$. One has $\mathcal{K}_X \simeq \mathcal{O}(-C_1 - C_2)$, so $C_1 + C_2$ is an anti-canonical (AC) divisor on X. Figure 3.5 shows the possible configurations of curves on Inoue-Hirzebruch surfaces with $b_2 \leqslant 3$.

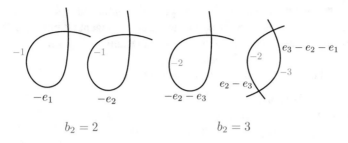

Fig. 3.5 Curves on Inoue-Hirzebruch with $b_2 \leqslant 3$

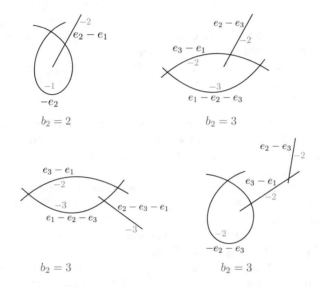

Fig. 3.6 Curves on intermediate surfaces with $b_2 \leqslant 3$

3.3.2.5   Intermediate Surfaces

A surface $X \in \mathrm{VII}^{\min}_{b_2>0}$ is called an *intermediate* surface, if it contains a single cycle C of rational curves, and also trees (at least one tree) of rational curves intersecting the cycle. The maximal reduced divisor of an intermediate surface is connected, and has $b_2(X)$ irreducible components. Figure 3.6 shows the possible configurations of curves on intermediate surfaces with $b_2 \leqslant 3$.

3.3.3   General Properties of Kato Surfaces

The Kato surfaces have the following remarkable properties:

1. Any Kato surface X has exactly $b_2(X)$ rational curves.

2. A Kato surface has either one or two cycles of rational curves.
3. Any Kato surface is a deformation in large (a degeneration) of a family of blown up primary Hopf surfaces. In particular any Kato surface with $b_2 = b$ is diffeomorphic to $(S^1 \times S^3)\#b\bar{\mathbb{P}}^2_{\mathbb{C}}$.
4. Any small deformation of a Kato surface is either

   - a Kato surface, or
   - a blown up Kato surface, or
   - a blown up primary Hopf surface.

5. Any Kato surface has a holomorphic foliation.

Remark 3.3.3 All Kato surfaces with fixed b_2 are deformation equivalent. However the intersection numbers of the holomorphic curves and the homology classes represented by analytic cycles are different, depending on the considered subclass. This phenomenon is not consistent with the intuition we have from algebraic and Kählerian geometry, and illustrates a well-known difficulty occurring in non-Kählerian complex geometry: the "explosion of volume" of analytic cycles in holomorphic families [Ba], [DlTe1].

Example 3.3.1 Let $X_0 \hookrightarrow \mathcal{X} \to B$ be the versal deformation of a half Inoue surface X_0 with $b_2 = 1$. B can be identified with the ball $B \subset \mathbb{C}^2$ such that X_0 is the fibre over 0. There exists a curve $Z \ni 0$ in B such that for any $z \in B \setminus Z$ the fibre X_z is a blown up primary Hopf surface, and for any $z \in Z \setminus \{0\}$ the fibre X_z is a generic Enoki surface. As a moving point $z \in B \setminus Z$ approaches a point $\zeta \in Z \setminus \{0\}$, the area of the exceptional curve $E_z \subset X_z$ tends to ∞, and the limit fibre X_ζ does not contain any 1-dimensional analytic cycle representing the class $e_1 = [E_z]$. The only curve of X_ζ is a homologically trivial singular rational curve C_ζ. Similarly, as a moving point $\zeta \in Z$ approaches 0, the area of $C_\zeta \subset X_\zeta$ tends to ∞, and the limit fibre X_0 does not contain any non-empty 1-dimensional analytic cycle which is homologically trivial. The only curve of X_0 is a singular rational curve representing the class $-e_1$ (see Fig. 3.7).

Fig. 3.7 A family of class VII surfaces with $b_2 = 1$. It contains blown up Hopf surfaces, Enoki surfaces and a half Inoue surface

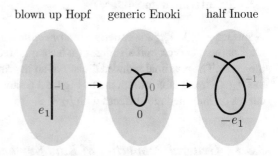

3.3.4   The Moduli Spaces of Framed Kato Surfaces with $b_2 = 1$ and $b_2 = 2$

In this section we describe the moduli spaces of Kato surfaces with $b_2 = 1$ and $b_2 = 2$. These examples show that, in principle, Kato surfaces can be classified explicitly, and should be regarded as the *known* surfaces in the class $\text{VII}^{\min}_{b_2>0}$ of minimal class VII surfaces with positive b_2.

An important tool in the classification of Kato surfaces is the trace invariant introduced by Dloussky [Dl1], [DlKo], which can be defined as follows. First, for a blown up Hopf surface X we denote by $\text{tr}(X)$ the trace of the differential at 0 of the holomorphic contraction $\mathfrak{c} : \mathbb{C}^2 \to \mathbb{C}^2$ defining the minimal model of X.

Let now X_0 be a Kato surface, and let

$$X_0 \hookrightarrow \mathcal{X} \to B$$

be a deformation of X_0 containing blown up primary Hopf surfaces, where $B \subset \mathbb{C}^n$ is the standard ball (see Sect. 3.3.3). The condition "X_z is blown up Hopf surface" defines a Zariski open set analytic $B_H \subset S$. The map $B_H \ni z \to \text{tr}(X_z)$ has a holomorphic extension at 0, and $\text{tr}(X_0)$ coincides with the value of this extension at 0. Denoting by $\text{VII}_K \subset \text{VII}^{\min}_{b_2>0}$ the class of Kato surfaces, we obtain a biholomorphic invariant

$$\text{tr} : \text{VII}_K \to D,$$

where $D \subset \mathbb{C}$ is the standard disc.

3.3.4.1   Moduli Spaces of Framed Class VII Surfaces

A (cohomologically) framed class VII surface is a pair (X, e), where X is a class VII surface, and $e \in B_X$ (see Remark 3.2.1). Framing class VII surfaces in this way is motivated by Remark 3.3.2, which shows that, for a Kato surface, fixing a class $e \in B_X$ determines a standard basis of $H^2(X, \mathbb{Z})$; this basis obtained by applying to e the powers of the canonical cyclic permutation c_X (see Remark 3.3.2).

Let b be a fixed positive integer. We denote by $\mathcal{M}_K(b)$ the moduli space (regarded as topological spaces) of framed Kato surfaces with $b_2 = b$. This space can be defined either as the underlying topological space of the analytic stack of framed Kato surfaces, or as the quotient of the space of Kato complex structures on the differentiable 4-manifold $(S^1 \times S^3) \# b\bar{\mathbb{P}}^2_{\mathbb{C}}$ by a suitable diffeomorphism group. We do not discuss here these technical definitions, but we note that one can also construct this moduli space using the correspondence between Kato surfaces and contracting germs explained in Sect. 3.3.1. The trace invariant defines a continuous map

$$\text{tr} : \mathcal{M}_K(b) \to D.$$

An important remark of Dloussky-Kohler states that any Kato surface with non-vanishing trace is an Enoki surface [DlKo, Remark p. 54]. In other words, putting $D^\bullet := D \setminus \{0\}$, we have

Proposition 3.3.4 *Let $b > 0$. The open subspace $\mathrm{tr}^{-1}(D^\bullet) \subset \mathcal{M}_K(b)$ coincides with the moduli space of framed (generic and special) Enoki surfaces with $b_2 = b$.*

This moduli space has been studied by Dloussky-Kohler [DlKo]. Theorem [DlKo, Theorem 1.23] can be reformulated as follows:

Theorem 3.3.5 *The space $\mathrm{tr}^{-1}(D^\bullet) \subset \mathcal{M}_K(b)$ can be identified with the quotient*

$$D^\bullet \times \mathbb{C}^b \Big/ \mathbb{C}^* \times \mu_b,$$

where $\mathbb{C}^* \times \mu_b$ acts on $\mathbb{C}^b$ by

$$(\zeta, r) \cdot \left(\tau, (z_j)_{0 \leqslant j \leqslant b-1}\right) := \left(\tau, (r^j \zeta z_j)_{0 \leqslant j \leqslant b-1}\right).$$

Via this identification, the trace map is given by $(\tau, (z_j)_{0 \leqslant j \leqslant b-1}) \mapsto \tau$.

For $z \in D^\bullet$ the fibre

$$\Phi_z^b := \mathrm{tr}^{-1}(z) \subset \mathcal{M}_K(b)$$

is the moduli space of framed Enoki surfaces with $b_2 = b$ and fixed trace z; it can be identified with the non-Hausdorff quotient

$$\mathfrak{P}^{b-1} := \mathbb{C}^b \Big/ \mathbb{C}^* \times \mu_b,$$

which decomposes as the union $\mathfrak{P}^{b-1} = (\mathbb{P}^{b-1}/\mu_b) \cup \{*\}$, where $P^{b-1} := \mathbb{P}^{b-1}/\mu_b$ is a projective variety and is *open* in $\mathfrak{P}^{b-1}$. Any neighbourhood of $*$ coincided with the whole $\mathfrak{P}^{b-1}$. Note that one has $P^0 = \{*\}$, $P^1 \simeq \mathbb{P}^1$, but P^k is singular for $k \geqslant 2$. In our pictures we will use the following symbols for the quotients $\mathfrak{P}^0$, $\mathfrak{P}^1$:

$$\mathfrak{P}^0: \; \begin{smallmatrix} \bullet \\ \bullet \end{smallmatrix} \qquad \mathfrak{P}^1: \; \textbf{\textcircled{\raisebox{0pt}{}}}$$

3.3.4.2 The Moduli Space of Kato Surfaces with $b_2 = 1$

The moduli space $\mathcal{M}_K(1)$ can be identified with the quotient

$$(D \times \mathbb{C}) \setminus \{(0, 0)\} \Big/ \mathbb{C}^*,$$

where $\mathbb{C}^*$ acts on $(D \times \mathbb{C}) \setminus \{(0,0)\}$ by $\zeta \cdot (\tau, z) := (\tau, \zeta z)$. This quotient can be further identified with the union

$$(D^\bullet \times \{0\}) \cup (D \times \{1\})$$

endowed with the topology determined by the following conditions:

- a basis of open neighbourhoods of a point $(\tau, 1) \in D \times \{1\}$ is

$$\{V \times \{1\} \mid V \text{ open neighbourhood of } \tau \text{ in } D\}.$$

- a basis of open neighbourhoods of a point $(\tau, 0) \in D^\bullet \times \{0\}$ is

$$\{V \times \{0, 1\} \mid V \text{ open neighbourhood of } \tau \text{ in } D^\bullet\}.$$

Via this identification, the trace map is given by $(\tau, t) \mapsto \tau$; its fibre over a point $\tau \in D^\bullet$ is the non-Hausdorff quotient $\mathfrak{P}^0$. The points of the form $(\tau, 1)$ (respectively $(\tau, 0)$) with $\tau \in D^\bullet$ correspond to generic (special) Enoki surfaces, and the point $(0, 1)$ corresponds to the (up to isomorphism) unique half Inoue surface with $b_2 = 1$. As Fig. 3.8 suggests, any small deformation of the half Inoue surface is a *generic* Enoki surface. This can be proved directly, without using our description of the moduli space $\mathcal{M}_K(1)$, taking into account that

- any special Enoki surface has an AC effective divisor.
- the existence of an AC effective divisor is a closed condition in holomorphic families.
- A half Inoue surface does have an AC effective divisor, so it cannot be the limit of a family of special Enoki surfaces.

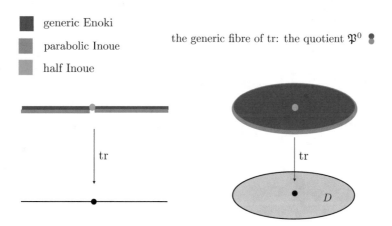

Fig. 3.8 The moduli space $\mathcal{M}_K(1)$ of framed Kato surfaces with $b_2 = 1$

3.3.4.3 The Moduli Space of Kato Surfaces with $b_2 = 2$

Let $D \subset \mathbb{C}$ be the standard disc.

Definition 3.3.6 The disc with a line of origins is the quotient

$$\mathfrak{D} := {}^{D \times \mathbb{C}}\!/_{\mathbb{C}},$$

where $\mathbb{C}$ acts on $D \times \mathbb{C}$ by

$$\zeta \cdot (z, u) = (z, u + \zeta z).$$

This $\mathbb{C}$-action on $D \times \mathbb{C}$ is fibrewise transitive over $D^\bullet$ and trivial over 0. The obvious map $\mathfrak{D} \to D$ is surjective; its fibre over a point $z \in D^\bullet$ is a singleton, whereas its fibre over $0 \in D$ is the line $\{[0, u] | u \in \mathbb{C}\}$. Therefore $\mathfrak{D}$ can be intuitively thought of as a disc with a 1-parameter family of mutually non-separable "origins" $0_u := [0, u]$. For a disc with a line of origins we will use the symbol illustrated in Fig. 3.9.

The open subspace $\mathrm{tr}^{-1}(D^\bullet) \subset \mathcal{M}_K(2)$ is described by Theorem 3.3.5, and the result is simple: this open subspace of $\mathcal{M}_K(2)$ is a trivial fibre bundle over $D^\bullet$ with the non-Hausdorff quotient $\mathfrak{P}^1$ as fibre (Fig. 3.10):

The closed subspace $\mathrm{tr}^{-1}(0) \subset \mathcal{M}_K(2)$ is more interesting; using the results of [Fa] and [OT] one can see that this subspace is homeomorphic to the space $\mathfrak{Y}$ obtained as follows:

- Consider the degenerate conic $Y \subset \mathbb{P}^2$ defined be the equation $X_1 X_2 = 0$. The involution $\iota : Y \to Y$ given by $[X_0, X_1, X_2] \to [X_0, X_2, X_1]$ interchanges the irreducible components $Y_\pm$ of Y and leaves its singularity $\mathfrak{c} = [1, 0, 0]$ fixed.

Fig. 3.9 A disc with a line of origins

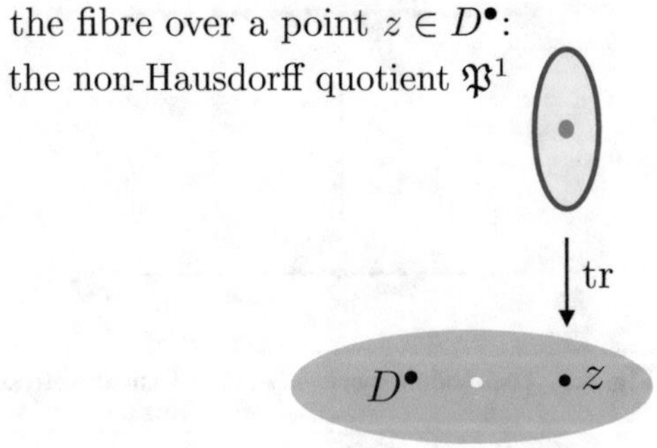

Fig. 3.10 The open subspace $\mathrm{tr}^{-1}(D^\bullet) \subset \mathcal{M}_K(2)$

the fibre over a point $z \in D^\bullet$:
the non-Hausdorff quotient $\mathfrak{P}^1$

- Choose ι-conjugate points $y_\pm \in Y_\pm \setminus \{c\}$, copies $D_\pm$ of the standard disc, and biholomorphic maps $f_\pm : D_\pm \to U_\pm$ on open neighbourhoods $U_\pm$ of $y_\pm$.
- Define

$$\mathfrak{Y} := (\mathfrak{D}_- \coprod \mathfrak{D}_+) \coprod_{(D_-^\bullet \coprod D_+)} (Y \setminus \{y_-, y_+\}),$$

where the expression on the right denotes the push-out associated with the inclusion $D_-^\bullet \coprod D_+^\bullet \hookrightarrow \mathfrak{D}_- \coprod \mathfrak{D}_+$ and the homeomorphism

$$D_-^\bullet \coprod D_+^\bullet \to (U^- \setminus \{y_-\}) \coprod (U^+ \setminus \{y_+\})$$

induced by $f^\pm$. In other words $\mathfrak{Y}$ is obtained from Y by replacing $U^\pm$ with a disc with a line origins $\mathfrak{D}_\pm$, and $y_\pm$ with the corresponding line of origins of $\mathfrak{D}_\pm$.

The singularity c of $\mathfrak{Y}$ corresponds to the (up to isomorphism) unique framed Inoue-Hirzebruch surface with $b_2 = 2$. There are two isomorphism classes of framed half Inoue surfaces with $b_2 = 2$, and they correspond to ι-conjugate points $v^\pm \in Y^\pm \setminus \{y^\pm, c\}$. The complement $\mathfrak{Y} \setminus \{c, v^-, v^+\}$ corresponds to the moduli space of framed intermediate surfaces with $b_2 = 2$, and the union of the two lines of origins corresponds to the subspace of intermediate surfaces admitting a non-trivial vector field. $\mathfrak{Y}$ contains two remarkable points, which are also ι-conjugate and correspond to intermediate surfaces admitting an AC effective divisor (Fig. 3.11).

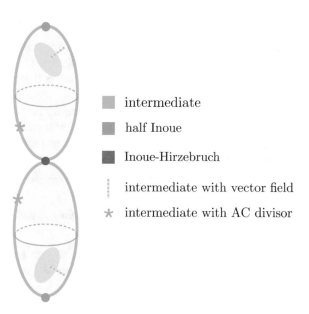

Fig. 3.11 The closed subspace $\mathrm{tr}^{-1}(0) \subset \mathcal{M}_K(2)$

intermediate

half Inoue

Inoue-Hirzebruch

intermediate with vector field

$*$ intermediate with AC divisor

Note any small deformation with non-trivial trace of a point $y \in \mathfrak{Y}$ is a *generic Enoki surface*. This can be proved directly, using a semicontinuity argument, as in the case $b_2 = 1$, by taking into account that there is no Kato surface with $b_2 = 2$ and vanishing trace that has both a non-trivial vector field and an AC effective divisor.

3.3.5   Standard Conjectures on Class VII Surfaces

Two important results show that existence of curves plays an important role in the classification of class VII surfaces:

The first theorem is due to Dloussky-Oeljeklaus-Toma [DOT], and gives a positive answer to a conjecture stated by Kato:

Theorem 3.3.7 *Any surface* $X \in \mathrm{VII}^{\min}_{b_2>0}$ *with* $b_2(X)$ *rational curves is a Kato surface.*

The second is an older result of Nakamura [Na3]:

Theorem 3.3.8 *Any surface* $X \in \mathrm{VII}^{\min}_{b_2>0}$ *with a cycle of rational curves is a degeneration of a 1-parameter family of blown up primary Hopf surfaces.*

These results suggest the conjectures:

Conjecture 1 Any surface $X \in \mathrm{VII}^{\min}_{b_2>0}$ has $b_2(X)$ rational curves.

Conjecture 2 Any surface $X \in \mathrm{VII}^{\min}_{b_2>0}$ has a cycle of rational curves.

By Theorem 3.3.7, Conjecture 1 will solve the classification problem up to biholomorphism (equivalent to the GSS conjecture).

By Theorem 3.3.8, Conjecture 2 will solve the classification problem up to deformation equivalence.

Nakamura's Theorem 3.3.8 states that any surface $X \in \mathrm{VII}^{\min}_{b_2>0}$ with a cycle of curves belongs to the known component of the moduli space of class VII surfaces; this moduli space contains both Kato surfaces and blown up primary Hopf surfaces, the latter being generic. In the next section we will see that, using a combination of complex geometric and gauge theoretical techniques, one can prove [Te2], [Te3], [Te8], [Te9]:

Theorem 3.3.9 *Any minimal class VII surface X with $1 \leqslant b_2(X) \leqslant 3$ contains a cycle of curves.*

Therefore any minimal class VII surface X with $1 \leqslant b_2(X) \leqslant 3$ belongs to the known component of the moduli space of class VII surfaces.

3.4 Gauge Theoretical Methods in the Classification of Class VII Surfaces

3.4.1 Instantons and Holomorphic Bundles on Complex Surfaces

Let (X, g) be a complex surface endowed with a Gauduchon metric (see [Gau] and Appendix in this article). A holomorphic rank 2 bundle $\mathcal{E}$ on X is called

- *stable*, if for every line bundle $\mathcal{L}$ and any sheaf monomorphism $0 \to \mathcal{L} \to \mathcal{E}$ one has

$$\deg(\mathcal{L}) < \frac{1}{2} \deg_g(\det(\mathcal{E})).$$

- *polystable*, if it is either stable, or isomorphic to a direct sum $\mathcal{L} \oplus \mathcal{M}$ of line bundles with $\deg_g(\mathcal{L}) = \deg_g(\mathcal{M})$.

We will consider moduli spaces of (poly)stable bundles with fixed determinant line bundle. Let (E, h) be a differentiable Hermitian rank 2-bundle on X, and $\mathcal{D}$ be a fixed holomorphic structure on the line bundle $D := \det(E)$. Denote by

$$\mathcal{M}_{\mathcal{D}}^{st}(E), \ \mathcal{M}_{\mathcal{D}}^{pst}(E)$$

the moduli sets of stable, respectively polystable holomorphic structures $\mathcal{E}$ on E inducing the fixed holomorphic structure $\mathcal{D}$ on $\det(E)$, modulo the complex gauge group $\mathcal{G}^{\mathbb{C}} := \Gamma(X, \mathrm{SL}(E))$. $\mathcal{M}_{\mathcal{D}}^{st}(E)$ has a natural complex space structure obtained using classical deformation theory. $\mathcal{M}_{\mathcal{D}}^{pst}(E)$ can be endowed with a natural topology induced by the Kobayashi-Hitchin correspondence we explain briefly below (see [Bu1], [LT]).

Let a be the Chern connection of the pair $(\mathcal{D}, \det(h))$, and $\mathcal{G} := \Gamma(X, \mathrm{SU}(E))$ be the gauge group of special unitary automorphisms of the Hermitian bundle (E, h). Denote by $\mathcal{A}(E)$ the space of unitary connections on (E, h) and by

$$\mathcal{M}_a^{ASD}(E) := \{A \in \mathcal{A}(E) \mid \det(A) = a, \ (F_A^0)^+ = 0\}/\mathcal{G}$$

the moduli space of projectively ASD unitary connections on (E, h) which induce a on D. We will also need the open subspace $\mathcal{M}_a^{ASD}(E)^* \subset \mathcal{M}_a^{ASD}(E)$ defined by the condition "A is irreducible"; it has the structure of a real analytic space.

The *Kobayashi-Hitchin correspondence* states that the map

$$A \mapsto \text{ the holomorphic structure on } E \text{ defined by } \bar{\partial}_A$$

induces a bijection $KH : \mathcal{M}_a^{\mathrm{ASD}}(E) \to \mathcal{M}_{\mathcal{D}}^{\mathrm{pst}}(E)$. More precisely we have a commutative diagram

$$
\begin{array}{ccc}
\mathcal{M}_a^{\mathrm{ASD}}(E)^* & \hookrightarrow & \mathcal{M}_a^{\mathrm{ASD}}(E) \\
{\scriptstyle KH^*}\downarrow \simeq & & \simeq \downarrow {\scriptstyle KH} \\
\mathcal{M}_{\mathcal{D}}^{\mathrm{st}}(E) & \hookrightarrow & \mathcal{M}_{\mathcal{D}}^{\mathrm{pst}}(E)
\end{array}
$$

where KH is a bijection and KH^* a real analytic isomorphism. We endow $\mathcal{M}_{\mathcal{D}}^{\mathrm{pst}}(E)$ with the topology induced by KH. In general $\mathcal{M}_{\mathcal{D}}^{\mathrm{pst}}(E)$ is not a complex space around the reduction locus $\mathcal{R} := \mathcal{M}_{\mathcal{D}}^{\mathrm{pst}}(E) \setminus \mathcal{M}_{\mathcal{D}}^{\mathrm{st}}(E)$. $\mathcal{R}$ can be identified with the subspace of reducible instantons in $\mathcal{M}_a^{\mathrm{ASD}}(E)$. If $c_1(D) \notin 2H^2(X, \mathbb{Z})$ this subspace is a union of tori of real dimension $b_1(X)$. The local structure of $\mathcal{M}_{\mathcal{D}}^{\mathrm{pst}}(E)$ around $\mathcal{R}$ can be described explicitly using the Kobayashi-Hitchin correspondence and Donaldson theory.

The Kobayashi-Hitchin correspondence has been first used by Donaldson as a tool to describe moduli spaces of instantons on algebraic surfaces. The "unknown" (the object to describe) was $\mathcal{M}_a^{\mathrm{ASD}}(E)$ and the computable object was $\mathcal{M}_{\mathcal{D}}^{\mathrm{pst}}(E)$. In our case the moduli space $\mathcal{M}_{\mathcal{D}}^{\mathrm{pst}}(E)$ cannot be regarded as a computable object. The problem is that, on non-algebraic surfaces, the appearance of *non-filtrable bundles* complicates the description of such a moduli space. We explain briefly this difficulty. Recall that a rank 2 holomorphic bundle $\mathcal{E}$ on X is called *filtrable* if there exists a sheaf epimorphism

$$\mathcal{E} \to \mathcal{F} \to 0$$

onto a torsion free coherent sheaf $\mathcal{F}$ of rank 1. A filtrable bundle $\mathcal{E}$ fits in a short exact sequence

$$0 \to \mathcal{M} \to \mathcal{E} \to \mathcal{N} \otimes \mathcal{I}_Z \to 0,$$

for line bundles $\mathcal{M}, \mathcal{N}$ and a 0-dimensional locally complete intersection $Z \subset X$. Therefore, filtrable rank 2 bundles are, in principle, classifiable. A non-filtrable bundle is stable with respect to *any* Gauduchon metric. There exists no classification method for non-filtrable bundles. Usually their presence is detected using deformation theory.

3.4.2 A Moduli Space of Instantons on Class VII Surfaces

Let now (X, g) be a class VII surface endowed with a Gauduchon metric, and (E, h) be a differentiable rank 2-bundle on X with $c_2(E) = 0$ and $\det(E) = K_X$, where K_X stands for the underlying C^∞ bundle of the canonical line bundle $\mathcal{K}_X$. Let a be the Chern connection of $(\mathcal{K}_X, \det(h))$. The fundamental objects used in our program to prove existence of curves on class VII surfaces are the moduli space

$$\mathcal{M} := \mathcal{M}_{\mathcal{K}}^{\mathrm{pst}}(E) \simeq \mathcal{M}_a^{\mathrm{ASD}}(E).$$

and its open subspace $\mathcal{M}^{\mathrm{st}} := \mathcal{M}_{\mathcal{K}}^{\mathrm{st}}(E) \simeq \mathcal{M}_a^{\mathrm{ASD}}(E)^*$ of stable bundles, which is a complex space of dimension $b := b_2(X)$. The rough idea of our strategy to use moduli spaces of holomorphic bundles is simple: prove that the same filtrable bundle can be written as an extension in two different ways. This will yield a non-trivial, non-isomorphic morphism of line bundles, whose vanishing locus will be a curve.

The first crucial property of the moduli space $\mathcal{M}$ is compactness, which holds in full generality [Te3]:

Theorem 3.4.1 (X, g) be a class VII surface endowed with a Gauduchon metric, and (E, h) be a differentiable rank 2-bundle on X with $c_2(E) = 0$ and $\det(E) = K_X$. Then $\mathcal{M}_{\mathcal{K}}^{\mathrm{pst}}(E)$ is compact.

The proof is due to N. Buchdahl and the author, and makes use of a combination of gauge theoretical and complex geometric arguments.

Suppose that $X \in \mathrm{VII}_{b_2>0}^{\mathrm{min}}$ is not an Enoki surface, and that g has been chosen such that $\deg_g(\mathcal{K}_X) < 0$ (this is also always possible, see [Te3]). Then $\mathcal{M}^{\mathrm{st}}$ is a smooth b-dimensional complex manifold. Moreover, the reduction locus $\mathcal{R}$ is a finite disjoint union of circles, and every circle $C \subset \mathcal{R}$ has a compact neighbourhood which is homeomorphic to the trivial bundle over C with fibre a cone (in the topological sense) over $\mathbb{P}_{\mathbb{C}}^{b-1}$, where $b := b_2(X)$.

From now on we will suppose for simplicity that $H_1(X, \mathbb{Z}) \simeq \mathbb{Z}$. This simplifying assumption allows us to parameterize the set of connected components of $\mathcal{R}$ in a simple way: $\mathcal{R}$ is a disjoint union of 2^{b-1} circles $C_{\{I, \bar{I}\}}$ indexed by unordered partitions $\{I, \bar{I}\}$ of the index set $\mathfrak{I} := \{1, \ldots, b\}$.

Note also that he moduli space $\mathcal{M}$ comes with a natural involution given by $\otimes \mathcal{L}_0$, where $[\mathcal{L}_0] \in \mathrm{Pic}^0(X)$ is the non-trivial square root of $[\mathcal{O}_X]$. The fixed points of this involution are called twisted reductions. The fixed point set is always finite and is contained in $\mathcal{M}^{\mathrm{st}}$; if $\pi_1(X, x_0) \simeq \mathbb{Z}$ this set has 2^{b-1} elements. The filtrable bundles in our moduli space can be classified as follows: Let $(e_1, \ldots, e_b)$ be a standard basis of $H^2(X, \mathbb{Z})$. Let $\mathcal{E}$ be rank 2 bundle on X with $\det(\mathcal{E}) = \mathcal{K}_X$, $c_2(\mathcal{E}) = 0$, and let $\mathcal{E} \to \mathcal{L}$ be an epimorphism onto a rank 1 torsion-free sheaf. One can prove [Te3] that $\mathcal{L}$ is locally free, and there exists a subset $I \subset \mathfrak{I}$ such that

$$c_1(\mathcal{L}) = e_I := \sum_{i \in I} e_i.$$

Therefore any filtrable bundle $\mathcal{E}$ in our moduli space fits in a short exact sequence

$$0 \to \mathcal{K}_X \otimes \mathcal{L}^\vee \to \mathcal{E} \to \mathcal{L} \to 0, \tag{3.6}$$

where $c_1(\mathcal{L}) = e_I$ for an index set $I \subset \mathfrak{I}$. We will denote by $\mathcal{M}_I^{\text{st}} \subset \mathcal{M}^{\text{st}}$ the subset of isomorphism classes of stable bundles which are extensions of type (3.6) with fixed $c_1(\mathcal{L}) = e_I$. One has

$$\mathcal{M}_\emptyset^{\text{st}} = \{[\mathcal{A}], [\mathcal{A}']\}$$

where $\mathcal{A}$ is the *canonical extension* of X, defined as the essentially unique non-trivial extension of the form

$$0 \to \mathcal{K}_X \to \mathcal{A} \to \mathcal{O}_X \to 0$$

(note that $h^1(\mathcal{K}_X) = 1$ by Serre duality) and $\mathcal{A}' := \mathcal{A} \otimes \mathcal{L}_0$.

If $I \neq \emptyset$ the closure $\overline{\mathcal{M}}_I^{\text{st}}$ of $\mathcal{M}_I^{\text{st}}$ in $\mathcal{M}$ contains the circle $C_{I,\bar{i}}$; moreover, if X has no curves in certain homology classes, the subset $\mathcal{M}_I^{\text{st}} \subset \mathcal{M}^{\text{st}}$ is a $\mathbb{P}_{\mathbb{C}}^{|I|-1}$-fibration over a punctured disc, and these fibrations are pairwise disjoint.

3.4.3 Existence of a Cycle on Class VII Surfaces with Small b_2

We present the main ideas of the proof of Theorem 3.3.9 for $b_2 = 1$, and we explain the general strategy for larger b_2. If the method generalizes for arbitrary b_2, we will have a proof of Conjecture 2, which would complete the classification of class VII surfaces up to deformation equivalence.

Let X be class VII surface. A cycle in X is an effective divisor which is either en elliptic curve, or a cycle of rational curves. Note that minimal class VII surfaces $X \in \text{VII}_{b_2>0}^{\text{min}}$ containing an elliptic curve have been classified [Na1, Na2, Na3], and we know that any such surface is a parabolic Inoue surface, so it also has a cycle of rational curves.

Proposition 3.4.2 *If the canonical extension $\mathcal{A}$ can be written as a line bundle extension in a different way (with a different kernel), then X has a cycle.*

In particular, if $\mathcal{A}$ belongs to $\mathcal{M}_I^{\text{st}}$ for $I \neq \emptyset$ or coincides with a twisted reduction, then X has a cycle.

Proof Suppose that $\mathcal{A}$ can be written as a line bundle extension in a different way, i.e. there exists an invertible subsheaf $\mathcal{L} \xhookrightarrow{j} \mathcal{A}$ such that

(i) $\mathcal{L} \neq \mathcal{K}_X$.
(ii) j is a bundle embedding, i.e. the linear map $\mathcal{L}(x) \to \mathcal{A}(x)$ induced by j is injective for any $x \in X$.

Since $\mathcal{L} \neq \mathcal{K}_X$ the composition $p \circ j$ is non-zero, because, if it were 0, we would have $\mathcal{L} \subsetneq \mathcal{K}_X$, so $\mathcal{L} = \mathcal{K}_X(-D)$ for an effective divisor D which is nonempty by (i). But this contradicts (ii) taking $x \in D$.

$$0 \longrightarrow \mathcal{K}_X \lhook\joinrel\longrightarrow A \xrightarrow{\ p\ } \mathcal{O}_X \longrightarrow 0$$

with maps $j \uparrow$ and $p \circ j$ from $\mathcal{L}$.

The composition $p \circ j$ cannot be an isomorphism either, because the canonical extension is non-split by definition. Therefore $\mathrm{im}(p \circ j) = \mathcal{O}_X(-D)$ where $D > 0$ is the vanishing divisor of $p \circ j$.

Restricting the diagram to D and taking into account (ii) we obtain $\mathcal{L}_D \simeq \mathcal{K}_D$. But $\mathcal{L} \simeq \mathcal{O}(-D)$. Therefore $\omega_D := \mathcal{K}_X(D)_D$ is trivial. This implies that D is either a cycle (when it is reduced) or it contains a cycle [Te3]. ∎

Proposition 3.4.2 shows that, in order to prove the existence of a cycle on X, it "suffices" to prove the following "remarkable incidence":

$$[\mathcal{A}] \in \{\text{twisted reductions}\} \cup \left(\bigcup_{I \neq \emptyset} \mathcal{M}_I^{\mathrm{st}}\right). \tag{RI}$$

3.4.3.1 The Existence of a Cycle for $b_2 = 1$

In the case $b_2 = 1$ one can prove that the connected component $\mathcal{M}_0$ of the circle $C_{\emptyset; \mathfrak{J}}$ in the moduli space $\mathcal{M}$ is a compact disc. The boundary of this disc is the circle $C_{\emptyset; \mathfrak{J}}$, its center b is a twisted reduction, and the punctured open disc $\mathcal{M} \setminus (C \cup \{b\})$ coincides with $\mathcal{M}_{\mathfrak{J}}^{\mathrm{st}}$. Therefore we have the following dichotomy: either $[\mathcal{A}] \in \mathcal{M}_0$, so the remarkable incidence (RI) holds, and the theorem is proved, or the connected component of $\mathcal{A}$ in $\mathcal{M}$ is a closed Riemann surface $Y \subset \mathcal{M}^{\mathrm{st}}$ which has at most two filtrable points. The latter possibility is ruled out by the following proposition [Te2]:

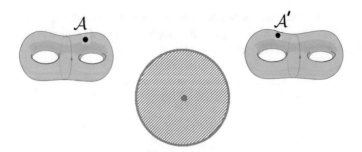

Proposition 3.4.3 *Suppose that X is a complex surface with $a(X) = 0$, E is a differentiable rank 2 bundle over X, Y is a closed Riemann surface and*

$$f : Y \to \mathcal{M}^{\text{simple}}(E), \ y \mapsto [\mathcal{E}_y]$$

is a holomorphic map to the moduli space of simple structures on E [LT]. Then all the bundles $\mathcal{E}_y$ contain a constant (independent of y) subsheaf $\mathcal{T}$ of rank 1 or 2. In particular the bundles $\mathcal{E}_y$ are either all filtrable or all non-filtrable.

The idea of the proof is to show that f has a classifying bundle $\mathcal{E}$ on $Y \times X$, and to regard $\mathcal{E}$ as a family of bundles on Y parameterized by X. If this family is generically stable, one obtains a meromorphic map $X \dashrightarrow \mathcal{M}^{\text{sst}}(Y)$ to a moduli space of semistable bundles on Y. Since Y is a Riemann surface, this moduli space is a projective variety.

Therefore, Proposition 3.4.3 shows that the appearance of an "unexpected" component in the moduli spaces leads to a contradiction, hence the remarkable incidence holds, which proves that X has a cycle.

3.4.3.2   The Strategy for Larger b_2

For larger b_2 the idea is to prove the following dichotomy (similar to the dichotomy used for $b_2 = 1$):

D: *Either the remarkable incidence holds (and then X has a cycle) or $\mathcal{M}^{\text{st}}$ contains an irreducible compact subspace $Y \subset \mathcal{M}^{\text{st}}$ of positive dimension with an open subspace $Y_0 \ni [\mathcal{A}]$ such that the points of $Y_0 \setminus \{[\mathcal{A}]\}$ correspond to non-filtrable bundles.*

The proof of this dichotomy for $b_2 \leq 3$ is explained in [Te3], [Te8], [Te9] and makes use of [Te5], [Te6]. On the other hand the following proposition shows that the existence of such a compact subspace $Y \subset \mathcal{M}^{\text{st}}$ leads to a contradiction:

Proposition 3.4.4 *Let X be minimal class VII surface with $b_2 > 0$ arbitrary. There does not exist any irreducible compact subspace $Y \subset \mathcal{M}^{\text{st}}$ of positive dimension* *with the following property:* $[\mathcal{A}] \in Y$ *and there exists an open neighbourhood of* Y_0 *of* $[\mathcal{A}]$ *in* Y *such that the points of* $Y_0 \setminus \{[\mathcal{A}]\}$ *correspond to non-filtrable bundles.*

The proof of Proposition 3.4.4 uses of a variation formula for determinant line bundles in non-Kählerian geometry [Te7].

3.5 Algebraic Deformations of Singular Contractions of Class VII Surfaces

3.5.1 From Local to Global Smoothability

Recall that, by the theorem Dloussky-Oeljeklaus-Toma [DOT] (Theorem 3.3.7 in this article), Conjecture 1 will solve the classification problem up to biholomorphism, and by the theorem of Nakamura [Na2] (Theorem 3.3.8 in this article), Conjecture 2 will solve the classification problem up to deformation equivalence. More precisely, if X has a cycle of curves, then X belongs to the known component of the moduli space of class VII surfaces, and this known component contains both Kato surfaces and blown up primary Hopf surfaces, the latter being generic. We have seen that, using Donaldson theory, one can prove Conjecture 2 at least for small b_2. Therefore an important question is: can one pass from "X has a cycle" to the stronger property "X is a Kato surface"? We know that the existence of curves cannot be obtained by "passing to the limit" because of the "explosion of area" phenomenon (see Remark 3.3.3).

A new approach have been suggested by G. Dloussky and the author in [DlTe2]: for a minimal class VII surface with a cycle of rational curves, use *algebraic* deformations of the singular surface obtained by contracting the cycle.

Let $X \in \mathrm{VII}^{\mathrm{min}}_{b_2 > 0}$ with a cycle $C = C_0 + \cdots + C_{r-1}$ of $r \geqslant 1$ rational curves. We know $1 \leqslant r \leqslant b_2(X)$. If $C^2 = 0$, then X is an Enoki surface, hence a Kato surface.

Suppose $C^2 < 0$. Contracting C one obtains a singular surface Y with an isolated singularity c. The germ (Y, c) is *a cusp*, in particular it is an elliptic, Gorenstein singularity.

Theorem 3.5.1 ([DlTe2]) *If the germ of singularity* (Y, c) *is smoothable, then the singular surface* Y *is globally smoothable. Moreover*

(1) If $r < b_2(X)$, *then any smooth deformation* Y' *of* Y *is a rational surface* Y' *with* $b_2(Y') = 10 + b_2(X) + C^2$.
(2) If $r = b_2(X)$, *then any smooth deformation* Y' *of* Y *is an Enriques surface.*

The main ingredients used in the proof are:

1. Let $\pi : X \to Y$ be the contraction map. One has

$$\omega_Y = \pi_*(\mathcal{K}_X(C)), \quad \Omega_Y^{\vee\vee} = \pi_*(\Omega_X).$$

2. Using Serre duality on Cohen-Macauley spaces and ideas coming from Nakamura, we proved the following vanishing theorem:

   Theorem 3.5.2 *With the notations above one has:* $H^2(Y, \Theta_Y) = 0$.

3. A lemma due to Manetti [Ma]:

   Proposition 3.5.3 *Let Y be a complex surface with finitely many isolated
   singularities $\{y_1, \ldots, y_s\}$. If $H^2(Y, \Theta_Y) = 0$ then the natural germ morphism
   $\mathrm{Def}(Y) \to \times_{i=1}^s \mathrm{Def}(Y, y_i)$ is smooth, in particular surjective.*

   Here we denoted by $\mathrm{Def}(Y)$ the base of a versal deformation of Y, and by
   $\mathrm{Def}(Y, y_i)$ the base of a versal deformation of the singularity (Y, y_i).

3.5.2 Local Smoothability: Looijenga's Conjecture

Our result shows that Y is globally smoothable if the cusp (Y, c) is smoothable.
Having this implication the natural question is: which cusps (Y, c) are smoothable?
The answer is given by Looijenga's conjecture, which has recently become a
theorem. In order to state this result we need a preparation.

Definition 3.5.4 Let $C = \sum_{i \in \mathbb{Z}_r} C_i$ be an oriented cycle of r rational curves
[DlTe2]. The type of C is given by

$$[c_0, \ldots, c_{r-1}] = \begin{cases} [-C_0^2, \ldots, -C_{r-1}^2] & \text{if } r \geq 2 \\ [2 - C_0^2] & \text{if } r = 1 \end{cases}.$$

The type of a cycle should be regarded as an element of $\mathbb{Z}^{\mathbb{Z}_r}/\mathbb{Z}_r$. In other words
one has

$$[c_0, c_1, \ldots, c_{r-1}] = [c_1, \ldots, c_{r-1}, c_0] = \cdots = [c_{r-1}, c_0, \ldots, c_{r-2}]$$

(see [DlTe2] for details). Note that for any cycle C one has

$$-C^2 = \sum_{i=0}^{r-1}(c_i - 2),$$

and for any cycle $C \subset X \in \mathrm{VII}_{b_2>0}^{\min}$ with $C^2 < 0$ it holds

$$\forall i \in \{0, \ldots, r-1\}, \ c_i \geqslant 2, \ \exists j \in \{0, \ldots, r-1\}, \ c_j \geqslant 3. \tag{3.7}$$

There exists an important involution on the set of "types" verifying conditions (3.7).
Put $s := \sum_{j=0}^{r-1}(c_j - 2) = -C^2$. The Hirzebruch-Zagier *dual type* $[d_0, \ldots, d_{s-1}]$
is constructed by replacing any element $c_j \geqslant 3$ in the original type by the (possibly

empty) sequence

$$\underbrace{2, \ldots, 2,}_{c_j - 3}$$

and any maximal (possibly empty) sequence of the form

$$\underbrace{2, \ldots, 2}_{l}$$

in the original type by the single element $l + 3$.

Example 3.5.1 The dual of the type $[3, 2, \ldots, 2]$ (of length r) is $[r + 2]$. If $r = 1$ we obtain a selfdual type: $[3]$.

The following conjecture of Looijenga [Loo] became a theorem (see [GHK], [Eng]):

Theorem 3.5.5 *The cusp* (Y, c) *is smoothable if and only if there exists a smooth rational surface with an anti-canonical cycle D whose type $[d_0, \ldots, d_{s-1}]$ is the Hirzebruch-Zagier dual of $[c_0, \ldots, c_{r-1}]$.*

Note that the condition "$[d_0, \ldots, d_{s-1}]$ is the type of an anti-canonical cycle in a rational surface" is equivalent to the condition "$[d_0, \ldots, d_{s-1}]$ is the type of an anti-canonical cycle in a blown up $\mathbb{P}^2_{\mathbb{C}}$", and can be checked algorithmically.

Example 3.5.2 Suppose that the type of C is $[3]$ (the type of a nodal curve with self-intersection -1). The dual type is $[3]$. A singular rational cubic $\Gamma \subset \mathbb{P}^2_C$ is anti-canonical, but has $\Gamma^2 = 9$. The proper transform $\tilde{\Gamma}$ of Γ in the blown up $\hat{\mathbb{P}}^2_C$ of $\mathbb{P}^2_C$ at 10 smooth points of Γ is an anti-canonical nodal curve of $\hat{\mathbb{P}}^2_C$, and its type is $[3]$. Therefore the smoothability condition given by Looijenga's conjecture is fulfilled, so (Y, c) is smoothable.

In a similar way one can prove [DlTe2, Theorem 5.7]:

Corollary 3.5.6 *The cusp* (Y, c) *is always smoothable if* $\sum_{i=0}^{r-1}(c_i - 2) \leqslant 10$.

Therefore

Theorem 3.5.7 ([DlTe2]) *Let X be a minimal class VII surface, $C \subset X$ be a cycle of r rational curves with $C^2 < 0$, $[c_0, \ldots, c_{r-1}]$ be its type, and (Y, c) be the singular contraction of (X, C). Then $r \leqslant b_2(X)$ and*

(1) Y is smoothable if and only if the dual type $[d_0, \ldots, d_{s-1}]$ of $[c_0, \ldots, c_{r-1}]$ is the type of an anti-canonical cycle in a smooth rational surface which admits $\mathbb{P}^2$ as minimal model. This condition is always satisfied when $\sum_{i=0}^{r-1}(c_i - 2) \leqslant 10$.

(2) If $r < b_2(X)$, then any smooth deformation Y' of Y is a rational surface with

$$b_2(Y') = 10 + b_2(X) + C^2 = 10 + r.$$

Fig. 3.12 Smooth
deformations of the singular
surface Y obtained by
contracting a cycle in a
minimal class VII surface

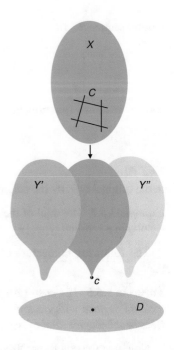

(3) If $r = b_2(X)$, then X is a half-Inoue surface, and any smooth deformation Y' of Y is an Enriques surface.

The condition $\sum_{i=0}^{r-1}(c_i - 2) \leqslant 10$ is automatically satisfied when $r < b_2(X) \leqslant 11$ so, in this range, the singular surface Y obtained starting with a pair (X, C) as above, is always smoothable by rational surfaces. Therefore any minimal class VII with a cycle C satisfying the conditions $C^2 < 0$, $r < b_2(X) \leqslant 11$ can be "connected" to a smooth rational surface using a two-step "geometric transition": contract, then deform (Fig. 3.12).

This reminds us of the *"Web Conjecture"* in the theory of Calabi-Yau 3-folds: *Any two deformation classes of CY 3-folds can be connected by a sequence of "geometric transitions"* consisting of

- birational contraction to a normal variety,
- deformation of the contraction.

Theorem 3.5.7 suggests a possible strategy for proving Conjecture 1 for minimal class VII surfaces with a cycle of rational curves and $b_2 \leqslant 11$: classify all families of smooth rational surfaces parameterized by $D^\bullet$, degenerating to a surface with a single singularity which is a cusp; prove than only contractions of the Kato surfaces may occur as limits.

Acknowledgements The author thanks Daniele Angella, Leandro Arosio, and Eleonora Di Nezza, the organizers of the "CIME School "Complex non-Kähler geometry", for the invitation to give a lecture series, and to submit a written version of my lectures for publication in the proceedings of the meeting. The author is grateful to Georges Dloussky for his constant help, encouragement and collaboration, and for his useful suggestions and comments on the text.

Appendix

The Picard Group and the Gauduchon Degree

Let X be a connected, compact n-dimensional complex manifold. Recall that the Picard group $\mathrm{Pic}(X)$ of X is the group of isomorphism classes of holomorphic line bundles on X, and can be identified with $H^1(X, \mathcal{O}_X^*)$. The exponential short exact sequence gives the cohomology exact sequence

$$0 \to H^1(X, \mathbb{Z}) \xrightarrow{2\pi i \cdot} H^1(X, \mathcal{O}_X) \to H^1(X, \mathcal{O}_X^*) = \mathrm{Pic}(X) \xrightarrow{c_1} \mathrm{NS}(X) \to 0,$$

where the Neron-Severi group $\mathrm{NS}(X)$ is defined by

$$\mathrm{NS}(X) := \ker(H^2(X, \mathbb{Z}) \to H^2(X, \mathcal{O}_X)).$$

A class $c \in H^2(X, \mathbb{Z})$ belongs to $\mathrm{NS}(X)$ if and only if it can be represented by a closed real 2-form of type $(1,1)$. Denoting by $\mathrm{Pic}^0(X)$ the kernel of the Chern class morphism $\mathrm{Pic}(X) \xrightarrow{c_1} \mathrm{NS}(X)$, we obtain a short exact sequence

$$\{1\} \to \mathrm{Pic}^0(X) \to \mathrm{Pic}(X) \xrightarrow{c_1} \mathrm{NS}(X) \to 0 \tag{3.8}$$

and an identification

$$\mathrm{Pic}^0(X) = {H^1(X, \mathcal{O}_X)}\big/{2\pi i H^1(X, \mathbb{Z})}. \tag{3.9}$$

Remark 1 The image of $2\pi i H^1(X, \mathbb{Z})$ in $H^1(X, \mathcal{O}_X)$ is closed, in particular the quotient $H^1(X, \mathcal{O}_X)/2\pi i H^1(X, \mathbb{Z})$ has the structure of an Abelian connected complex Lie group.

Proof The obvious embedding $2\pi i H^1(X, \mathbb{Z}) \hookrightarrow H^1(X, \mathcal{O}_X)$ factorizes as

$$2\pi i H^1(X, \mathbb{Z}) \hookrightarrow H^1(X, i\mathbb{R}) \to H^1(X, \mathcal{O}_X).$$

The coefficients formula shows that $2\pi i H^1(X, \mathbb{Z})$ is a lattice in $H^1(X, i\mathbb{R})$, in particular it is closed in this real vector space. On the other hand, using de Rham and Dolbeault theorems, it is easy to prove that the $\mathbb{R}$-linear morphism $H^1(X, i\mathbb{R}) \to$

$H^1(X, \mathcal{O}_X)$ is injective. Therefore $2\pi i H^1(X, \mathbb{Z})$ is closed in an $\mathbb{R}$-linear subspace of $H^1(X, \mathcal{O}_X)$, so it is closed in $H^1(X, \mathcal{O}_X)$. ∎

Using (3.8) it follows that

Remark 2 Pic(X) has a natural structure of an Abelian complex Lie group, and Pic$^0(X)$ is the connected component of its unit element.

For a class $c \in \mathrm{NS}(X)$ we put Pic$^c(X) := \{[\mathcal{L}] \in \mathrm{Pic}(X)|\ c_1(\mathcal{L}) = c\}$, and

$$\mathrm{Pic}^T(X) := \{[\mathcal{L}] \in \mathrm{Pic}(X)|\ c_1(\mathcal{L}) \in \mathrm{Tors}\}.$$

Recall that a Gauduchon metric on X is a Hermitian metric g on X whose Kähler form ω_g satisfies $dd^c\omega_g^{n-1} = 0$. An important theorem of Gauduchon states that any conformal class of Hermitian metrics on X contains a Gauduchon metric (which is unique up to constant factor if $n \geq 2$), so there is no obstruction to the existence of Gauduchon metrics.

The degree map associated with a Gauduchon metric g on X is the group morphism

$$\deg_g : \mathrm{Pic}(X) \to \mathbb{R}$$

defined by

$$\deg_g(\mathcal{L}) := \int_X \frac{i}{2\pi} F_{A_h} \wedge \omega_g^{n-1}, \qquad (3.10)$$

where h is a Hermitian metric on $\mathcal{L}$, and $F_{A_h} \in iA^{1,1}(X)$ is the curvature of the Chern connection A_h associated with the pair $(\mathcal{L}, h)$. Changing h will modify the Chern form $c_1(\mathcal{L}, h) := \frac{i}{2\pi} F_{A_h}$ by a dd^c-exact form. Since g is Gauduchon, the right hand term of (3.10) is independent of h, so $\deg_g$ is well defined.

The Kobayashi-Hitchin Correspondence for Line Bundles

Let (X, g) be a connected, compact complex manifold endowed with a Gauduchon metric. We denote by $\Lambda_X^{p,q}$ the bundle of (p, q)-forms on X, and by $\Lambda_g : \Lambda_X^{p,q} \to \Lambda_X^{p-1,q-1}$ the adjoint of the wedge product operator $\omega_g \wedge \cdot : \Lambda_X^{p-1,q-1} \to \Lambda_X^{p,q}$. The same symbol Λ_g will be used for the induced operator $A^{p,q}(X) \to A^{p-1,q-1}(X)$ between spaces of global forms. Denoting by $\mathrm{vol}_g = \frac{1}{n!}\omega_g^n$ the volume form on X, and using the identity $\Lambda_g\omega_g = n$, we obtain easily the identity

$$\alpha \wedge \omega_g^{n-1} = (n-1)!(\Lambda_g\alpha)\mathrm{vol}_g \ \forall \alpha \in A^2(X, \mathbb{C}). \qquad (3.11)$$

Definition 3 Let $\mathcal{L}$ be a holomorphic line bundle on X. A Hermitian metric h on $\mathcal{L}$ is called Hermitian-Einstein if the real function $i\Lambda_g F_{A_h}$ is constant. If this is the case, this constant is called the Einstein constant of h and is denoted c_h.

Applying (3.11) to $iF_{A_h} = 2\pi c_1(\mathcal{L}, h)$ we obtain the following formula

$$c_h = \frac{2\pi}{(n-1)!\mathrm{Vol}_g(X)}\deg_g(\mathcal{L}) \tag{3.12}$$

for the Einstein constant of a Hermitian-Einstein connection on $\mathcal{L}$. Therefore a Hermitian metric h on $\mathcal{L}$ is Hermitian-Einstein if and only if

$$i\Lambda_g F_{A_h} = \frac{2\pi}{(n-1)!\mathrm{Vol}_g(X)}\deg_g(\mathcal{L}).$$

Proposition 4 *Let $\mathcal{L}$ be a holomorphic line bundle on X. Then $\mathcal{L}$ admits a Hermitian-Einstein metric h, which is unique up to constant factor.*

Proof Let h_0 be an arbitrary Hermitian metric on $\mathcal{L}$, and let $u \in C^\infty(X, \mathbb{R})$. The metric $h = e^u h_0$ is Hermite-Einstein if and only if

$$\Lambda_g(i\bar\partial\partial u + iF_{A_{h_0}}) = \frac{2\pi}{(n-1)!\mathrm{Vol}_g(X)}\deg_g(\mathcal{L}),$$

i.e. if and only if u is a solution of the elliptic equation

$$i\Lambda_g\bar\partial\partial u = \frac{2\pi}{(n-1)!\mathrm{Vol}_g(X)}\deg_g(\mathcal{L}) - i\Lambda_g F_{A_{h_0}}.$$

The definition of $\deg_g(\mathcal{L})$ gives

$$\int_X \left(\frac{2\pi}{(n-1)!\mathrm{Vol}_g(X)}\deg_g(\mathcal{L}) - i\Lambda_g F_{A_{h_0}}\right)\mathrm{vol}_g = 0,$$

so the result follows from Lemma 5 below (see [LT, Corollary 1.2.9]). ∎

Lemma 5 *Let (X, g) be a connected, compact complex manifold endowed with a Gauduchon metric. Denote by $\mathbb{R}$ the line of constant real functions on X. The operator $P = i\Lambda_g\bar\partial\partial : A^0(X, \mathbb{R}) \to A^0(X, \mathbb{R})$ has the following properties:*

(1) $\ker(P) = \mathbb{R}$.
(2) $\mathrm{im}(P) = \mathbb{R}^\perp$, *where the symbol* $\perp$ *stands for the orthogonal complement with respect to the L^2-inner product.*

Proof

(1) Note first that P is a second order elliptic operator which vanishes on locally defined constant functions (has no 0-order terms in local coordinates).

The maximum principle applies and shows that a function $u \in \ker(P)$ is constant around any local maximum. Therefore the non-empty closed subset $u^{-1}(u_{\max}) \subset X$ where u attains its global maximum $u_{\max}$ is open in X. Since X is connected, it follows $u^{-1}(u_{\max}) = X$, so u is constant.

(2) Since g is Gauduchon it follows easily that $\mathbb{R} \subset \mathrm{im}(P)^{\perp}$. On the other hand, the symbol of P is self-adjoint, so the index of P vanishes. Taking into account (1) it follows that $\dim(\mathrm{im}(P)^{\perp}) = 1$, so the inclusion $\mathbb{R} \subset \mathrm{im}(P)^{\perp}$ is an equality.

■

Proposition 4 has an important interpretation in terms of moduli spaces. In order to explain this interpretation, we will change the point of view and we will consider variable unitary connections on a fixed differentiable Hermitian line bundle. Let (L, h) be a differentiable Hermitian line bundle on X, and let $\mathcal{A}(L, h)$ be the set of unitary connection on (L, h). This set is an affine space over the linear space $A^1(X, i\mathbb{R})$ of $i\mathbb{R}$-valued 1-forms on X, so it has a natural Fréchet topology. The gauge real group $\mathcal{C}^\infty(X, S^1)$ acts on $\mathcal{A}(L, h)$ in the obvious way: denoting by

$$d_A : A^0(L) \to A^1(L)$$

the linear connection associated with A, the gauge action on $\mathcal{A}(L, h)$ satisfies the identity:

$$d_{f \cdot A} = f \circ d_A \circ f^{-1} = d_A - df f^{-1}.$$

The quotient $\mathcal{M}(L, h) := \mathcal{A}(L, h)/\mathcal{C}^\infty(X, S^1)$, endowed with the quotient topology, is called the moduli space of unitary connections on (L, h). It is an infinite dimensional Hausdorff space [Te4].

Definition 6 Let (L, h) be a differentiable Hermitian line bundle on X with $c_1(L) \in \mathrm{NS}(X)$. A unitary connection A on (L, h) is called Hermitian-Einstein if the curvature form F_{A_h} has type $(1,1)$ and $i\Lambda_g F_A$ is constant.

The Hermite-Einstein condition is gauge invariant so, denoting by $\mathcal{A}^{\mathrm{HE}}(L, h) \subset \mathcal{A}(L, h)$ the subspace of Hermitian-Einstein on (L, h), we obtain a closed subspace

$$\mathcal{M}^{\mathrm{HE}}(L, h) \to \mathcal{M}(L, h)$$

called the moduli space of Hermitian-Einstein on (L, h).

Let $\bar{\partial}_A : A^0(L) \to A^{0,1}(L)$ be the second component of the first order differential operator $d_A : A^0(L) \to A^1(L) = A^{1,0}(L) \oplus A^{0,1}(L)$. The first condition in Definition 6 means that $\bar{\partial}_A^2 = 0$, so $\bar{\partial}_A$ defines a holomorphic structure $\mathcal{L}_A$ on L; the corresponding sheaf of holomorphic sections is given by

$$\mathcal{L}_A(U) = \{s \in \Gamma(U, L) | \; \bar{\partial}_A s = 0\}$$

for open sets $U \subset X$. The assignment $A \mapsto \mathcal{L}_A$ induces a well defined map

$$K H_{g,L,h} : \mathcal{M}^{\mathrm{HE}}(L, h) \to \mathrm{Pic}^{c_1(L)}(X)$$

called the Kobayashi-Hitchin correspondence associated with (L, h). Proposition 4 can be reformulated as follows:

Corollary 7 *Let (L, h) be a differentiable Hermitian line bundle on X with $c_1(L) \in$ NS(X). The Kobayashi-Hitchin correspondence*

$$K H_{g,L,h} : \mathcal{M}^{\mathrm{HE}}(L, h) \to \mathrm{Pic}^{c_1(L)}(X)$$

is a homeomorphism.

For a class $c \in$ NS(X) let (L_c, h_c) be a differentiable Hermitian line bundle of Chern class c. The classification theorem for differentiable S^1-bundles shows that (L_c, h_c) is well-defined up to unitary isomorphism.

Remark 8 Corollary 7 gives a homeomorphism

$$K H_g : \mathcal{M}^{\mathrm{HE}} := \coprod_{c \in \mathrm{NS}(X)} \mathcal{M}^{\mathrm{HE}}(L_c, h_c) \to \mathrm{Pic}(X).$$

The moduli space $\mathcal{M}^{\mathrm{HE}}$ has a natural group structure defined by tensor product of Hermite-Einstein connections and, with respect to this structure, $K H_g$ is an isomorphism of real Lie groups.

Moduli Spaces of Flat S^1-Connections

Recall that a Hermitian line bundle (L, h) on a compact differentiable manifold X admits a unitary flat connection if and only if $c_1(L) \in$ Tors. This follows easily using the cohomology exact sequence associated with the short exact sequence of constant sheaves on X

$$\{1\} \to \mathbb{Z} \xrightarrow{2\pi i \cdot} i\mathbb{R} \xrightarrow{\exp} S^1 \to \{1\}.$$

Therefore the moduli space $\mathcal{M}_{\mathrm{fl}}$ of flat S^1-connections on X decomposes as

$$\mathcal{M}_{\mathrm{fl}} = \coprod_{c \in \mathrm{Tors}} \mathcal{M}_{\mathrm{fl}}(L_c, h_c),$$

where (L_c, h_c) is a Hermitian line bundle of Chern class c, and $\mathcal{M}_{\mathrm{fl}}(L_c, h_c)$ denotes the moduli space of flat unitary connections on (L_c, h_c). The classical classification theorem for flat connections gives an identification

$$\mathcal{M}_{\mathrm{fl}} \xrightarrow[\mathrm{hol}]{\simeq} \mathrm{Hom}(\pi_1(X, \mathbb{Z}), S^1) = \mathrm{Hom}(H_1(X, \mathbb{Z}), S^1)$$

given by the map hol which assigns to any flat connection its holonomy representation. Note that $\mathcal{M}_{\mathrm{fl}}$ has a natural Lie group structure, and fits in the short exact sequence

$$\{1\} \rightarrow \mathcal{M}_{\mathrm{fl}}^0 \hookrightarrow \mathcal{M}_{\mathrm{fl}} \xrightarrow{c_1} \mathrm{Tors} \rightarrow 0,$$

where $\mathcal{M}_{\mathrm{fl}}^0$ is the moduli space of flat unitary connections on the trivial Hermitian line bundle.

Remark 9 Let X be a compact differentiable manifold. The moduli space $\mathcal{M}_{\mathrm{fl}}^0$ is canonically isomorphic to the quotient $H^1(X, i\mathbb{R})/H^1(X, 2\pi i\mathbb{Z})$-torsor, so it is a real torus of dimension $b_1(X)$. In particular $\mathcal{M}_{\mathrm{fl}}$ is compact, and the connected component of its unit element is the torus $\mathcal{M}_{\mathrm{fl}}^0$.

Let now X be a compact complex manifold, and g be a Gauduchon metric on X. Any flat connection on a Hermitian line bundle on X is obviously Hermite-Einstein, so we get an obvious inclusion $\mathcal{M}_{\mathrm{fl}} \hookrightarrow \mathcal{M}^{\mathrm{HE}}$ which identifies $\mathcal{M}_{\mathrm{fl}}$ with a subgroup of $\mathcal{M}^{\mathrm{HE}}$.

Remark 10 The images

$$\mathrm{Pic}_{\mathrm{ufl}}(X) := K H_g(\mathcal{M}_{\mathrm{fl}}), \ \mathrm{Pic}_{\mathrm{ufl}}^0(X) := K H_g(\mathcal{M}_{\mathrm{fl}}^0)$$

of $\mathcal{M}_{\mathrm{fl}}$ ($\mathcal{M}_{\mathrm{fl}}^0$) in $\mathrm{Pic}(X)$ (respectively $\mathrm{Pic}^0(X)$) are independent of g; the first (respectively second) image coincides with the subgroup of isomorphism classes of (topologically trivial) holomorphic line bundles $\mathcal{L}$ on X admitting a Hermitian metric h with A_h flat.

Corollary 11 *Let X be a complex surface, and g be a Gauduchon metric on X. One has*

$$\mathrm{Pic}_{\mathrm{ufl}}(X) = \ker \left(\deg_g \big|_{\mathrm{Pic}^T(X)} : \mathrm{Pic}^T(X) \rightarrow \mathbb{R}\right).$$

In particular, the kernel $\ker \left(\deg_g \big|_{\mathrm{Pic}^T(X)} : \mathrm{Pic}^T(X) \rightarrow \mathbb{R}\right)$ *is independent of the Gauduchon metric g, and is a compact Lie group of real dimension $b_1(X)$.*

Proof The inclusion $\mathrm{Pic}_{\mathrm{ufl}}(X) \subset \ker \left(\deg_g \big|_{\mathrm{Pic}^T(X)} : \mathrm{Pic}^T(X) \rightarrow \mathbb{R}\right)$ is obvious and holds for manifolds of arbitrary dimension. Conversely, let $[\mathcal{L}] \in \mathrm{Pic}^T(X)$ with $\deg_g(\mathcal{L}) = 0$. Remark 8 shows that $\mathcal{L} \simeq \mathcal{L}_A$ for a Hermite-Einstein connection $A \in \mathcal{A}^{\mathrm{HE}}(L_c)$, where $c := c_1(\mathcal{L}) \in \mathrm{Tors}$. Since $\deg_g(\mathcal{L}) = 0$, formula (3.12) gives $i\Lambda_g F_A = 0$. Taking into account that iF_A is a (1,1)-form, it follows that $\frac{i}{2\pi}F_A$ is an anti-selfdual 2-form on the compact, oriented Riemannian 4-manifold (X, g).

Since this form is also closed, it follows that it is harmonic, so it coincides with the harmonic representative of the Chern class $c_1^{\mathrm{DR}}(L)$ in de Rham cohomology. But this de Rham class vanishes, because $c_1(L) \in \mathrm{Tors}$. Therefore $\frac{i}{2\pi} F_A = 0$, which shows that A is flat, so $[A] \in \mathcal{M}_{\mathrm{fl}}$, and $[\mathcal{L}] \in KH_g(\mathcal{M}_{\mathrm{fl}}) = \mathrm{Pic}_{\mathrm{ufl}}(X)$ as claimed. The other claims follow from Remarks 9 and 10. $\blacksquare$

The real dimension of $\mathrm{Pic}^0(X)$ is $2q(X)$, where $q(X) := \dim(H^1(X, \mathcal{O}_X))$ is the irregularity of X. By Remark 9 the real dimension of $\mathrm{Pic}_{\mathrm{ufl}}^0(X)$ is $b_1(X)$. Taking into account Corollary 11 we obtain:

Corollary 12 *Let X be a complex surface, and let $q(X)$ be its irregularity.*

(1) One has $2q(X) - 1 \leqslant b_1(X) \leqslant 2q(X)$.
(2) If $b_1(X)$ is even, then $b_1(X) = 2q(X)$, $\mathrm{Pic}_{\mathrm{ufl}}(X) = \mathrm{Pic}^T(X)$ and the degree map associated with any Gauduchon metric on X is a topological invariant.
(3) If $b_1(X)$ is odd, then $b_1(X) = 2q(X) - 1$, $\mathrm{Pic}_{\mathrm{ufl}}(X)$ has real codimension 1 in $\mathrm{Pic}^T(X)$ and the degree map associated with any Gauduchon metric on X induces an isomorphism

$$\mathrm{Pic}^0(X) \Big/ \mathrm{Pic}_{\mathrm{ufl}}^0(X) = \mathrm{Pic}^T(X) \Big/ \mathrm{Pic}_{\mathrm{ufl}}(X) \xrightarrow{\simeq} \mathbb{R}.$$

The identity

$$b_1(X) = \begin{cases} 2q(X) & \text{if } b_1(X) \text{ is even,} \\ 2q(X) - 1 & \text{if } b_1(X) \text{ is odd,} \end{cases}$$

is well known [BHPV]. Our proof uses gauge theoretical methods, and gives a geometric interpretation of this identity.

References

[Ba] D. Barlet, Convexité de l'espace des cycles. Bull. Soc. Math. France **106**, 373–397 (1978)
[BHPV] W. Barth, K. Hulek, Ch. Peters, A. Van de Ven, *Compact Complex Surfaces* (Springer, Berlin, 2004)
[Bu1] N. Buchdahl, Hermitian-Einstein connections and stable vector bundles over compact complex surfaces. Math. Ann. **280**, 625–648 (1988)
[Bu2] N. Buchdahl, Algebraic deformations of compact Kähler surfaces. Math. Z. **253**, 453–459 (2006)
[Bu3] N. Buchdahl, Algebraic deformations of compact Kähler surfaces II. Math. Z. **258**, 493–498 (2008)
[Dl1] G. Dloussky, Structure des surfaces de Kato. Mémoires de la Société Mathématique de France, tome **14**, 1–120 (1984)
[Dl2] G. Dloussky, Sur la classification des germes d'applications holomorphes contractantes. Math. Ann. **289**(4), 649–661
[DlKo] G. Dloussky, F. Kohler, Classification of singular germs of mappings and deformations of compact surfaces of class VII$_0$. Ann. Polon. Math. **LXX**, 49–83 (1998)

[DlTe1] G. Dloussky, A. Teleman, Infinite bubbling in non-Kählerian geometry. Math. Ann. **353**(4), 1283–1314 (2012)

[DlTe2] G. Dloussky, A. Teleman, Smooth deformations of singular contractions of class VII surfaces. https://arxiv.org/pdf/1803.07631.pdf

[DOT] G. Dloussky, K. Oeljeklaus, M. Toma, Class VII_0 surfaces with b_2 curves. Tohoku Math. J. **55**, 283–309 (2003)

[En] I. Enoki, Surfaces of class VII_0 with curves. Tohoku Math. J. **33**, 453–492 (1981)

[Eng] Ph. Engel, A proof of Looijenga's conjecture via integral-affine geometry. Ph.D. thesis, Columbia University, 2015. ISBN: 978-1321-69596-0. arXiv:1409.7676

[Fa] Ch. Favre, Classification of 2-dimensional contracting rigid germs and Kato surfaces. I. J. Math. Pures Appl. **79**(5), 475–514 (2000)

[Gau] P. Gauduchon, Sur la 1-forme de torsion d'une variété hermitienne compacte. Math. Ann. **267**, 495–518 (1984)

[GHK] M. Gross, P. Hacking S. Keel, Mirror symmetry for log Calabi-Yau surfaces I. Publ. Math. IHES **122**(1), 65–168 (2015)

[In] M. Inoue, New surfaces with no meromorphic functions. Proc. Int. Congr. Math. Vancouver **1974**, 423–426 (1976)

[Ka1] M. Kato, Compact complex manifolds containing "global" spherical shells. Proc. Jpn. Acad. **53**(1), 15–16 (1977)

[Ka2] M. Kato, Compact complex manifolds containing "global" spherical shells. I, in *Proceedings of the International Symposium on Algebraic Geometry (Kyoto Univ., Kyoto, 1977), Kinokuniya Book Store, Tokyo* (1978), pp. 45–84

[Ka3] M. Kato, On a certain class of nonalgebraic non-Kähler compact complex manifolds, in *Recent Progress of Algebraic Geometry in Japan*. North-Holland Mathematics Studies, vol. 73 (North-Holland, Amsterdam, 1983), pp. 28–50

[Loo] E. Looijenga, Rational surfaces with an anti-canonical cycle. Ann. Math. **114**(2), 267–322 (1981)

[LT] M. Lübke, A. Teleman, *The Kobayashi-Hitchin Correspondence* (World Scientific Publishing Co Pte Ltd, Singapore, 1995), 250 pp.

[Ma] M. Manetti, Normal degenerations of the complex projective plane. J. Reine Angew. Math. **419**, 89–118 (1991)

[Na1] I. Nakamura, On surfaces of class VII_0 surfaces with curves. Invent. Math. **78**, 393–443 (1984)

[Na2] I. Nakamura, Towards classification of non-Kählerian surfaces. Sugaku Expositions **2**(2), 209–229 (1989)

[Na3] I. Nakamura, On surfaces of class VII_0 surfaces with curves II. Tohoku Math. J. **42**(4), 475–516 (1990)

[OT] K. Oeljeklaus, M. Toma, Logarithmic moduli spaces for surfaces of class VII. Math. Ann. **341**(2), 323–345 (2008)

[Pl] R. Plantiko, Stable bundles with torsion Chern classes on non-Kählerian elliptic surfaces. Manuscripta Math. **87**, 527–543 (1995)

[Siu] Y.T. Siu, Every K3-surface is Kähler. Invent. Math. **73**, 139–150 (1983)

[Te1] A. Teleman, Projectively flat surfaces and Bogomolov's theorem on class VII_0-surfaces. Int. J. Math. **05**(02), 253–264 (1994)

[Te2] A. Teleman, Donaldson Theory on non-Kählerian surfaces and class VII surfaces with $b_2 = 1$. Invent. Math. **162**, 493–521 (2005)

[Te3] A. Teleman, Instantons and holomorphic curves on class VII surfaces. Ann. Math. **172**, 1749–1804 (2010)

[Te4] A. Teleman, *Introduction áá la théorie de jauge*, Cours Spécialisés, SMF (2012)

[Te5] A. Teleman, Instanton moduli spaces on non-Kählerian surfaces. Holomorphic models around the reduction loci. J. Geom. Phys. **91**, 66–87 (2015)

[Te6] A. Teleman, Analytic cycles in flip passages and in instanton moduli spaces over non-Kählerian surfaces. Int. J. Math. **27**(07), 1640009 (2016)

[Te7] A. Teleman, A variation formula for the determinant line bundle. Compact subspaces of moduli spaces of stable bundles over class VII surfaces, in *Geometry, Analysis and Probability: In Honor of Jean-Michel Bismut.* Progress in Mathematics (Birkhäuser, Basel, 2017)

[Te8] A. Teleman, Towards the Classification of Class VII Surfaces, in *Complex and Symplectic Geometry.* Springer INdAM Series, vol. 21 (Springer, Cham, 2017)

[Te9] A. Teleman, Donaldson theory in non-Kählerian geometry, in *Modern Geometry: A Celebration of the Work of Simon Donaldson.* Proceedings of Symposia in Pure Mathematics, vol. 99 (2018), pp. 363–392

Chapter 4
Intersection of Quadrics in $\mathbb{C}^n$, Moment-Angle Manifolds, Complex Manifolds and Convex Polytopes

Alberto Verjovsky

Abstract These are notes for the *CIME school on Complex non-Kähler geometry* from July 9th to July 13th of 2018 in Cetraro, Italy. It is an overview of different properties of a class of non-Kähler compact complex manifolds called LVMB manifolds, obtained as the Hausdorff space of leaves of systems of commuting complex linear equations in an open set in complex projective space $\mathbb{P}_{\mathbb{C}}^{n-1}$.

4.1 Introduction

The origin of the so-called LVM manifolds is the paper [DV97] by Santiago López de Medrano and the author of these notes. There they define and study a new infinite family of compact complex manifolds (a finite number of diffeomorphism classes for each dimension) which, except for a series of cases corresponding to complex tori, are not symplectic. The construction is based on the following principle discovered by André Heafliger:

> If $\mathcal{F}$ is a holomorphic foliation of complex codimension m on a complex manifold M with $m \leqslant n = dim_{\mathbb{C}} M$ and Σ is a C^∞ manifold of real dimension $2m$ which is transversal to $\mathcal{F}$ then Σ is a complex manifold. Indeed it suffices to provide Σ with a holomorphic atlas from transversals to the plaques of a foliation atlas of $\mathcal{F}$.

The essential point is that one can obtain *non-algebraic* complex manifolds as the space of leaves of holomorphic foliations of complex *algebraic* manifolds, as long as the space of leaves is Hausdorff. In particular the foliation could be given by a holomorphic action of a complex Lie group. In fact the construction in [DV97] uses an explicit linear action of $\mathbb{C}$ in $\mathbb{C}^n$ ($n \geqslant 3$) which descends to a projective linear action on complex projective space $\mathbb{P}_{\mathbb{C}}^{n-1}$ and there is an open set $\mathcal{V} \subset \mathbb{P}_{\mathbb{C}}^{n-1}$ which is invariant under the action and such that every leaf (orbit) of the action

A. Verjovsky (✉)

Instituto de Matemáticas, Universidad Nacional Autónoma de México (UNAM), México City, Mexico

e-mail: alberto@matcuer.unam.mx

© Springer Nature Switzerland AG 2019

D. Angella et al. (eds.), *Complex Non-Kähler Geometry*, Lecture Notes in Mathematics 2246, https://doi.org/10.1007/978-3-030-25883-2_4

163

is an immersed copy of $\mathbb{C}$ or $\mathbb{C}^*$; furthermore, the space of leaves of the foliation by orbits of $\mathcal{V}$ is compact and Hausdorff and therefore it is a compact complex manifold. In some sense the set $\mathcal{V}$ is the union of "semi-stable orbits" (or Siegel leaves) of the action in the sense of *Geometric Invariant Theory* (GIT) and is the complement of a union of projective subspaces of different dimensions. In fact $\mathcal{V}$ is the image under the canonical projection $\mathbb{C}^n \to \mathbb{P}_{\mathbb{C}}^{n-1}$ of the set of orbits in $\mathbb{C}^n$ that do not accumulate to the origin (a sort of Kempf-Ness condition). In a very pretty paper [CZ07] Stéphanie Cupit-Foutou and Dan Zaffran describe how to construct the generalized family of LVMB manifolds from certain Geometric Invariant Theory (GIT) quotients.

They show that Bosio's generalization parallels exactly the extension obtained by Mumford's GIT to the more general GIT developed by Białynicki-Birula and Świecicka.

The article [DV97] is a continuation of the foundational papers by Girbau et al. [GHS83] about the deformations of foliations which are transversally holomorphic. In fact, André Haefliger used these results to study in [HA85] the deformations of Hopf manifolds which are realized as the space of leaves of a foliation.

Another continuation of that work was obtained by Jean-Jacques Loeb and Marcel Nicolau which uses the foliation in order to describe the deformations of the Calabi–Eckmann manifolds [LN96].

The initial construction in [DV97] can be extended to the case of projective linear actions of $\mathbb{C}^m$ for any positive integer m on $\mathbb{P}_{\mathbb{C}}^n$ as long as $n > 2m$. Then, under two assumptions related to the $n \times m$ complex matrix $\mathbf{\Lambda}$ of eigenvalues of the linear flows which determine the action one obtains new compact manifolds. These assumptions are that $\mathbf{\Lambda}$ be admissible i.e., it satisfies the *weak hyperbolicity and Siegel conditions*. This was achieved by Laurent Meersseman who studies in detail several aspects of the compact manifolds in [ME00]. These compact complex manifolds $N_{\mathbf{\Lambda}}$ are now known as LVM manifolds. A very interesting property of these manifolds when $m > 1$ is their very rich topology. For instance, any finite abelian group is a summand of the homology group of one of these manifolds. In particular some of the manifolds have arbitrarily large torsion in its homology groups. In [BO01], Frédéric Bosio gives a generalization of the construction of LVM manifolds. The idea is to relax the weak hyperbolicity and Siegel conditions $\mathbf{\Lambda}$ and to look for all the subsets $\mathcal{S}$ of $\mathbb{C}^n$ such that action (4.1) in Sect. 4.2 is free and proper. The manifolds that are either LVM manifolds or the generalization by Frédéric Bosio are now known as LVMB manifolds. The manifolds $N_{\mathbf{\Lambda}}$ are obtained as the orbit space of a free action of the circle on an odd-dimensional manifold $M_1(\mathbf{\Lambda})$ contained in the sphere $\mathbb{S}^{2n-1}$ which is the intersection of homogeneous quadratic equations and called *moment-angle manifold*. Santiago López de Medrano has studied deeply these intersection of quadrics in several papers by himself and some collaborators [GL13, GL05, GL14, DM88, DM89, DM14, DM17] in particular the paper [GL13] by Samuel Gitler and Santiago López de Medrano has been a great advance to understand the topology of moment-angle manifolds.

The LVM manifolds are not symplectic (except when $2m + 1 = n$ when the manifolds are compact complex tori) however under an arbitrarily small deformation of Λ the manifolds N_Λ fiber à la Seifert-Orlik over a toric manifold (or orbifold) with fiber a compact complex torus. This is due to the following fact:

The complex manifolds N_Λ of complex dimension $n - m - 1$ admit a locally-free holomorphic action of $\mathbb{C}^m$ (recall that $n > 2m$); although N_Λ is not Kähler, the foliation $\mathcal{G}_\Lambda$ on N_Λ by the leaves of the action is transversally Kähler, in particular $\mathcal{G}_\Lambda$ is a Riemannian foliation and thus admits a transverse invariant volume form. In particular either has a Zariski open set of noncompact leaves or else all are compact complex tori.

If Λ satisfies a rationality condition called condition **K** in Definition 4.12 then all the leaves of $\mathcal{G}_\Lambda$ are compact, in fact they are complex tori $\mathbb{C}^m / \Gamma$ ($\Gamma \cong \mathbb{Z}^{2m}$) and the quotient is Hausdorff. Hence it is a compact complex manifold (or an orbifold).

Furthermore, the rationality conditions **K** in Definition 4.12 imply that the transversal Kähler form is "integral" (a sort of transversal Kodaira embedding condition) which makes this quotient an algebraic manifold or variety with quotient singularities of dimension $n - 2m - 1$. In fact this quotient admits an action of $(\mathbb{C}^*)^{n-2m-1}$ with a principal dense orbit so it is a toric manifold $X(\Delta)$ where $\Delta = \Delta_\Lambda$ is the corresponding fan which depends on Λ.

The reciprocal is true as shown by the author and Meersseman [MV04]: If $X(\Delta)$ is a toric variety with at most singularities which are quotients then there exists an admissible configuration Λ which satisfies conditions **K** in Definition 4.12 and therefore any toric variety with at most quotient singularities is obtained by the quotient of a LVM manifold by a holomorphic locally-free action of a compact complex torus. In this paper one uses Delzant construction over a rational simple convex polytope which is naturally associated to the convex hull $\mathcal{H}(\Lambda)$ of the configuration. When the leaves of $\mathcal{G}_\Lambda$ are not compact the leaf space is not Hausdorff and one has a "noncommutative" complex manifold in the sense of Alain Connes [CO85, CO94]. This happens when the convex polytope $\mathcal{H}_\Lambda$ is not rational and a convex polytope associated to the foliation $\mathcal{G}_\Lambda$ is non-rational. There are important reasons to consider *nonrational* polytopes. For instance, toric varieties corresponding to simple rational polytopes are rigid (i.e., they cannot be deformed) whereas simple rational polytopes can be perturbed simply by moving the vertices to non-rational simple polytopes. The problem of associating to a non-rational polytope a geometric space of some kind is an old one and emerges in different subjects, including symplectic geometry, via the convexity theorem and the Delzant construction. In fact it also is connected with the combinatorics of convex polytopes see for instance Stanley [ST83] where a link between rational simplicial polytopes and the geometry and topology of toric varieties is explained following earlier work of McMullen [MA93] and Stanley [ST96]. There are important reasons to consider *nonrational* simple polytopes and its and give them an interpretation in relation to toric geometry. In this respect the article by Prato [PR01] is the first work that addresses this problem via symplectic geometry and she defines the notion of *quasifolds*, which is a generalization of the notion of orbifolds and associates to a non-rational simple polytope a quasifold. In a joint paper [BP01] Elisa Prato and Fiammetta Battaglia generalize the notion of toric variety and associate to each non-rational simplicial polytope a Kähler quasifold and compute the Betti numbers (see also [BP01]). The paper by Battaglia and Zaffran [BZ15] uses also the leaf space of the foliation $\mathcal{G}_\Lambda$ of the manifolds N_Λ to have either toric orbifolds in the rational case or quasifolds in the non-rational case. Thus the papers [BZ15, BP01, BA01, BP01, PR01] are foundational papers in the theory of non-rational polytopes.

In [KLMV14] a different interpretation as *non-commutative toric varieties* is given of the pair $(N_\Lambda, \mathcal{G}_\Lambda)$ in the case Λ does not satisfy the rational condition (**K**). Non-commutative toric varieties are to toric varieties what non-commutative tori are to tori and, as such, they can be interpreted in multiple ways: As (noncommutative) topological spaces, they are C^*-algebras associated to dense foliations, that is to say, deformations of the commutative C^*-algebras associated to tori in the spirit of Alain Connes.

However, while non-commutative tori correspond to linear foliations (deformations) on classical tori, non-commutative toric varieties correspond to the holomorphic foliation $\mathcal{G}_\Lambda$) on N_Λ.

The manifolds N_Λ are certain intersections of real quadrics in complex projective spaces of a very explicit nature. The homotopy type of LVM-manifolds is described by moment angle complexes.

The paper of Bosio and Meersseman [BM06] is a beautiful paper with many ideas and interconnection of several branches of mathematics. In fact the title and subject of the present notes is very much inspired on this paper.

They do a deep study of the properties of LVM manifolds and also made significant advances in the study of the topology of the intersection of k homogeneous quadrics.

In particular, the question of whether they are always connected sums of sphere products was considered: they produced new examples for any k which are so, but also showed how to construct many more cases where they are not.

Independently in [DJ91] Michael W. Davis and Janusz Januszkiewicz had introduced new constructions, part of which essentially coincide with those above, where the main objective was the study of some important quotients of them (different from the ones mentioned above) which they called *toric manifolds* (in contrast with toric *varieties* that are algebraic). These toric manifolds are topological analogues of toric varieties in algebraic geometry. They are even dimensional manifolds with an effective action of an n-dimensional compact torus $(\mathbb{S}^1)^n$, there is a kind of "moment map" and the orbit space is a simple convex polytope. One can do combinatorics on the quotient polytope to obtain information on the manifold above. For example one can compute the Euler characteristic and describe the cohomology ring of the manifold in terms of the polytope. The paper by Davis and Januszkiewicz originated an important development through the work of many authors, for which we refer the reader to the book of Buchstaber and Panov [BP02]. A line of research derived from [DJ91] is the paper [BBCG10] where a far-reaching generalization is made and a general splitting formula is derived that provides a very good geometric tool for understanding the relations among the homology groups of different spaces.

There is a principal circle bundle $p : M_1(\Lambda) \to N_\Lambda$ over each manifold N_Λ. The manifold $M_1(\Lambda)$ is a smooth manifold of real odd dimension $2n - 2m - 1$ called *moment-angle manifold.* The manifold $M_1(\Lambda)$ admits an action of the torus $(\mathbb{S}^1)^n$. The orbit space of this action is a convex polytope of dimension $3n - 2m - 1$ thus there is a "moment map". In addition $M_1(\Lambda)$ has in a contact structure and in many cases is an open book with a very interesting structure [BLV17].

In the present notes we present various results and properties of LVMB manifolds:

1. Complex analytic proprieties
2. The relation between these manifolds and toric manifolds and orbifolds with quotient singularities.
3. Their topology and geometric structures

The main body of the results presented in these notes are in part contained in the papers [BLV17, BM06, DV97, ME00, MV04, MV08].

4.2 Singular Holomorphic Foliations of $\mathbb{C}^n$ and $\mathbb{P}^{n-1}_\mathbb{C}$ Given by Linear Holomorphic Actions of $\mathbb{C}^m$ on $\mathbb{C}^n$ $(n > 2m)$

Let M be a complex manifold of complex dimension n and $0 \leqslant p \leqslant n$.

Definition 4.1 A holomorphic foliation $\mathcal{F}$ of complex dimension p (or complex codimension $n - p$) is given by a *foliated atlas* $(U_\alpha, \Phi_\alpha)_{\alpha \in \mathcal{I}}$ where U_α are open in M, $\{U_i\}_{i \in \mathcal{I}}$ is an open covering of M and $\Phi_i : U_i \to V_i \subset \mathbb{C}^{n-p} \times \mathbb{C}^p = \mathbb{C}^n$ are homeomorphisms such that for overlapping pairs U_i, U_j the transition functions $\Phi_{ij} = \Phi_j \Phi_i^{-1} : \Phi_i(U_i \cap U_j) \to \Phi_j(U_i \cap U_j)$ are of the form:

$$\Phi_{ij}(x, y) = (\Phi^1_{ij}(x), \Phi^2_{ij}(x, y)) \quad x \in \mathbb{C}^{n-p}, \quad y \in \mathbb{C}^p \tag{A}$$

where Φ^1_{ij} and Φ^2_{ij} are holomorphic and Φ^1_{ij} is a local biholomorphism between open sets of $\mathbb{C}^{n-p}$ and Φ^2_{ij} is a local holomorphic submersion from an open set in $\mathbb{C}^n$ onto an open set of $\mathbb{C}^p$.

Definition 4.2 The atlas $(U_\alpha, \Phi_\alpha)_{\alpha \in \mathcal{I}}$ is called a *holomorphic foliation atlas* and the maps Φ_α are called *holomorphic flow boxes* or *holomorphic foliation charts*. The sets of the form $\Phi_\alpha^{-1}(\{x\} \times \mathbb{C}^p)$, $x \in \mathbb{C}^{n-p}$, i.e., the set of points whose coordinates (X, Y) with $X = (x_1, \cdots, x_{n-p}) \in \mathbb{C}^{n-p}$, $Y = (y_1, \cdots, y_p) \in \mathbb{C}^p$ satisfy $X = C$ for some constant vector $C \in \mathbb{C}^{n-p}$ are called *plaques*. Condition (A) says that the plaques glue together to form complex submanifolds called *leaves*, which are immersed in M (not necessarily properly immersed). If $(U_\alpha, \varphi_\alpha)_{\alpha \in \mathcal{I}}$ is a complex atlas as in Definition 4.1 the leaves are immersed p-dimensional holomorphic submanifolds of W.

The family of biholomorphisms $\{\Phi_{ij}^1\}_{i \in I}$ defines a groupoid called the *transverse holonomy groupoid*. It can be used to define noncommutative toric varieties [BZ15, KLMV14].

Let m and n be two positive natural numbers such that $n > 2m$. Let $\Lambda := (\Lambda_1, \cdots, \Lambda_n)$ be an n-tuple of vectors in $\mathbb{C}^m$ where $\Lambda_i = (\lambda_i^1, \cdots, \lambda_i^m)$ for $i = 1, \cdots, n$.

To the configuration $(\Lambda_1, \cdots, \Lambda_n)$ we can associate the linear (singular) foliation of $\mathbb{C}^n$ generated by the m holomorphic linear commuting vector fields $(1 \leqslant j \leqslant m)$

$$\mathbb{C}^n \ni (z_1, \cdots, z_n) \longmapsto \sum_{i=1}^n \lambda_i^j z_i \frac{\partial}{\partial z_i}$$

$$\frac{d\mathbf{Z}}{dT} = \begin{bmatrix} \lambda_1^j & 0 & 0 & \dots & 0 \\ 0 & \lambda_2^j & 0 & \dots & 0 \\ & & \dots \dots \\ 0 & 0 & & \dots & \lambda_n^j \end{bmatrix} \mathbf{Z}, \qquad \text{(System of linear equations)}$$

$$\mathbf{Z} = \begin{bmatrix} z_1 \\ \vdots \\ z_n \end{bmatrix}, \quad j = 1, \cdots, m, \quad T \in \mathbb{C}$$

Let us start with the construction of an infinite family of compact complex manifolds. Let m be a positive integer and n and integer such that $n > 2m$.

Definition 4.3 Let $\Lambda = (\Lambda_1, \ldots, \Lambda_n)$ be a configuration of n vectors in $\mathbb{C}^m$. Let $\mathcal{H}(\Lambda_1, \cdots, \Lambda_n)$ be the convex hull of $(\Lambda_1, \cdots, \Lambda_n)$.
We say that Λ is *admissible* if:

(1) **(SC)** The Siegel condition: 0 belongs to the convex hull $\mathcal{H}(\Lambda) := \mathcal{H}(\Lambda_1, \ldots, \Lambda_n)$ of $(\Lambda_1, \ldots, \Lambda_n)$ in $\mathbb{C}^m \simeq \mathbb{R}^{2m}$.
(2) **(WH)** The weak hyperbolicity condition: for every $2m$-tuple of integers $i_1, \cdots, i_{2m}$ such that $1 \leqslant i_1 < \cdots < i_{2m} \leqslant n$ we have $0 \notin \mathcal{H}(\Lambda_{i_1}, \cdots, \Lambda_{i_{2m}})$

This definition can be reformulated geometrically in the following way: the convex polytope $\mathcal{H}(\Lambda_1, \cdots, \Lambda_n)$ contains 0, but neither external nor internal facet of this polytope (that is to say hyperplane passing through $2m$ vertices) contains 0. An admissible configuration satisfies the following regularity property (Fig. 4.1).

Lemma 4.1 Let $\Lambda_i' = (\Lambda_i, 1) \in \mathbb{C}^{m+1}$, for $i \in \{1, \cdots, n\}$. For all set of integers $J \subset \{1, \cdots, n\}$ such that $0 \in \mathcal{H}((\Lambda_j')_{j \in J})$ the complex rank of the matrix whose columns are the vectors $(\Lambda_j')_{j \in J}$ is equal to $m + 1$, therefore it is of maximal rank.

Fig. 4.1 Quadrilateral in $\mathbb{C}$ corresponding to $m = 1$ and $n = 4$ given by the single equation in $\mathbb{C}^4$ given by the diagonal matrix $\Lambda_1 =$ diagonal$(\lambda_1, \lambda_2, \lambda_3, \lambda_4)$

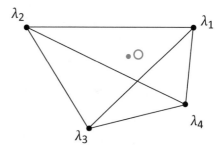

One considers the holomorphic (singular) foliation $\mathcal{F}$ in projective space $\mathbb{P}_{\mathbb{C}}^{n-1}$ given by the orbits of the linear action of $\mathbb{C}^m$ on $\mathbb{C}^n$ induced by the linear vector fields (4.1).

$$(T, [z]) \in \mathbb{C}^m \times \mathbb{P}_{\mathbb{C}}^{n-1} \longmapsto [z_1 \cdot \exp\langle\Lambda_1, T\rangle, \ldots, z_n \cdot \exp\langle\Lambda_n, T\rangle] \in \mathbb{P}_{\mathbb{C}}^{n-1} \quad (4.1)$$

where $T = (t_1, \cdots, t_m) \in \mathbb{C}^m$, $[z_1, \cdots, z_n]$ are projective coordinates and $\langle -, - \rangle$ is inner product $\langle Z, W \rangle = \sum\limits_{i=1}^{n} z_i w_i$.

One can lift this foliation to a foliation $\tilde{\mathcal{F}}$ in $\mathbb{C}^n$ given by the linear action

$$(T, z) \in \mathbb{C}^m \times \mathbb{C}^n \longmapsto (z_1 \cdot \exp\langle\Lambda_1, T\rangle, \ldots, z_n \cdot \exp\langle\Lambda_n, T\rangle) \in \mathbb{C}^n. \quad (B)$$

The so-defined foliation is singular, in particular 0 is a singular point. The behavior in the neighborhood of 0 determines two different sorts of leaves.

Definition 4.4 (Poincaré and Siegel Leaves) Let L be a leaf of the previous foliation. If 0 belongs to the closure of L, we say that L is a *Poincaré* leaf. In the opposite case, we talk of a *Siegel leaf*.

If L is a Siegel leaf then the distance from that leaf to the origin is positive and one can show that there exists a unique point $\mathbf{z} = (z_1, \cdots, z_n) \in L$ which minimizes the distance to the origin and this point satisfies

$$\sum_{i=1}^{n} \Lambda_i |z_i|^2 = 0 \quad (4.2)$$

This is because the leaf L_W through the point $W = (w_1, \cdots, w_n)$ in the Siegel domain is the Riemann surface

$$L_W = \left\{ (w_1 \cdot \exp\langle\Lambda_1, T\rangle, \ldots, w_n \cdot \exp\langle\Lambda_n, T\rangle) \in \mathbb{C}^n \mid T \in \mathbb{C}^m \right\}$$

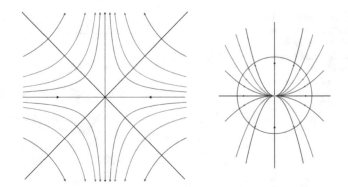

Fig. 4.2 Siegel and Poincaré leaves

and to minimize the (square of the) distance to the origin we see that Lagrange multipliers imply that the complex line from the origin to a point that minimizes the square of the distance must be orthogonal to the orbit at the point (Fig. 4.2).

One has the following dichotomy:

1. If $0 \notin \mathcal{H}(\Lambda_1, \ldots, \Lambda_n)$ then every leaf is of Poincaré type.
2. The set of Siegel leaves is nonempty if and only if $0 \in \mathcal{H}(\Lambda_1, \ldots, \Lambda_n)$

For $z = (z_1, \cdots, z_n) \in \mathbb{C}^n$ let $I_z \subset \{1, \cdots, n\}$ defined as follows $I_z = \{j : z_j \neq 0\}$ and let $\Lambda_{I_z} = \{\Lambda_j : j \in I_z\}$. One defines:

$$\mathcal{S} = \mathcal{S}_\Lambda$$

$$= \{z \in \mathbb{C}^n \mid 0 \in \mathcal{H}(\Lambda_{I_z})\}, \quad \mathcal{S} \text{ as the complement of subspaces in } \mathbb{C}^n$$

$$(4.3)$$

Definition 4.5 We define $\mathcal{V} = \mathcal{V}(\Lambda) \subset \mathbb{P}^{n-1}$ to be the image of $\mathcal{S}$ in $\mathbb{P}^{n-1}$ under the canonical projection $\pi : \mathbb{C}^n - \{0\} \to \mathbb{P}^{n-1}$.

Let

$$\mathcal{T} = \mathcal{T}(\Lambda) = \{z \in \mathbb{C}^n \mid z \neq 0, \sum_{i=1}^{n} \Lambda_i |z_i|^2 = 0\} \quad (C)$$

then $\mathcal{T}$ is the set of points that realize the minimum distance in each Siegel leaf. Then $\mathcal{T}$ meets ever Siegel leaf in exactly one point and it meets each leaf transversally. We have that $\overline{\mathcal{T}} = \mathcal{T} \cup \{0\}$ is a singular manifold with an isolated singularity at the origin.

Let

$$N = N_\Lambda = \{[z] \in \mathbb{P}^{n-1} \quad |$$

$$\sum_{i=1}^{n} \Lambda_i |z_i|^2 = 0\} \quad \text{Equations of LVM manifolds} \tag{D}$$

One can verify that S is the union of the Siegel leaves and that S is an open set of the form $S = \mathbb{C}^n - E$ where E is an analytic set, whose different components correspond to subspaces of $\mathbb{C}^n$ where some coordinates vanish.

The leaf space of the foliation restricted to S, that we call M, or M_Λ if we want to emphasize Λ, is identified with $\mathcal{T}$.

Since S contains $(\mathbb{C}^*)^n$ we see that S is dense in $\mathbb{C}^n$.

The weak hyperbolicity condition implies that the system of quadrics given by the preceding equations which define $\mathcal{T}$ and N have maximal rank in every point thanks to Lemma 4.1.

The Siegel condition implies that $\mathcal{T}$ and N are nonempty. One can show also that $\tilde{\mathcal{F}}$ is regular in S and that $\mathcal{T}$ is a smooth manifold transverse to the restriction of $\tilde{\mathcal{F}}$ to S. In other words the quotient space of $\tilde{\mathcal{F}}$ restricted to S can be identified with $\mathcal{T}$.

As mentioned before, the open set S is a deleted complex cone in $\mathbb{C}^n$: i.e. if $Z \in S$ then $\lambda Z \in S$ for all $\lambda \in \mathbb{C}^*$. Therefore $V = \pi(S)$ (Definition 4.5) is an open set of $\mathbb{P}_\mathbb{C}^{n-1}$. Then $\pi(\mathcal{T})$ is a smooth manifold of dimension equal to the codimension of $\mathcal{F}$ and transversal to the leaves.

By the following Lemma by André Haefliger $\pi(\mathcal{T})$ is a complex manifold:

Lemma 4.2 (A. Haefliger) *Let $\mathfrak{M}$ be a complex manifold of complex dimension $n \geqslant 2$ and $\mathcal{F}$ a holomorphic foliation of $\mathfrak{M}$ of codimension $m \geqslant 1$ with $n \geqslant m$. Let $\mathfrak{N} \subset \mathfrak{M}$ be a smooth manifold of real dimension $2m$ which is transversal to the leaves of $\mathcal{F}$. Then $\mathfrak{N}$ is in a natural way a complex manifold of complex dimension m.*

Proof In fact if $V \subset \mathfrak{M}$ is an open subset of $\mathfrak{N}$ which is contained the domain U_α of the foliation chart $\Phi_\alpha : U_\alpha \to \mathbb{C}^m \times \mathbb{C}^{m-n}$ of $\mathcal{F}$ then if $\hat{\Phi}_\alpha = \Phi_\alpha{\restriction}V$ is the restriction of Φ_α to V and $\pi_1 : \mathbb{C}^m \times \mathbb{C}^{n-m}$ is projection onto the first factor then $\Psi_\alpha = \pi_1 \circ \hat{\Phi}_\alpha$ is a holomorphic coordinate chart of $\mathfrak{M}$. Condition (A) in Definition 4.1 implies that the coordinate changes are holomorphic. □

Remark 4.1 Bogomolov has conjectured that every compact complex manifold W can be obtained by this process for a singular holomorphic foliation of projective space and W transversal to the foliation outside of the singularities. More precisely he asks: *can one embed every compact complex manifold as a C^∞ smooth subvariety that is transverse to an algebraic foliation on a complex projective algebraic variety?*

In this respect, Demailly and Gaussier [DG17] have shown an embedding theorem for compact almost complex manifolds into complex algebraic varieties.

They show that every almost complex structure can be realized by the transverse structure to an algebraic distribution on an affine algebraic variety, namely an algebraic subbundle of the tangent bundle.

By Haefliger's Lemma 4.2 above, $\mathcal{T}$ defined in (C) has the structure of a (non-compact) complex manifold which we call M.

Also N can be identified with the quotient space of $\mathcal{F}$ restricted to V (Definition 4.5) and therefore it inherits a complex structure. Let us denote this complex manifold by N. The complex dimension of M is $n - m$ and of N is $n - m - 1$.

The natural projection $M \to N$, induced by the projection $\pi : \mathbb{C}^n \setminus \{0\} \to \mathbb{P}_{\mathbb{C}}^{n-1}$, is in fact a principal $\mathbb{C}^*$ fibration. Let M_1 denote the total space of the associated circle fibration It has the same homotopy type as M but it has the advantage of being compact.

Let us observe that M_1 can be identified with the transverse intersection of the cone $\mathcal{T}$ (with the vertex at the origin deleted) and the unit sphere $\mathbb{S}^{2n-1}$ in $\mathbb{C}^n$. For this reason we make the following definition

Definition 4.6 Let

$$M_1 = M_1(\Lambda) = \{z = (z_1, \cdots, z_n) \in \mathbb{C}^n \quad | \quad \sum_{i=1}^{n} \Lambda_i |z_i|^2 = 0, \ \sum_{i=1}^{n} |z_i|^2 = 1\}.$$

$$(4.4)$$

Then $M_1(\Lambda)$ is called the **moment-angle** manifold corresponding to Λ.

Remark 4.2 Let Λ be an admissible configuration. Then N_Λ and $N_{(A\Lambda+B)}$ (with $A \in \mathrm{GL}_m(\mathbb{C})$ and $B \in \mathbb{C}^m$) are biholomorphic (provided that $(A\Lambda + B)$ is admissible and provided that the corresponding sets S are the same).

Remark 4.3 The manifold N is naturally equipped with the principal $\mathbb{C}^*$-bundle $\mathcal{T} \to N$.

Remark 4.4 The natural projection $M_1 \to N$ is a $\mathbb{S}^1$-principal bundle. It is in fact the unit bundle associated to the bundle $\mathcal{T} \to N$.

Then, the differentiable embedding of N into the projective space described yields an embedding of fibre bundles

$$\begin{array}{ccc} M_1 & \longrightarrow & \mathbb{S}^{2n-1} \\ \downarrow & & \downarrow \\ N & \longrightarrow & \mathbb{C}P^{n-1} \end{array}$$

Let us denote by ω the pull-back of the Fubini-Study Kählerian form by this embedding. The form ω is thus a closed real two-form on N which represents the Euler class of the bundle $M_1 \to N$.

Definition 4.7 We call ω the *canonical Euler form* of the bundle $M_1 \to N$.

Definition 4.8 Let $1 \leqslant i \leqslant n$. We say that Λ_i (or more briefly i) is *indispensable* if $(\Lambda_j)_{j \in \{i\}^c}$ is not admissible. Let $I \subset \{1, \ldots, n\}$. We say that $(\Lambda_i)_{i \in I}$ (or more briefly I) is *removable* if $(\Lambda_j)_{j \in I^c}$ is still admissible.

Remark 4.5 Let $I \subset \{1, \ldots, n\}$ of cardinal p. If I is removable, then the configuration $(\Lambda_i)_{i \in I^c}$ gives rise to a holomorphic LVM submanifold of $N(\Lambda_1, \ldots, \Lambda_n)$ of codimension p.

Remark 4.6 We write $\mathcal{S}_\Lambda$, N_Λ, $M_1(\Lambda)$ etc., if we want to emphasize the configuration Λ. However many times we omit Λ if it is clearly understood and no confusion is possible.

Another characterization of $\mathcal{S}$ is the following:

$$\mathcal{S} = \{z \in \mathbb{C}^n \mid 0 \text{ is not in the closure of the leaf of } \tilde{\mathcal{F}} \text{ through } z\} \qquad (4.5)$$

in other words $\mathcal{S}$ is the union of the Siegel Leaves and it open and invariant under the action of $\mathbb{C}^m$.

Remark 4.7 The space of Siegel leaves $\mathcal{S}$ has the same homotopy type as M and therefore also as M_1. This is because $\mathcal{S}$ is a deleted complex cone over the origen. It is a union of disjoint copies of $\mathbb{C} - \{0\} = \mathbb{C}^*$ corresponding to complex lines through the origin. Therefore it retracts to M_1 which is the union of unit circles in each line.

Remark 4.8 The linear holomorphic action of $(\mathbb{C}^*)^m$ commutes with the diagonal action (by diagonal matrices) hence $(\mathbb{C}^*)^n$ acts on M.

Definition 4.9 (LVM Manifolds) If Λ is an admissible configuration the manifold $N = N_\Lambda$ given by formula (D) above is called a LVM manifold corresponding to Λ. It is a compact complex manifold and $dim_{\mathbb{C}} N_\Lambda = n - m - 1$.

4.3 Examples

4.3.1 Elliptic Curves

We consider $m = 1$, $n = 3$ i.e., one linear equation in $\mathbb{C}^3$. In $\mathbb{C}$ consider a non-degenerate triangle with vertices λ_1, λ_2 and λ_3. Let $\Lambda = (\lambda_1, \lambda_2, \lambda_3)$. Suppose that the origin is in the interior of this triangle. Then the open set of Siegel leaves $\mathcal{S}_\Lambda \subset \mathbb{C}^3$ is the complement of the three coordinate hyperplanes $z_1 = 0$, $z_2 = 0$ and $z_3 = 0$.

The set in $\mathcal{T} \subset \mathbb{C}^3 - \{0\}$ given by the equation:

$$\lambda_1 |z_1|^2 + \lambda_2 |z_2|^2 + \lambda_3 |z_3|^2 = 0 \qquad (4.6)$$

is the transversal as in formula (C) above and it meets every leaf in $\mathcal{S}_\Lambda$ in exactly one point. So that the space of leaves in $\mathcal{S}_\Lambda$ can be identified with the set given by Eq. (4.6).

The set $\mathcal{T}$ is a complex cone with the origin deleted so that if $Z \in \mathcal{T}$ also $cZ \in \mathcal{T}$ for all $c \in \mathbb{C}^*$.

We see that $N_\Lambda := N_{(\lambda_1, \lambda_s, \lambda_3)}$ is the projectivization of $\mathcal{T}$ and therefore N_Λ can be identified as the quotient set of points satisfying the following two equations:

$$\begin{cases} \lambda_1 |z_1|^2 + \lambda_2 |z_2|^2 + \lambda_3 |z_3|^2 = 0 \\ |z_1|^2 + |z_2|^2 + |z_3|^2 = 1 \end{cases}$$

modulo the natural action of the circle given by

$$(z_1, z_2, z_3) \mapsto (\mu z_1, \mu z_2, \mu z_3), \ |\mu| = 1, \ (z_1, z_2, z_3) \in N.$$

The set of points satisfying the two equations is $\mathbb{T}^3 = \mathbb{S}^1 \times \mathbb{S}^1 \times \mathbb{S}^1$.

Hence one has a free action of $\mathbb{C}^*$ and the quotient $N_\Lambda := M/\mathbb{C}^*$, then a complex, compact manifold of dimension one. In fact N_Λ is an elliptic curve. To identify the corresponding complex structure, observe that in this case $\mathcal{S} = (\mathbb{C}^*)^3$. The mapping $exp : \mathbb{C}^3 \to \mathcal{S} = (\mathbb{C}^*)^3$ given by $exp(\zeta_1, \zeta_2, \zeta_3) = (e^{\zeta_1}, e^{\zeta_2}, e^{\zeta_3})$ can be used to identify $N_{(\lambda_1, \lambda_2, \lambda_3)}$ with the quotient of $\mathbb{C}$ by the lattice generated by $\lambda_3 - \lambda_2$ and $\lambda_1 - \lambda_2$. So we have that

$N_{(\lambda_1, \lambda_2, \lambda_3)}$ *is biholomorphically equivalent to the elliptic curve with modulus* $\frac{\lambda_3 - \lambda_2}{\lambda_1 - \lambda_2}$.

Observe that in this case we obtain all complex structures on the torus. By choosing adequately the order of the λ_i we obtain a mapping from the Siegel domain to the Siegel upper half-plane in $\mathbb{C}$. Therefore *Any elliptic curve is obtained this way.*

4.3.2 Compact Complex Tori

(i) If $n = 2m + 1$, the convex hull $\{\Lambda_i\}_{i \in \{1, \cdots, 2m+1\}}$ is a simplex in $\mathbb{C}^m \simeq \mathbb{R}^{2m}$.

In fact if one removes one the Λ's then 0 is not in the complex hull of the remaining. In other words $\mathcal{S}$ is equal to $(\mathbb{C}^*)^n$ and one can show that N is a complex torus.

Remark 4.9 Every compact complex torus is obtained by this process. In particular, if $n = 3$ and $m = 1$ we obtain every elliptic curve.

4.3.3 Hopf Manifolds

(ii) If $m = 1$ let us define for $n \geqslant 4$

$$\Lambda_1 = 1 \qquad \Lambda_2 = i \qquad \Lambda_3 = \ldots = \Lambda_n = -1 - i \, .$$

It is easy to verify that under these conditions $\mathcal{S}$ is equal to $(\mathbb{C}^*)^2 \times \mathbb{C}^{n-2} \setminus \{0\}$. Consider the two real equations that are used to define $\mathcal{T}$:

$$\begin{cases} |z_1|^2 = |z_3|^2 + \ldots + |z_n|^2 \\[2mm] |z_2|^2 = |z_3|^2 + \ldots + |z_n|^2. \end{cases}$$

If we fix the modules of z_1 and z_2 (by the definition of $\mathcal{S}$ they cannot be 0) the above equations imply that these modules are equal and that $(z_3, \ldots, z_n)$ belong to a sphere $\mathbb{S}^{2n-5}$. Therefore these equations define a manifold which is diffeomorphic to $\mathbb{S}^{2n-5} \times S^1 \times \mathbb{S}^1 \times \mathbb{R}_*^+$. The manifold M_1 obtained as the intersection of $\mathcal{T}$ and the unit sphere of $\mathbb{C}^n$ is diffeomorphic to $\mathbb{S}^{2n-5} \times \mathbb{S}^1 \times \mathbb{S}^1$ and N is diffeomorphic to $\mathbb{S}^1 \times \mathbb{S}^{2n-5}$. In particular for $n = 4$, on has all the linear Hopf surfaces.

4.3.4 Calabi–Eckmann Manifolds

(iii) Let $m = 1$, $n = 5$ and

$$\Lambda_1 = 1 \qquad \Lambda_2 = \Lambda_3 = i \qquad \Lambda_4 = \Lambda_5 = -1 - i \, .$$

An argument similar to the previous one shows that N is diffeomorphic to $\mathbb{S}^3 \times \mathbb{S}^3$. One obtains an example of a Calabi–Eckmann manifold that is not a Kähler manifold.

Remark 4.10 In general one obtains complex structures in products of odd dimensional spheres $\mathbb{S}^{2r+1} \times \mathbb{S}^{2l+1}$ like in the classical Calabi–Eckmann manifolds. In fact: *Every Calabi–Eckmann manifold is obtained by this process.*

4.3.5 Connected Sums

(iv) S. López de Medrano has shown that for the pentagon in the picture below M_1 is diffeomorphic to the connected sum of five copies of $\mathbb{S}^3 \times \mathbb{S}^4$. The complex manifold N is the quotient of this connected sum under a non-trivial action of $\mathbb{S}^1$ (Fig. 4.3).

Fig. 4.3 Pentagon in $\mathbb{C}$. The number n_i is the multiplicity of λ_i

λ_2 $(n_2=1)$

$(n_3=1)$ λ_3

$(n_1=1)$ λ_1

0

$(n_4=1)$ λ_4

λ_5 $(n_5=1)$

Fig. 4.4 Polygon, the number n_i is the multiplicity of λ_i

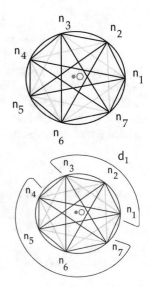

When $m = 1$ it can be assumed Λ is one of the following normal forms: Take $n = n_1 + \cdots + n_{2\ell+1}$ a partition of n into an odd number of positive integers. Consider the configuration consisting of the vertices of a regular polygon with $(2\ell + 1)$ vertices, where the i-th vertex in the cyclic order appears with multiplicity n_i.

The topology of M_1 and N can be completely described in terms of the numbers $d_i = n_i + \cdots + n_{i+\ell-1}$, i.e., the sums of ℓ consecutive n_i in the cyclic order of the partition:

For $\ell = 1$: $M_1 = \mathbb{S}^{2n_1-1} \times \mathbb{S}^{2n_2-1} \times \mathbb{S}^{2n_3-1}$. For $\ell > 1$: $M_1 = \#_{j=1}^{2\ell+1} \left(\mathbb{S}^{2d_i-1} \times \mathbb{S}^{2n-2d_i-2} \right)$. See Theorem 4.1 below (Fig. 4.4).

To describe the topology of N we will use the following known facts about the topology of M_1: First observe that the smooth topological type of M_1 (as well as that of N) does not change if we vary continuously the parameters Λ as long as we do not violate condition (WH) in Definition 4.3 in the process. It is shown in [DM89] that the parameters Λ can always be so deformed until they occupy the vertices of a regular k-gon in the unit circle, where $k = 2l + 1$ is an odd integer, every vertex being occupied by one or more of the λ_i.

Therefore the topology of M_1 (and that of N also) is totally described by this final configuration, which can be specified by the multiplicities of those vertices, that is, by the partition

$$n = n_1 + \cdots + n_k.$$

Observe that different partitions give different open sets $\mathcal{S}$ and therefore also different reduced deformation spaces. It is clear that if we permute cyclically the numbers n_i we obtain again the same manifolds and deformation spaces, but it follows from the next result that the cyclic order is relevant for their description. It is shown in [DM89] that the topology of M_1 is given as follows:

Let $d_i = n_i + n_{i+1} + \cdots + n_{i+l-1}$, for $i = 1, \ldots, k$ (the subscripts being taken modulo k). Let also $d = min\{d_1, \ldots, d_k\}$.

These numbers determine the topology of M_1:

Theorem 4.1

(1) If $k = 1$ then $M_1 = \emptyset$.
(2) If $k = 3$ then $M_1 = S^{2n_1-1} \times S^{2n_2-1} \times S^{2n_3-1}$.
(3) If $k = 2l + 1 > 3$ then $\ell > 1$ and M_1 is diffeomorphic to the connected sum of the manifolds $S^{2d_i-1} \times S^{2n-2d_i-2}$ ($i = 1, \ldots, k$) i.e., $M_1 = \#_{j=1}^{2\ell+1} \left(S^{2d_i-1} \times S^{2n-2d_i-2} \right)$.

The proof of parts (1) and (2) is quite direct, while the proof of part (3) is long and complicated [DM89]. In what follows we shall only use the fact that the integral homology groups of M_1 coincide with those of the above described connected sum and the fact that M_1 is $(2d - 2)$-connected. The homology calculations (and part (2) of Theorem 4.1) were first obtained by Wall [WA80]. Thus our results will be independent of [DM89] and will provide a simplified proof of some of the cases of Theorem 4.1.

4.3.6 Some Examples of LVM

In the cases where M_1 is not simply connected (i.e., when $k = 3$ and $d = n_1 = 1$), the complex structure on N can be described in terms of the defining parameters by identifying it with previous descriptions of these known manifolds, for instance when $n = k = 3$ the manifold N_Λ is diffeomorphic to the torus $\mathbb{S}^1 \times \mathbb{S}^1$. When M_1 is simply connected we obtain new complex structures on manifolds. An intermediate situation is given by the cases $k = 3$, with $n_1 = 2$, n_2 and n_3 even, where one can show, using the fact that each $\mathbb{C}^{n_i}$ can be considered as a quaternionic vector space, that M_1 is diffeomorphic $\mathbb{S}^3 \times \mathbb{S}^{2n_2-1} \times \mathbb{S}^{2n_3-1}$ and the action of the circle acts on the first factor so N is diffeomorphic to $\mathbb{P}^1_{\mathbb{C}} \times \mathbb{S}^{2n_2-1} \times \mathbb{S}^{2n_3-1}$. It is easy to see that in some cases N can be identified with the product of $\mathbb{P}^1_{\mathbb{C}}$ with one of the Loeb-Nicolau complex structures on $\mathbb{S}^{2n_2-1} \times \mathbb{S}^{2n_3-1}$. But in other cases there is no simple way to establish such an identification, and it is plausible that these give new complex structures.

When $k = 3, n_1 = d > 2$ we definitely get a manifold which is not a product, but a twisted fibration over $\mathbb{P}^{n_1-1}_{\mathbb{C}}$. In fact, N clearly fibers over $\mathbb{P}^{n_1-1}_{\mathbb{C}}$ with fiber $\mathbb{S}^{2n_2-1} \times \mathbb{S}^{2n_3-1}$. This fibration does have a section (recall that we are assuming that $n_1 = d$ is not bigger than the other n_i) which is homotopic to the map $\mathbb{P}^{n_1-1}_{\mathbb{C}} \to N$ constructed in Lemma 4.3 in Sect. 4.4 below. But, by Remark 4.12 at the end of the same Sect. 4.4 below, the normal bundle of $\mathbb{P}^{n_1-1}_{\mathbb{C}}$ in N is stably equivalent to the normal bundle of $\mathbb{P}^{n_1-1}_{\mathbb{C}}$ in $\mathbb{P}^{n-1}_{\mathbb{C}}$. By computing the Pontryagin classes of this bundle one shows that it is not trivial. We therefore have:

Theorem 4.2 *When $3 \leq n_1 \leq n_2 \leq n_3$ there is a non-trivial $(\mathbb{S}^{2n_2-1} \times \mathbb{S}^{2n_3-1})$-fibration over $\mathbb{P}^{n_1-1}_{\mathbb{C}}$ with an $(n-2)$-dimensional space of complex structures.*

When $k > 3$ we get new complex structures on manifolds. We will give the complete description of the underlying real smooth manifold only in the case where all $n_i = 1$ (so $n = k = 2l + 1$), where the computations and arguments are simpler. To do this we can assume as before that the λ_i are the n-th roots of unity: $\lambda_i = \rho^i$, ρ a primitive root.

One has that $\ell > 1$ and M_1 is a parallelizable $(2n - 3)$-manifold with nontrivial homology only in the two middle dimensions $n - 2$ and $n - 1$ only, where it is free of rank n:

$$H_{n-2}(M_1) = H_{n-1}(M_1) = \mathbb{Z}^n$$

It follows from the Gysin sequence of the fibration $M_1 \to N$ (and from the order of its Euler class in Remark 4.11 found in Sect. 4.4 below) that N has homology only in dimensions $2i, i = 1, \ldots, n - 2$ where it is free of rank 1, and in dimension $2l - 1$ where it is free of rank $2l$.

On the other hand, M_1 is the boundary of a manifold Q constructed as follows: Let

$$Z = \{z \in \mathbb{C}^n = \mathbb{C}^{2l+1} \mid \Sigma \Re(\lambda_i) z_i \bar{z}_i = 0, \ \Sigma z_i \bar{z}_i = 1\}.$$

Z is diffeomorphic to the $4l$ manifold $S^{2l-1} \times S^{2l+1}$ (since the defining quadratic form has index $2l$) and is the union of two manifolds with boundary

$$Q^{\pm} = \{z \in \mathbb{C}^n \mid \Sigma \Re(\lambda_i) z_i \bar{z}_i = 0, \ \pm \Sigma \Im(\lambda_i) z_i \bar{z}_i \geq 0, \ \Sigma z_i \bar{z}_i = 1\}$$

whose intersection is M_1.

The involution of $\mathbb{C}^n$ which interchanges the coordinates z_i and z_{n-i} preserves Z and M_1, and interchanges Q^+ with Q^-. Therefore these two are diffeomorphic and M_1 is an equator of Z.

Let $Q = Q^+$. It follows now easily from the Mayer-Vietoris sequence of the triple $(S^{2l-1} \times S^{2l+1}, Q, Q^-)$ that $H_i(Q) = 0$ for $i \neq 2l - 1, 2l$, in which case it is free of rank $l + 1$ and l, respectively, and that $H_i(M_1) \to H_i(Q)$ is always surjective. Q is also simply connected by Van Kampen's Theorem. The Hurewicz and Whitehead Theorems now show that all homology classes in Q can be represented by spheres which for dimensional reasons can be assumed to be embedded in M by Whitney's Imbedding Theorem. (This is enough to show, using the h-Cobordism Theorem, that M_1 is a connected sum, as described in Theorem 4.1. It is shown in [DM89] that these facts are true in general, by a detailed description of all homology classes in M_1).

The S^1 scalar action leaves Q invariant, so the quotient $R = Q/S^1$ is a compact manifold with boundary $\partial R = N$. Now the fibration $Q \to R$ again embeds in a diagram

$$S^{2d-1} \to M_1 \to S^{2d-1} \to S^{2n-1}$$

$$\downarrow \qquad \downarrow \qquad \downarrow \qquad \downarrow$$

$$\mathbb{P}^{d-1}_{\mathbb{C}} \to N \to \mathbb{P}^{d-1}_{\mathbb{C}} \to \mathbb{P}^{n-1}_{\mathbb{C}}$$

See Lemma 4.3 in Sect. 4.4 below.

It follows now from the cohomology Gysin sequence of the fibration $Q \to R$ that $H_{2i}(R) = \mathbb{Z}, i = 0, \ldots, l - 1$ and $H_{2l-1}(R) = \mathbb{Z}^l$, all other homology groups being trivial.

Now we can embed, by Lemma 4.3 in Sect. 4.4 below, $\mathbb{P}^{l-1}_{\mathbb{C}}$ in R representing all even dimensional homology classes, and l disjoint $(2l - 1)$-spheres with trivial normal bundle representing the generators of the corresponding homology group of R (since all these classes come from Q and are therefore spherical, and their normal bundles are again stably equivalent to the trivial normal bundle of S^{2l-1} in $\mathbb{P}^{n-1}_{\mathbb{C}}$). Taking a tubular neighborhood of these manifolds and joining them by tubes we get a manifold with boundary R' whose inclusion in R induces isomorphisms in homology groups. It follows from the h-Cobordism Theorem [MI65] that $N = \partial R$ is diffeomorphic to $\partial R'$ which is a connected sum of simple manifolds. These are l copies of $S^{2l-1} \times S^{2l-1}$ and the boundary of the tubular neighborhood of $\mathbb{P}^{l-1}_{\mathbb{C}}$ in R. By the remark at the end of Lemma we know that the normal bundle of this inclusion is stably equivalent to the normal bundle of $\mathbb{P}^{l-1}_{\mathbb{C}}$ in $\mathbb{P}^{2l}_{\mathbb{C}}$. We have therefore proved the following

Theorem 4.3 *For every $l > 1$ there is a $(2l - 1)$-dimensional space of complex structures on the connected sum of $\mathbb{P}^{l-1}_{\mathbb{C}} \tilde{\times} S^{2l}$ and l copies of $S^{2l-1} \times S^{2l-1}$, where $\mathbb{P}^{l-1}_{\mathbb{C}} \tilde{\times} S^{2l}$ denotes the total space of the S^{2l}-bundle over $\mathbb{P}^{l-1}_{\mathbb{C}}$ stably equivalent to the spherical normal bundle of $\mathbb{P}^{l-1}_{\mathbb{C}}$ in $\mathbb{P}^{2l}_{\mathbb{C}}$.*

Observe that for $l = 2$ we get a manifold which is similar, but not equal, to the one constructed by Kato and Yamada [KA86], where the first summand is a product instead of a nontrivial bundle. Both manifolds had been considered before, from the point of view of group actions, by Goldstein and Lininger (see [GL71]).

In general, these complex structures are very symmetric, in the sense that we can still find holomorphic actions of large groups on them (see [DM88]). In particular, there is an action of the complex, noncompact, $(n-2)$-torus $(\mathbb{C}^*)^{n-2}$ on them with a dense orbit. In this sense, our manifolds behave as toric varieties.

4.4 For $m = 1$ and $n > 3$ the Manifolds N Are Not Symplectic

Theorem 4.4 *For $n > 3$, the manifold $N = N_{\Lambda}$ is a compact, complex manifold that does not admit a symplectic structure.*

Proof In fact it follows from the classification given by Theorem 4.1 that the manifold depends on the polygon of k vertices and for $k = 1$ the manifold M_1 is empty and M_1 is a nontrivial circle bundle over N. In general we have that M_1 lies in the sphere $\mathbb{S}^{2n-1}$ and that N sits inside the complex projective space $\mathbb{P}_{\mathbb{C}}^{n-1}$ (but not as a holomorphic submanifold), so we have an inclusion of $\mathbb{S}^1$-bundles:

$$
\begin{array}{ccc}
M_1 & \longhookrightarrow & \mathbb{S}^{2n-1} \\
{\scriptstyle \pi_1}\downarrow & & \downarrow{\scriptstyle \pi_2} \\
N & \longhookrightarrow & \mathbb{P}_{\mathbb{C}}^{n-1}
\end{array}
$$

where π_1 and π_2 are the restrictions of the canonical map $\pi : \mathbb{C}^n \setminus \{0\} \to \mathbb{P}^{n-1}$ to M_1 and $\mathbb{S}^{2n-1}$ respectively.

We will prove first that the inclusion of N can be deformed down in $\mathbb{P}_{\mathbb{C}}^{n-1}$ into a projective subspace of low dimension $d - 1$, but not lower. We will prove first the following:

Lemma 4.3 *The above inclusion of $\mathbb{S}^1$-bundles embeds homotopically in the following sequence of bundle maps:*

$$
\begin{array}{ccccccc}
\mathbb{S}^{2d-1} & \to & M_1 & \to & \mathbb{S}^{2d-1} & \to & \mathbb{S}^{2n-1} \\
\downarrow & & \downarrow & & \downarrow & & \downarrow \\
\mathbb{P}_{\mathbb{C}}^{d-1} & \to & N & \to & \mathbb{P}_{\mathbb{C}}^{d-1} & \to & \mathbb{P}_{\mathbb{C}}^{n-1}
\end{array}
\qquad \text{(Diagram)}
$$

where the composition of the bottom arrows is homotopic to the natural inclusion.

Proof of Theorem 4.3 If we put d coordinates $z_i = 0$ we obtain a new manifold $M_1(\Lambda')$ where Λ' is a configuration of eigenvalues that is concentrated in $l + 1$

consecutive vertices of the regular $(2l + 1)$-gon. This configuration being in the Poincaré domain, it follows that the above manifold is empty.   □

This means that the original $M_1(\Lambda)$ does not intersect a linear subspace of C^n of codimension d and that correspondingly N does not intersect an d-codimensional projective subspace of $\mathbb{P}_{\mathbb{C}}^{n-1}$. Then the inclusion of N in $\mathbb{P}_{\mathbb{C}}^{n-1}$ can be deformed into a complementary projective subspace of dimension $d - 1$, which gives the middle bundle map.

Now, M_1 being $(2d - 2)$-connected (by Theorem 4.1), it follows that $M_1 \to N$ is a universal S^1-bundle for spaces of dimension less than $2d - 1$ (see [ST51, p. 19]) and therefore the Hopf bundle over $\mathbb{P}_{\mathbb{C}}^{d-1}$ admits a classifying map into it, which gives the first map in the bottom row. The composition of the bottom maps also classifies this Hopf bundle and is therefore homotopic to the natural inclusion, so the Lemma is proved. From the description of M_1 it follows that M_1 is simply connected, except for the cases $k = 3$, $d = n_1 = 1$. In these cases the S^1-action on $M_1 = S^1 \times S^{2n_2-1} \times S^{2n_3-1}$ can be concentrated on the first factor, and therefore N is diffeomorphic to $S^{2n_2-1} \times S^{2n_3-1}$. Unless $n_2 = n_3 = 1$ we have that $H^2(N) = 0$ and N is not symplectic.

In all the other cases we have that $d > 1$ and M_1 is 2-connected. From the cohomology Gysin sequence of the fibration $M_1 \to N$ it follows that $H^2(N) = \mathbb{Z}$ generated by the Euler class e. However, it follows from the Lemma the following:

Remark 4.11

$$e^{d-1} \neq 0, \qquad e^d = 0$$

so this class does not go up to the top cohomology group $H^{2n-4}(N, \mathbb{Z})$, and it follows again that N is not symplectic, and Theorem 4.4 is proved.   □

Nevertheless, observe that N is a real algebraic submanifold of $\mathbb{P}_{\mathbb{C}}^{n-1}$ since it is the regular zero set of the (non holomorphic) function $g : \mathbb{P}_{\mathbb{C}}^{n-1} \to \mathbb{R}^2$ defined by

$$g([z_1, \ldots, z_n]) = \frac{\Sigma \lambda_i z_i \bar{z}_i}{\Sigma z_i \bar{z}_i}$$

Remark 4.12 This implies that the normal bundle of N in $\mathbb{P}_{\mathbb{C}}^{n-1}$ is trivial. Observe also that the map $\mathbb{P}_{\mathbb{C}}^{d-1} \to N$ in the Lemma is homotopic to an embedding, whose normal bundle is then stably equivalent to the normal bundle of $\mathbb{P}_{\mathbb{C}}^{d-1}$ in $\mathbb{P}_{\mathbb{C}}^{n-1}$.

4.4.1 Compact Complex Tori Are the Only Kähler LVM Manifolds

Let k denote the number of indispensable points (remember Definition 4.8 above). By Carathéodory's Theorem $k \leqslant 2m + 1$ and the maximum is attained only when $n = 2m + 1$. One has:

Lemma 4.4

(1) $S = (\mathbb{C}^*)^k \times (\mathbb{C}^{n-k} \setminus A)$ with A an analytic set of codimension at least two in every point.

(2) This decomposition descends to a decomposition $M_1 = (\mathbb{S}^1)^k \times M_0$ where M_0 is a real compact manifold which is 2-connected.

Sketch of the Proof Let $S = \mathbb{C}^n \setminus E$, where E is a union of subspaces (see (4.3))

$$E = \{z \in \mathbb{C}^n \mid 0 \notin \mathcal{H}(\Lambda_{I_z})\}.$$

The components of codimension one are given by indices corresponding to indispensable points in the configuration. This proves the first part. Since A is of complex codimension at least 2 in every point $(\mathbb{C}^{n-k} \setminus A)$ is 2-connected, hence M_0 is 2-connected, since they have the same homotopy type. $\square$

In examples 4.3.3, 4.3.4 and 4.3.5 above one obtains compact complex manifolds which are not symplectic because the second de Rham cohomology group is trivial. This is in fact a general property of the manifolds we obtain:

Theorem 4.5 *Let Λ be an admissible configuration as in Definition 4.3 and N_Λ the corresponding compact complex manifold. The the following are equivalent:*

(1) $\mathcal{H}(\Lambda)$ is a simplex

(2) N_Λ is symplectic.

(3) N_Λ is Kähler.

(4) N_Λ is a complex torus.

(5) $n = 2m + 1$.

Sketch of the Proof It is easy to prove the equivalence of (3) and (4): If N is a complex torus, one must have $S = (\mathbb{C}^*)^n$ hence all the Λ_i must be indispensable and in this case the convex hull must be a simplex and $n = 2m + 1$. If the convex hull is a simplex then, as in Example 4.3.2, N is a compact complex torus.

The most difficult part is that (1) implies (4). One proves that by contradiction. Suppose $n > 2m + 1$. As in the examples one must study the de Rham cohomology of N and to prove that it is incompatible with the existence of a symplectic form.

We consider two cases:

First Case There exists indispensable points. From here one can deduce that the fibration $M \to N$ is trivial. Hence the decomposition $M_1 = (\mathbb{S}^1)^k \times M_0$ of the previous Lemma gives a decomposition $N = (\mathbb{S}^1)^{k-1} \times M_0$.

In other words if N has a symplectic structure it must be supported by $(\mathbb{S}^1)^{k-1}$. The maximal power of this symplectic form must be a volume form in N but that is only possible only if $k - 1$ is equal $2n - 2m - 2$ which is the real dimension of N.

Second Case If there are not indispensable points then M is 2-connected and the fibration $M \to N$ is not topologically trivial. Therefore the second de Rham cohomology group of N is generated by the Euler class of that fibration. Analyzing carefully this fibration one shows that the Euler class is trivial. Therefore this class is not symplectic, the proof is similar to that of Theorem 4.4.   □

4.5 Meromorphic Functions on the Manifold N_Λ

Many analytic properties of LVM manifolds are related to the arithmetic properties of the configuration Λ. One nice example of this fact is given by the following

Theorem 4.6 ([ME00, Theorem 4]) *Let N be a LVM manifold without indispensable points. Then the algebraic dimension of N_Λ is equal to the dimension over $\mathbb{Q}$ of the $\mathbb{Q}$-vector space of rational solutions of the system (S):*

$$
\begin{cases}
\displaystyle\sum_{i=1}^{n} s_i \Lambda_i = 0 \\[2mm]
\displaystyle\sum_{i=1}^{n} s_i = 0
\end{cases}
\tag{S}
$$

The idea of the proof is very simple. If f is a meromorphic function on N, then it can be lifted to a meromorphic function $\tilde{f}$ of S which is constant along the leaves of $\tilde{\mathcal{F}}$. Since we have assumed that there is not an indispensable point Lemma 4.4 implies that S is obtained from $\mathbb{C}^n$ by removing an analytic subspace of codimension at least two at every point. Therefore $\tilde{f}$ can be extended to all $\mathbb{C}^n$ by Levi's Extension Theorem (see [BHPV04, p. 26]). Furthermore $\tilde{f}$ must be invariant by the action given by (4.1) in Sect. 4.2. In particular $\tilde{f}$ must be invariant by the standard action of $\mathbb{C}^*$ on $\mathbb{C}^n \setminus \{0\}$, and descends to $\mathbb{P}_{\mathbb{C}}^{n-1}$. Therefore $\tilde{f}$ is a rational function. We can show that the fact that $\tilde{f}$ is constant along the leaves of $\mathcal{F}$ implies that an algebraic basis of these rational functions is given by the monomials

$$
z_1^{s_1} \cdot \ldots \cdot z_n^{s_n},
$$

where $(s_1, \ldots, s_n)$ is a rational basis of the vector space of solutions of system (S).

Example 4.1 Let $n = 5$ et $m = 1$, and:

$$
\Lambda_1 = 1 \qquad \Lambda_2 = i \qquad \Lambda_3 = -1 - i \qquad \Lambda_4 = \frac{3}{2}i + 1 \qquad \Lambda_5 = -i - \frac{1}{2}
$$

One verifies immediately that there are not indispensable points. The complex dimension of N is 3 and its algebraic dimension is according to the preceding Theorem 4.6. Indeed

$$f(z) = \frac{z_1^5 z_2^5 z_3^2}{z_4^6 z_5^6}, \qquad g(z) = \frac{z_1 z_2^2}{z_3 z_4^2}$$

are meromorphic functions which are algebraically independent on N and in addition every meromorphic function on N depends algebraically on f and g.

Recall that a connected Moishezon manifold M is a compact complex manifold such that the field of meromorphic functions has transcendence degree equal the complex dimension of the manifold.

It is shown in [ME00] that when Theorem 4.6 applies the algebraic dimension of N is at most $n - 2m - 1$ therefore the dimension is strictly inferior to its dimension $n - m - 1$. In other words: if there are not indispensable points N is not Moishezon. This happens if and only if Λ is a simplex. Hence we have the following:

Theorem 4.7 ([ME00, Theorem 3]) *The following are equivalent:*

  (i) *N is Moishezon.*
 (ii) *N is projective.*
(iii) *N is a complex projective torus.*

Sketch of the Proof We follow the proof given by Frédéric Bosio in [BO01, pp. 1276–1277]. If I is a subset of $\{1, \ldots, n\}$ such that 0 is in the convex envelope of $(\Lambda_i)_{i \in I}$, then the restriction of action (4.2) to the complex vector subspace of $\mathbb{C}^n$ given by the equations

$$z_j = 0 \qquad \text{pour } j \notin I$$

defines also a LVM manifold that we denote N_I. Then this is a complex submanifold of N. One can verify that if $n > 2m + 1$, i.e., there are points that can be eliminated, one can find always submanifolds which have indispensable and in fact we can fine such a submanifold with odd first Betti number. But if N is Moishezon then all of its complex subvarieties are also Moishezon and therefore must have first Betti number even.   □

Remark 4.13 Exactly this last argument implies that N is not Kähler if $n > 2m + 1$.

4.6 Deformation Theory

4.6.1 Small Deformations

We will state without a proof a theorem of stability of LVM manifolds under small deformations. Let Λ be an admissible configuration and N the associated LVM manifold. For $\epsilon > 0$, let $(\Lambda_t)_{-\epsilon < t < \epsilon}$ be a small smooth perturbation of Λ (i.e., a smooth function from $(-\epsilon, \epsilon)$ to $(\mathbb{C}^m)^n$ such that $\Lambda_0 = \Lambda = (\Lambda_1, \dots, \Lambda_n)$).

Since the Siegel and weak hyperbolicity conditions are open in $(\mathbb{C}^m)^n$, if ϵ is sufficiently small all the configurations (Λ^t) are admissible. The manifold (ϵ, ϵ), $\bigcup_{t \in (\epsilon, \epsilon)} N_t \subset \mathbb{P}_{\mathbb{C}}^{n-1} \times \mathbb{R}$ admits an obvious submersion over $(-\epsilon, \epsilon)$ with compact fibers. Ehresmann' Lemma implies that all the N_t are diffeomorphic, however they are not necessarily biholomorphic it is enough, for instance, to start with a configuration Λ which verifies la condition (**K**) in Definition 4.12, and to perturb it in $(\mathbb{C}^m)^n$)in order to obtain $(\Lambda)^t$ which verify (H) in 4.12. This way one obtains a non-trivial family of de LVM manifold N_Λ parametrized by the interval $(-\epsilon, \epsilon)$.

On the other hand, if Λ et Λ' are two admissible configurations such that Λ' is obtained from Λ' by a complex affine transformation of $\mathbb{C}^m$, i.e., there exists a complex affine transformation A of $\mathbb{C}^m$ such that $\Lambda_i' = A(\Lambda_i)$ for all i, one sees immediately since $A(\mathcal{S}_\Lambda) = \mathcal{S}'$ and A sends a Siegel leaf of the system corresponding to Λ to a leaf corresponding to Λ'.

Definition 4.10 Let Λ be an admissible configuration and N_Λ the corresponding LVM manifold. One calls *space of parameters* of N_Λ the set of equivalence classes on an open connected neighborhood of Λ in $(\mathbb{C}^m)^n$ consisting of equivalence classes of admissible configurations under the equivalence $\cong$ given by $\Lambda \cong \Lambda'$ if and only if there exists a complex affine transformation A such that $A(\Lambda) = A(\Lambda')$.

The weak hyperbolicity condition implies that Λ affinely generates the space $\mathbb{C}^m$ [MV04, Lemma 1.1]. Up to renumbering the vectors on can assume that $(\Lambda_1, \dots, \Lambda_{m+1})$ are affinely independent. Given a sufficiently small open connected set of configurations in $(\mathbb{C}^m)^n$ containing Λ, one sees that every element in that open set can be transformed in a unique way to a configuration where the first $m + 1$ vectors coincide with those of Λ. Therefore:

Lemma 4.5 *Let D be an space of parameters for N_Λ. Then D can be identified with an open connected subset of $(\mathbb{C}^m)^{n-m-1}$.*

Under these conditions on can construct a holomorphic family $\mathcal{D}$ of deformations of N_Λ parametrized by D. It is enough to consider the quotient of $\mathcal{S} \times D$ under the action in formula (4.1) with parameters in D.

Theorem 4.8 ([ME00, Theorem 11]) *Let D be an space of parameters of the LVM manifold N_Λ corresponding to the configuration Λ. Let $\mathcal{D} \to D$ be the associated family of deformation. Then*

(i) *If $\mathcal{S}$ is at least 3-connected, the family $\mathcal{D}$ is a versal family of deformations of N_Λ.*

(ii) *If $\mathcal{S}$ is at least 4-connected and $\Lambda_i \neq \Lambda_j$ if $i \neq j$, the family $\mathcal{D}$ is universal*

Hence under rather restrictive conditions we have that all the small deformations of N_Λ are obtained by just perturbing the configuration Λ. However this is not the general case: the Hopf surfaces don't admit a universal family.

4.6.2   Rigidity and Versality for $m = 1$

We consider the configuration corresponding to the regular polygon with $n = 2l + 1$ vertices (see Sect. 4.3.5). Let $n = n_1 + \cdots + n_k$ be an ordered partition of n with $d \geq 4$. Let $\Lambda = (\lambda_1, \ldots, \lambda_k)$, $\lambda_i \in \mathbb{C}$ be the admissible configuration where the multiplicity of λ_i is n_i.

Recall that the complex structure on $N(\Lambda)$ does not vary within the affine equivalence class of Λ. We show now that the converse is true in most of the cases. These include in particular all cases with $k > 5$. It is plausible that the result is true in general.

Theorem 4.9 *Let $n = n_1 + \cdots + n_k$ be an ordered partition of n with $d \neq 2$. Then any two collections of eigenvalues corresponding to this partition give holomorphically equivalent manifolds N if, and only if, they are affinely equivalent.*

Proof The sufficiency of the condition was observed above. For the necessity, if $d = 1$ we are in the Calabi–Eckmann case, and this was shown by Loeb and Nicolau [LN96, Proposition 12]. For $d > 2$ we follow their argument:

Let $V = \mathcal{S}/\mathbb{C}^*$ which is an open subset of $\mathbb{P}^{n-1}_\mathbb{C}$. Then the complement of V in $\mathbb{P}^{n-1}_\mathbb{C}$ is a union of projective subspaces whose smallest codimension is d. By the results of Scheja [SC61] we have that

$$H^i(V, \mathcal{O}) = H^i(\mathbb{P}^{n-1}_\mathbb{C}, \mathcal{O}) \ for \ i \leq d - 2$$

where $\mathcal{O}_X$ denotes the sheaf of holomorphic functions on a manifold. The second cohomology groups were computed by Serre and are $\mathbb{C}$ in dimension 0 and trivial otherwise (see e.g. [GH78, p. 118]).

Now, let $\mathcal{O}^{inv}$ be the kernel of the map $\mathcal{O} \to \mathcal{O}$ given by the Lie derivative along the vector field ξ which generates the $\mathbb{C}$ action on V, so we have an exact sequence of sheaves:

$$0 \longrightarrow \mathcal{O}^{inv} \to \mathcal{O} \xrightarrow{L_\xi} \mathcal{O} \longrightarrow 0$$

The associated cohomology exact sequence shows that, for $d \geq 3$, $H^1(V, \mathcal{O}^{inv}) = \mathbb{C}$, but this group can be identified with $H^1(N, \mathcal{O})$. Therefore this group is also $\mathbb{C}$ and since it classifies the principal $\mathbb{C}$-bundles over N, any two non-trivial principal $\mathbb{C}$-bundles over N differ by a scalar factor.

Let N_1, N_2 be two such manifolds which are holomorphically equivalent and consider a biholomorphism $\phi : N_1 \to N_2$. Over each N_i there is a principal $\mathbb{C}$-bundle $V_i \to N_i$, where the total space V_i is in both cases V, but is foliated in two different ways by the projectivized leaves of each system. We have to lift ϕ to an equivalence of the principal $\mathbb{C}$-bundles V_i, which amounts to finding an equivalence between V_1 and $\phi^* V_2$. Now V_1 and $\phi^* V_2$ are non-trivial $\mathbb{C}$-bundles (otherwise they would have sections, N_i would embed holomorphically in $\mathbb{P}_{\mathbb{C}}^{n-1}$ and would be a Kähler manifold, recall [WE73, p. 182]). By the previous computation these differ by a scalar factor and there is an equivalence between V_1 and V_2 preserving the leaves of the foliations. By Hartog's Theorem this equivalence extends to one of $\mathbb{P}_{\mathbb{C}}^{n-1}$ into itself which must then necessarily be linear since the group of biholomorphisms of $\mathbb{P}_{\mathbb{C}}^{n-1}$ is the corresponding projective linear group. But then it follows easily that the corresponding eigenvalues must be affinely equivalent, and Theorem 4.9 is proved. □

Theorem 4.9 says that when $d \neq 2$ the reduced deformation space of N injects into its universal deformation space. For $d = 1$ the question of whether the reduced deformation space is universal or not depends on the existence of resonances among the λ_i (see [HA85, LN96]). For $d \geq 4$ the situation is simpler and only depends on the condition that all the λ_i be different:

Theorem 4.10 *Let* $n = n_1 + \cdots + n_k$ *be an ordered partition of n with* $d \geq 4$. *Let* Λ *be a collection of eigenvalues corresponding to this partition and assume that all* λ_i *are different. Then the corresponding reduced deformation space of* $N(\Lambda)$ *is universal.*

Proof Following again [LN96] we consider the exact sequences of sheaves over V:

$$0 \to \Theta^{inv} \to \Theta \xrightarrow{L_\xi} \Theta \to 0$$

$$0 \to \mathcal{O}^{inv}\xi \to \Theta^{inv} \to \Theta_b \to 0$$

where Θ denotes the sheaf of holomorphic vector fields on a manifold and Θ^{inv} and Θ_b are defined by these sequences. Now again by Scheja [SC61] we have

$$H^i(V, \Theta) = H^i(\mathbb{P}_{\mathbb{C}}^{n-1}, \Theta) \ for \ i \leq d - 2$$

$H^0(\mathbb{P}_{\mathbb{C}}^{n-1}, \Theta)$ is the space of holomorphic global vector fields on $\mathbb{P}_{\mathbb{C}}^{n-1}$ (all of which are linear) and can be identified with the space of $n \times n$ matrices modulo the scalar ones. For $i > 0$, $H^i(\mathbb{P}_{\mathbb{C}}^{n-1}, \Theta) = 0$. □

The first sequence above gives a cohomology exact sequence for $d \geqslant 4$:

$$0 \to H^0(\Theta^{inv}) \to H^0(\Theta) \xrightarrow{L_\xi} H^0(\Theta) \to H^1(\Theta^{inv}) \to 0.$$

Since ξ corresponds to the diagonal matrix with entries λ_i and these are different, the kernel and cokernel of L_ξ can be identified with the space of diagonal matrices modulo the scalar ones, so $H^1(\Theta^{inv})$ is a space of dimension $n - 1$. The class of ξ in this vector space is non-zero.

From the exact sequences of sheaves we have the diagram:

$$\begin{array}{ccccccc}
H^0(\Theta_b) & \to & H^1(\mathcal{O}^{inv}) & \to & H^1(\Theta^{inv}) & \to & H^1(\Theta_b) \to 0 \\
 & & \uparrow \cong & & \uparrow & & \\
 & & H^0(\mathcal{O}) & \to & H^0(\Theta) & &
\end{array}$$

where the two middle horizontal maps are induced by multiplication by ξ. Since the lower one is injective by the above remark, it follows that so is the upper one and that $H^1(\Theta_b)$ is of dimension $n - 2$.

Now it is easy to see that $H^i(\Theta_b)$ is isomorphic to $H^i(N, \Theta)$. It follows that $H^1(N, \Theta)$ is of dimension $n-2$ and is the tangent space to the universal deformation space of N. Since we have shown that the reduced deformation space is smooth, has dimension $n - 2$ and injects into this universal space, it follows that it is itself a universal deformation space and Theorem 4.10 is proved.   □

Observe that in Theorem 4.10 if $n = 2l + 1$ then for $l \geq 4$ the space of complex structures is the universal deformation space for any of its members.

4.6.3 Global Deformation Theory of LVM Manifolds

Here the deformation theory of equivariant LVM manifolds is explained and then together with the reconstruction theorem we conclude that this implies the existence of the moduli stack of torics.

Let Λ be an admissible configuration. We want to describe the set $\mathcal{M}_\Lambda$ of G-biholomorphism classes of LVM manifolds $N_{\Lambda'}$ such that $\mathcal{S}_{\Lambda'}$ is equal to $\mathcal{S}_\Lambda$ up to a permutation of coordinates in $\mathbb{C}^n$.

We assume that Λ satisfies (4.21) and

$$\Lambda_i \text{ is indispensable} \iff i \leqslant k \tag{4.7}$$

that is, the k indispensable points are the first k vectors of the configuration. In the same way, every class $[N_{\Lambda'}]$ of $\mathcal{M}_\Lambda$ can be represented by a configuration Λ' satisfying (4.21), (4.7) and

$$\mathcal{S} := \mathcal{S}_\Lambda = \mathcal{S}_{\Lambda'}. \tag{4.8}$$

Remark 4.14 Condition (4.8) is equivalent to K_Λ being combinatorially equivalent to $K_{\Lambda'}$ with same numbering (4.31). Observe that because of our convention (4.7), having the same numbering implies having the same number of indispensable points.

Now, observe that, because of (4.21), there exists an affine transformation T of $\mathbb{C}^m$ sending Λ onto a configuration (which we still denote by Λ) whose first $m+1$ vectors satisfies

$$\Lambda_1 = ie_1, \ \Lambda_2 - \Lambda_1 = e_1, \ \ldots, \ \Lambda_{m+1} - \Lambda_1 = e_m,$$

$$\text{where } (e_1, \ldots, e_m) \text{ is the canonical basis of } \mathbb{C}^m. \tag{4.9}$$

It is straightforward to check that this does not change N_Λ up to G-biholomorphism. In the same way, each class of $\mathcal{M}_\Lambda$ can be represented by an element Λ' satisfying (4.21), (4.7), (4.8) and (4.9). We call *S-normalized configuration* such a configuration.

Let $\mathcal{T}_\Lambda$ be the set of *S*-normalized configurations. This is an open and connected set in $(\mathbb{C}^m)^{n-m-1}$.

Assume now that $N_{\Lambda'}$ is G-biholomorphic to N_Λ. Then, G_Λ and $G_{\Lambda'}$ as subgroups of $\text{Aut}(N_\Lambda)$, respectively $\text{Aut}(N_{\Lambda'})$ are isomorphic Lie groups. Hence, their universal cover are isomorphic as Lie groups, that is, using the presentation given in Proposition 4.6, there exists a matrix M in $\text{GL}_{n-m-1}(\mathbb{C})$ which sends the lattice of G_Λ bijectively onto that of $G_{\Lambda'}$. Using notations (4.22) and (4.23), this means that there exists a matrix P in $\text{SL}_{n-1}(\mathbb{Z})$ such that

$$M(Id, B_\Lambda A_\Lambda^{-1}) = (Id, B_{\Lambda'} A_{\Lambda'}^{-1})P. \tag{4.10}$$

Decomposing P as

$$P = \begin{pmatrix} P_1 & P_2 \\ Q_1 & Q_2 \end{pmatrix} \tag{4.11}$$

with P_1 a square matrix of size $n - m - 1$ and Q_2 a square matrix of size m, we obtain

$$MB_\Lambda A_\Lambda^{-1} = (P_1 + B_{\Lambda'} A_{\Lambda'}^{-1} Q_1)B_\Lambda A_\Lambda^{-1} = P_2 + B_{\Lambda'} A_{\Lambda'}^{-1} Q_2. \tag{4.12}$$

Because of (4.9), this means that

$${}^tB_\Lambda = ({}^tP_2 + {}^tQ_2\,{}^tB_{\Lambda'})({}^tP_1 + {}^tQ_1\,{}^tB_{\Lambda'})^{-1} \tag{4.13}$$

that is

Proposition 4.1 *Let Λ and Λ' be two S-normalized configurations. Then N_Λ and $N_{\Lambda'}$ are G-biholomorphic if and only if Λ and Λ' satisfies (4.13).*

Thus, $\mathcal{M}_\Lambda$ is the quotient of $\mathcal{T}_\Lambda$ by the action of $\mathrm{SL}_{n-1}(\mathbb{Z})$ described in (4.13). We claim

Proposition 4.2 *If the number k of indispensable points is less than $m + 1$, then the moduli space $\mathcal{M}(X)$ is an orbifold.*

Proof From the previous description, it is enough to prove that the stabilizers of action (4.13) are finite. Let f be a G-biholomorphism of N_Λ. Set

$$S_1 = \{w \in \mathbb{C}^{n-m-1} \mid (1, \ldots, 1, w) \in S\}. \tag{4.14}$$

$\square$

Observe that (4.14) is a covering of the quotient N_1 of $S \cap \{z_1 \cdots z_{m+1} \neq 0\}$ by the action (4.1). Indeed, we have a commutative diagram

$$
\begin{array}{ccccc}
(\mathbb{C}^*)^{n-m-1} & \longrightarrow & S_1 & \longrightarrow & S \\
\downarrow & & \downarrow & & \downarrow \\
G_\Lambda & \longrightarrow & N_1 & \longrightarrow & N_\Lambda
\end{array}
\tag{4.15}
$$

where the horizontal maps are inclusions and the first two vertical ones are coverings.

Then, up to composing with a permutation of $\mathbb{C}^n$, we may assume that f sends N_1 onto itself. Because of assumption (4.7), the set (4.14) is a 2-connected open subset of $\mathbb{C}^{n-m-1}$, hence the restriction of f to N_1, say f_1, lifts to a biholomorphic map F_1 of (4.14). More precisely, S_1 is equal to $\mathbb{C}^{n-m-1}$ minus a finite union of codimension 2 vector subspaces, hence by Hartogs, F_1 extends as a biholomorphism of $\mathbb{C}^{n-m-1}$.

On the other hand, the restriction of f to G_Λ preserves G_Λ and lifts as a biholomorphism $\tilde{F}$ of its universal covering $\mathbb{C}^{n-m-1}$. And we have a commutative diagram

$$
\begin{array}{ccc}
\mathbb{C}^{n-m-1} & \xrightarrow{\exp(2i\pi -)} & (\mathbb{C}^*)^{n-m-1} \\
\tilde{F} \downarrow & & \downarrow F_1 \\
\mathbb{C}^{n-m-1} & \xrightarrow{\exp(2i\pi -)} & (\mathbb{C}^*)^{n-m-1}
\end{array}
\tag{4.16}
$$

But, since the linear map $\tilde{F} = M$ must preserve the abelian subgroup of Proposition 4.6, using (4.9) and (4.10), we have

$$\tilde{F}(z + e_i) = \tilde{F}(z) + P_1 e_i := \tilde{F}(z) + \sum_{j=1}^{n-m-1} a_{ij} e_j \tag{4.17}$$

that is Q_1 is equal to 0. But through (4.16), this implies that

$$F_1(w) = \left(w_1^{a_{1j}} \cdots w_{n-m-1}^{a_{n-m-1\,j}}\right)_{j=1}^{n-m-1} \tag{4.18}$$

Now, recall that F_1 is a biholomorphism of the whole $\mathbb{C}^{n-m-1}$, so must send a coordinate hyperplane onto another one without ramifying. This shows that $P_1 = (a_{ij})$ is a matrix of permutation. Hence every stabilizer is a subgroup of the group of permutations with $n - m - 1$ elements, so is finite.

Example 4.2 (Tori) Let $n = 2m + 1$, then there are $2m + 1$ indispensable points, $\mathcal{S}$ is $(\mathbb{C}^*)^n$ and N is a compact complex torus of dimension m [ME00, Theorem 1]. The associate polytope K is reduced to a point and $N = G$. The moduli space $\mathcal{M}$ is equal to the moduli space of compact complex tori of dimension m, which is not an orbifold for $m > 1$.

Example Hopf Surfaces Let $n = 4$ and $m = 1$, then there are two indispensable points and $\mathcal{S}$ is $(\mathbb{C}^*)^2 \times \mathbb{C}^2 \setminus \{(0,0)\}$. A $\mathcal{S}$-admissible configuration is given by a couple complex numbers (λ_3, λ_4) belonging to

$$\{z \in \mathbb{C} \mid \Re z < 0 \text{ and } \Re z < \Im z\}. \tag{4.19}$$

The manifold N_Λ is equal to the diagonal Hopf surface obtained by taking the quotient of $\mathbb{C}^2 \setminus \{(0,0)\}$ by the group generated by

$$(z, w) \longmapsto (\exp 2i\pi(\lambda_3 - \lambda_1) \cdot z, (\exp 2i\pi(\lambda_4 - \lambda_1) \cdot w) \tag{4.20}$$

Two points (λ_3, λ_4) and (λ_3', λ_4') with coordinates in (4.19) are equivalent if and only if their difference is in the lattice $\mathbb{Z} \oplus \mathbb{Z}$ or if the difference of (λ_3, λ_4) by the switched (λ_4', λ_3') is in this lattice. The isotropy group of a point is $\mathbb{Z}_2$ for the diagonal $\lambda_3 = \lambda_4$ and is zero elsewhere. The moduli space is an orbifold.

Observe that not all Hopf surfaces are obtained as LVM-manifolds, but only the linear diagonal ones. Now, they coincide with the set of Hopf surfaces that are equivariant compactifications of $(\mathbb{C}^*)^2$.

(a) Generalized Hopf Manifolds

When $n_1 = n_2 = 1$ the manifold N is diffeomorphic to $\mathbb{S}^1 \times \mathbb{S}^{2n_3-1}$. Here the mapping $exp : \mathbb{C}^2 \times (\mathbb{C}^{n_3}\setminus 0) \to \mathcal{S} = (\mathbb{C}^*)^2 \times (\mathbb{C}^{n_3}\setminus 0)$ given by $exp(\zeta_1, \zeta_2, \zeta) = (e^{\zeta_1}, e^{\zeta_2}, \zeta)$ can be used to identify N with the quotient of $\mathbb{C}^{n_3}\setminus 0$ by the action of $\mathbb{Z}$ defined by the multipliers

$$\alpha_i = \exp\left(2\pi i \frac{\lambda_{2+i} - \lambda_2}{\lambda_1 - \lambda_2}\right), \quad i = 1, \ldots, n_3.$$

In this case we obtain all complex structures on $\mathbb{S}^1 \times \mathbb{S}^{2n_3-1}$ having $\mathbb{C}^n \backslash 0$ as universal cover when there is no resonance among the α_i. But in the resonant case we do not obtain all such complex structures since we do not obtain the non-linear resonant cases of Haefliger (see [HA85] for the notion of resonance).

(b) Generalized Calabi–Eckmann Manifolds

When $n_1 = 1$ and n_2, n_3 are both greater than 1 we have seen that the manifold N is diffeomorphic to $\mathbb{S}^{2n_2-1} \times \mathbb{S}^{2n_3-1}$. Here the mapping $exp : \mathbb{C} \times (\mathbb{C}^{n_2} \backslash 0) \times (\mathbb{C}^{n_3} \backslash 0) \to \mathcal{S} = \mathbb{C}^* \times (\mathbb{C}^{n_2} \backslash 0) \times (\mathbb{C}^{n_3} \backslash 0)$ given by $exp(\zeta, \zeta_1, \zeta_2) = (e^\zeta, \zeta_1, \zeta_2)$ can be used to identify N with the quotient of $(\mathbb{C}^{n_2} \backslash 0) \times (\mathbb{C}^{n_3} \backslash 0)$ by the action of $\mathbb{C}$ defined by the linear differential equation with eigenvalues $\lambda'_i = 2\pi i (\lambda_i - \lambda_1)$, $i = 2, \ldots, n$. This is exactly the construction of the Loeb-Nicolau complex structure corresponding to a linear system of equations of Poincaré type [LN96].

Observe that in their construction only the quotients of the eigenvalues of the system are relevant for the definition of the complex structure on N, so once again only the quotients $\frac{\lambda_i - \lambda_k}{\lambda_j - \lambda_k}$ of our original eigenvalues count.

Again we obtain all their examples of complex structures on $\mathbb{S}^{2n_2-1}\mathbb{S}^{2n_3-1}$ when there is no resonance among the λ'_i. But, once more, in the resonant case we do not obtain all their complex structures since we do not obtain the non-linear resonant examples (see [HA85]).

Observe that in all the cases considered in this section only the quotients $\frac{\lambda_i - \lambda_k}{\lambda_j - \lambda_k}$ are relevant in the description of the complex structure of N (in accordance with Theorem 4.9 that affinely equivalent configurations of eigenvalues with the same $\mathcal{S}$ give the same complex structure) and that they are actually *moduli* of that complex structure.

4.7 LVM Manifolds as Equivariant Compactifications

Theorem 4.6 has a deeper explanation related to the structure of N_Λ and the arithmetic properties of Λ. In fact $\mathcal{S}$ contains always $(\mathbb{C}^*)^n$ as an open and dense subset *invariant under the foliation* $\tilde{\mathcal{F}}$. If we pass to the quotient under the action of (4.1) one obtains that N_Λ has as an open subset G_Λ, which is the quotient of $(\mathbb{C}^*)^n$ by $\tilde{\mathcal{F}}$. Since $\tilde{\mathcal{F}}$ is defined by the action (4.1) and this action commutes with the group structure of the multiplicative group de $(\mathbb{C}^*)^n$, it follows that G_Λ itself is a connected commutative complex Lie group. In other words:

Theorem 4.11 ([LM02]) N_Λ *is the equivariant compactification of a complex commutative Lie group* G_Λ.

Remark 4.15 In some sense, this theorem is the principal reason of the interconnection between LVMB manifolds, toric varieties, convex polytopes and moment-angle manifolds.

Definition 4.11 A connected complex Lie group G is called *Cousin group* (or toroidal group in [KO64]) if any holomorphic function on it is constant [AK01].

Proposition 4.3 *Cousin groups are commutative. Moreover, they are quotients of a complex vector space $\mathbb{V}$ by a discrete additive subgroup of $\mathbb{V}$ [AK01].*

Proposition 4.4 *Any commutative connected complex Lie group G can be written in a unique way as a product $G = C \times \mathbb{C}^l \times (\mathbb{C}^*)^r$ where C is a Cousin group $(l, r \geqslant 0)$.*

Proposition 4.5 *A commutative complex Lie group is Cousin if and only if it does not have nontrivial characters.*

Observe that $(\mathbb{C}^*)^n$ acts by multiplication on the space of Siegel leaves $\mathcal{S}_\Lambda$ with an open and dense orbit, making it a toric variety. This action commutes with projectivization and with (4.1), making of N_Λ an equivariant compactification of an abelian Lie group, say G_Λ. A straightforward computation shows the following [ME98, p.27]

Proposition 4.6 *Assume that*

$$\mathrm{rank}_\mathbb{C} \begin{pmatrix} \Lambda_1 & \dots & \Lambda_{m+1} \\ 1 & \dots & 1 \end{pmatrix} = m + 1. \tag{4.21}$$

Then G_Λ is isomorphic to the quotient of $\mathbb{C}^{n-m-1}$ by the $\mathbb{Z}^{n-1}$ abelian subgroup generated by $(Id, B_\Lambda A_\Lambda^{-1})$ where

$$A_\Lambda = {}^t(\Lambda_2 - \Lambda_1, \dots, \Lambda_{m+1} - \Lambda_1) \tag{4.22}$$

and

$$B_\Lambda = {}^t(\Lambda_{m+2} - \Lambda_1, \dots, \Lambda_{n-1} - \Lambda_1). \tag{4.23}$$

Remark 4.16 It is easy to prove that

$$\mathrm{rank}_\mathbb{C} \begin{pmatrix} \Lambda_1 & \dots & \Lambda_n \\ 1 & \dots & 1 \end{pmatrix} = m + 1.$$

(cf. [MV04, Lemma 1.1] in the LVM case). Hence, up to a permutation, condition (4.21) is always fulfilled.

We say that N_Λ and $N_{\Lambda'}$ are *G-biholomorphic* if they are $(G_\Lambda, G_{\Lambda'})$-equivariantly biholomorphic.

Remark 4.17 When $\mathcal{S}$ is $(\mathbb{C}^*)^n$, one has $N = G$ is a compact complex Lie group and therefore a compact complex torus. This is a direct proof of the example presented in 4.2.

The structure of groups like G_Λ is well-known [60]. Since we know that the dimension of G_Λ is equal to that of N we obtain that G_Λ is the quotient of $(\mathbb{C}^*)^{n-m-1}$ by a discrete multiplicative subgroup Γ. The group G_Λ is sometimes called a semi torus i.e., there exists a short equivariant exact sequence

$$0 \longrightarrow (\mathbb{C}^*)^{n-m-1} \longrightarrow G_\Lambda \longrightarrow T \longrightarrow 0$$

where T is an appropriate compact complex torus of dimension $n - m - 1$.

Furthermore, the group G_Λ is isomorphic to $(\mathbb{C}^*)^a \times C$, where $a \geqslant 0$ and C is a Cousin group. Compact Cousin groups are just complex tori. However there are non-compact Cousin group, for instance If $C = (\mathbb{C}^*)^{n-m-a-1}/\Gamma_0$, and Γ_0 is a "sufficiently generic" discrete subgroup in order to have that any holomorphic function on $(\mathbb{C}^*)^{n-m-a-1}$ which is invariant under Γ_0 must be constant then any holomorphic function on the quotient is constant.

In our case, any holomorphic function on G extends to a meromorphic function on N_Λ. Then Theorem 4.6 shows is the following:

Proposition 4.7 *If N_Λ does not have an indispensable point, then the algebraic dimension of N_Λ is equal to the dimension a of the factor $\mathbb{C}^*$ in the associated decomposition $G = (\mathbb{C}^*)^a \times C$.*

Hence we obtain the following.

Corollary 4.1 ([ME00, Proposition IV.1]) *Let N_Λ be an LVM manifold which is the equivariant compactification of the connected complex abelian Lie group G_Λ. Suppose N_Λ is without indispensable points. Then one has an equivalence:*

(i) N_Λ does not have non-constant meromorphic functions
(ii) G_Λ is a Cousin group, i.e., every holomorphic function on G_Λ is constant
(iii) System (S) has no solution in the rationals.

4.8 Toric Varieties and Generalized Calabi–Eckmann Fibrations

Let $\Lambda = (\Lambda_1, \cdots, \Lambda_n)$ be a configuration which is admissible i.e., it satisfies both the Siegel and weak hyperbolicity conditions as before.

Recall again the system of equations:

$$\begin{cases} \displaystyle\sum_{i=1}^{n} s_i \Lambda_i = 0 \\[2em] \displaystyle\sum_{i=1}^{n} s_i = 0 \end{cases} \tag{S}$$

Definition 4.12 We say that the configuration Λ satisfies condition (**K**) if the dimension over $\mathbb{Q}$ of the vector space of rational solutions of the system (S) above is maximal, in other words is of dimension $n - 2m - 1$.

Observe that any linear diagonal holomorphic vector field

$$\xi = \sum_{i=1}^{n} \alpha_i z_i \frac{\partial}{\partial z_i}$$

on $\mathbb{C}^n$ projects onto a holomorphic vector field on N. In particular, let

$$\Lambda_i = (\lambda_i^1, \dots, \lambda_i^m) \qquad\qquad 1 \leqslant i \leqslant n$$

and define the m commuting vector fields on S

$$\eta_i(z) = \left\langle \operatorname{Re} \Lambda_i \sum_{j=1}^{n} z_j \frac{\partial}{\partial z_j} \right\rangle = \sum_{j=1}^{n} \Re(\lambda_j^i) z_j \frac{\partial}{\partial z_j} \qquad 1 \leqslant i \leqslant m \qquad\qquad \text{(VF)}$$

for i between 1 and m. The composition of the (holomorphic) flows of these vector fields gives an action of $\mathbb{C}^m$ on N. The following theorem is proven in [ME00], as a generalization of a result of Loeb and Nicolau [LN99].

Theorem 4.12 ([ME00, Theorem 7]) *The projection onto N of the vector fields $(\eta_1, \dots, \eta_m)$ gives on N a regular holomorphic foliation $\mathcal{G}$ of dimension m. Moreover, the foliation $\mathcal{G}$ is transversely Kählerian with respect to ω, the canonical Euler form of the bundle $M_1 \to N$ (see Definition 4.7).*

Recall that transversely Kählerian means that

1. $\mathcal{G}$ is the kernel of ω.
2. ω is closed and real.
3. The quadratic form $h(-, -) = \omega(J-, -) + i\omega(-, -)$ (where J denotes the almost complex structure of N) defines a hermitian metric on the normal bundle to the foliation.

Remark 4.18 We note that this theorem gives non-trivial examples of transversely Kählerian foliations on compact complex manifolds.

4.8.1 Canonical Transversally Kähler Foliations

The aim of this subsection is to study the quotient space of N by $\mathcal{G}$. Observe that it can be obtained as the quotient space of S by the action induced by the action (B).

Going back to the abelian group G_Λ of Theorem 4.11, we see that its Lie algebra is generated by the linear vector fields $z_i \partial/\partial z_i$ for $i = 1, \dots, n$. Due to the quotient by (B), they only generate a vector space of dimension $n - m - 1$ as needed. Amongst these $n - m - 1$ linearly independent vector fields, we can find m of them which extend to S_Λ without zeros and which generates a locally free action of $\mathbb{C}^m$ onto S_Λ. For example, we can take the commuting vector fields in formula (VF) above.

Definition 4.13 (Canonical Foliation) We denote by $\mathcal{G} = \mathcal{G}_\Lambda$ the foliation induced by this action. This is the *canonical foliation* of N_Λ.

It is easy to check that $\mathcal{G}$ is independent of the choice of vector fields. Indeed, changing the vector fields just means changing the parametrization of $\mathcal{G}$, that is changing the $\mathbb{C}^m$-action by taking a different basis of $\mathbb{C}^m$.

Pull back the Fubini-Study form of $\mathbb{P}^{n-1}$ to the embedding (4.27). This is the *canonical Euler form* ω of Λ, as defined in Definition 4.7. It is a representative of the Euler class of a particular $\mathbb{S}^1$-bundle associated to N_Λ, hence the name. Then $\mathcal{G}$ is transversely Kähler with transverse Kähler form ω. For our purposes, we will not focus on ω but on the ray $\mathbb{R}^{>0}\omega$ it generates Recall that Λ fulfills condition (**K**) if (4.39) admits a basis of solutions with integer coordinates; and that Λ fulfills condition (H) if (S) does not admit any solution with integer coordinates. If condition **K** in Definition 4.12 is fulfilled, then $\mathcal{G}$ is a foliation by compact complex tori and the quotient space is a projective toric orbifold, see [MV04] which contains a thorough study of this case.

We just note here that, even if condition (**K**) in Definition 4.12 is not satisfied, the foliation $\mathcal{G}$ has some compact orbits. Indeed, let I be a vertex of K_Λ. Then, by (4.30), 0 belongs to $\mathcal{H}(\Lambda_{I^c})$, so by [MV04, Lemma 1.1],

$$\text{rank}_\mathbb{C} \begin{pmatrix} \Lambda_{i_1^c} & \cdots & \Lambda_{i_{2m+1}^c} \\ 1 & \cdots & 1 \end{pmatrix} = m + 1. \tag{4.24}$$

Hence, up to performing a permutation, we may assume at the same time (4.21) and

$$I \cap \{1, \ldots, m+1\} = \emptyset. \tag{4.25}$$

Definition 4.14 An n-dimensional toric variety W (possibly singular) is an algebraic variety with an open and dense subset biholomorphic to $(\mathbb{C}^*)^n$ such that the natural action of $(\mathbb{C}^*)^n$ extends to a holomorphic action on all of W. In other words: a toric variety of complex dimension n is an algebraic variety which is an equivariant compactification of the abelian algebraic torus $(\mathbb{C}^*)^n$.

We have

Proposition 4.8 *For each vertex I of K_Λ, the corresponding submanifold N_I is a compact complex torus of dimension m and is a leaf of $\mathcal{F}$. Moreover, assume that Λ satisfies (4.21) and (4.25). Then, letting B_I denote the matrix obtained from (4.23) by erasing the rows $\Lambda_i - \Lambda_1$ for $i \in I$, the torus N_I is isomorphic to the torus of lattice $(Id, B_I A_\Lambda^{-1})$.*

The following theorem is the fundamental connection between toric varieties with at most quotient singularities (i.e., quasi-regular varieties) and LVM manifolds.

Theorem 4.13 ([MV04, Theorem A]) *Let N be one of our manifolds corresponding to a configuration which satisfies condition (K) in Definition 4.12. Then N is a Seifert-Orlik fibration in complex tori of dimension m over a quasi-regular, projective, toric variety of dimension $n - 2m - 1$. More precisely: Let Λ be an admissible configuration satisfying condition (K) Then*

(1) *The leaves of the foliation $\mathcal{G}$ of N_Λ are compact complex tori of dimension m.*
(2) *The quotient space of N_Λ by $\mathcal{G}$ is a projective toric variety of dimension $n - 2m - 1$. We denote it by $X(\Delta)$, where Δ is the corresponding fan.*
(3) *The toric variety $X(\Delta)$ comes equipped with an equivariant orbifold structure.*
(4) *The natural projection $\pi : N \to X(\Delta)$ is a holomorphic principal Seifert bundle, with compact complex tori of dimension m as fibers.*
(5) *The transversely Kählerian form ω of N projects onto a Kählerian (singular at the singular locus of $X(\Delta)$ as a variety) form $\tilde{\omega}$ of $X(\Delta)$.*

Moreover, condition (K) is optimal with respect to these properties in the sense that the foliation $\mathcal{G}$ of a configuration which does not satisfy it has non-compact leaves, so item 1 is not verified.

4.9   Idea of the Proof of Theorem 4.13

4.9.1   Toy Example

We start with an example that shows the close relationship between LVM manifolds and Calabi–Eckmann manifolds.

Example 4.3 Consider the admissible configuration given by

$$\Lambda_1 = \Lambda_2 = 1 \qquad \Lambda_3 = i \qquad \Lambda_4 = -1 - i.$$

We have seen in Sect. 4.3.3 that the manifold corresponding N is diffeomorphic to $\mathbb{S}^1 \times \mathbb{S}^3$, i.e., N is a primary Hopf surface. One knows [BHPV04, Chapter V, Proposition 8.18] that such a surface contains either exactly two elliptic curves or else it is an elliptic fibration over $\mathbb{P}^1_{\mathbb{C}}$. In addition these two cases are distinguished by their algebraic dimensions: in the first case the algebraic dimension is 0 and in the second it is 1.

Although Theorem 4.6 does not apply directly here since the configuration has two indispensable points one still has that the algebraic dimension of N is *greater or equal to a*, the number of rational solutions which are $\mathbb{Q}$-linearly independent of system (S). Here $a = 1$, the solutions of system (S) are generated by

$$s_1 = 1 \qquad s_2 = -1 \qquad s_3 = 0 \qquad s_4 = 0$$

It follows that the algebraic dimension of N is equal to one and that N fibers over $\mathbb{P}^1_{\mathbb{C}}$ with fiber an elliptic curve.

Let us now consider $\Lambda_j = a_j + ib_j$ and the following action of $\mathbb{C}^2$ in $\mathbb{P}^3_{\mathbb{C}}$:

$$((t_1, t_2), [z]) \in (\mathbb{C}^*)^2 \times \mathbb{P}^3_{\mathbb{C}} \longmapsto [z_i \cdot t_1^{a_i} \cdot t_2^{b_i}]_{i=1}^{i=4} = [z_1 t_1, z_2 t_1, z_3 t_2, z_4/(t_1 t_2)] \in \mathbb{P}^3_{\mathbb{C}} \tag{E}$$

Let us restrict this action to $\mathcal{V}$ (the projection to $\mathbb{P}^3_{\mathbb{C}}$ of the open set of Siegel leaves of the system Definition 4.5)

$$\mathcal{V} = \{[z] \in \mathbb{P}^3_{\mathbb{C}} \mid (z_1, \ldots, z_4) \in (\mathbb{C}^2 \setminus \{(0, 0)\}) \times (\mathbb{C}^*)^2\}$$

The projection

$$[z_1, \ldots, z_4] \in \mathcal{V} \longmapsto [z_1, z_2] \in \mathbb{P}^1_{\mathbb{C}}$$

is invariant under the action (E), hence the quotient of $\mathcal{V}$ under the action can be identified with $\mathbb{P}^1_{\mathbb{C}}$.

Consider now the action (4.1) of $\mathbb{C}$ on $\mathcal{V}$ (Definition 4.5). This action commutes with the action of of $(\mathbb{C}^*)^2$, hence it respects the projection. In fact, the inclusion

$$T \in \mathbb{C} \longmapsto (t_1 = \exp T, t_2 = \exp iT) \in (\mathbb{C}^*)^2$$

intertwines the two actions: N is given by the action (E) of $(\mathbb{C}^*)^2$ restricted to couples (t_1, t_2) on its image. So one has the commutative diagram:

$$\begin{array}{ccc} \mathcal{V} & \overset{Id}{\longrightarrow} & \mathcal{V} \\ \downarrow & & \downarrow \\ N & \overset{P}{\longrightarrow} & \mathbb{P}^{n-1}_{\mathbb{C}} \end{array}$$

On the other hand the fibers of the projection in the righthand side are biholomorphic to $(\mathbb{C}^*)^2$, and the fibers in the lefthand side are biholomorphic to $\mathbb{C}$. The the fibers of p are given by the quotient of $(\mathbb{C}^*)^2$ $\mathbb{C}$ where $\mathbb{C}$ acts on $(\mathbb{C}^*)^2$ by the inclusion defined above. A direct calculation shows that the fibers are elliptic curves isomorphic to the quotient of $\mathbb{C}^*$ by the group generated by the homothety $z \to \exp 2\pi \cdot z$. In this way we obtain the elliptic fibration of N over $\mathbb{P}^1_{\mathbb{C}}$.

Everything in the preceding example can be generalized to the case of any manifold $N = N_{\Lambda}$ where Λ verifies condition (K) in Definition 4.12. Let

$$\Lambda_j = a_j + ib_j \qquad 1 \leqslant j \leqslant n$$

and consider the action of $\mathbb{C}^{2m}$ on $\mathbb{P}_{\mathbb{C}}^{n-1}$ given by the formula:

$$(R, S, [z]) \in \mathbb{C}^m \times \mathbb{C}^m \times \mathbb{P}_{\mathbb{C}}^{n-1} \longmapsto [z_j \cdot \exp \langle a_j, R \rangle \cdot \exp \langle b_j, S \rangle]_{j=1}^n \in \mathbb{P}_{\mathbb{C}}^{n-1}$$

Since N verifies condition (**K**) in Definition 4.12, up to replacing the configuration.
Λ by $A(\Lambda)$ where A is an appropriate *real* affine transformation of $\mathbb{R}^{2m} \simeq \mathbb{C}^m$ one can assume that the real and imaginary parts of each Λ_j are vectors belonging to the lattice $\mathbb{Z}^m$ [MV04, Lemma 2.4]:

This means that the preceding action of $\mathbb{C}^{2m}$ on $\mathbb{P}_{\mathbb{C}}^{n-1}$ is equivalent to an algebraic action of $(\mathbb{C}^*)^{2m}$ on $\mathbb{P}_{\mathbb{C}}^{n-1}$

$$(t, s, [z]) \in (\mathbb{C}^*)^m \times (\mathbb{C}^*)^m \times \mathbb{P}_{\mathbb{C}}^{n-1} \longmapsto [z_1 \cdot t^{a_1} \cdot s^{b_1}, \ldots, z_n \cdot t^{a_n} \cdot s^{b_n}] \in \mathbb{P}_{\mathbb{C}}^{n-1}$$
$$(4.26)$$

where a_j and b_j belong to $\mathbb{Z}^m$. Here t^{a_j} (respectively s^{b_j}) means $t_1^{a_j^1} \cdot \ldots \cdot t_m^{a_j^m}$ (respectively $s_1^{b_j^1} \cdot \ldots \cdot s_m^{b_j^m}$).

When one restricts the action (4.26) to $\mathcal{V}$ (see Definition 4.5), one can show that one obtains as quotient a projective toric variety X. In fact this procedure is precisely the construction of toric varieties as **GIT** (Geometric Invariant Theory) quotients by de David Cox in [CO95]. One simply verifies that the open set $\mathcal{V} \subset \mathbb{P}_{\mathbb{C}}^{n-1}$ corresponds to the semi-stable points for the natural linearization of $\mathbb{C}^n \to \mathbb{P}_{\mathbb{C}}^{n-1}$ [MV04, Lemma 2.12]. Since in our case the quotient is a geometric quotient (the orbit space which is Hausdorff) and not a quotient where one identifies instead the closure of the orbits, one deduces that the semi-stable points are in fact stable and, via [CO95], that the quotient is a projective quasi-regular toric variety i.e., it possesses at worst quotient singularities,

Let i be the inclusion: $T \in \mathbb{C}^m \xrightarrow{i} (\exp T, \exp iT) \in (\mathbb{C}^*)^m \times (\mathbb{C}^*)^m$.

Like in the toy example one can restrict the action (4.26) to the pairs in $(\mathbb{C}^*)^m \times (\mathbb{C}^*)^m$ that are in the image of i. This way one obtains an *algebraic* action of $\mathbb{C}^m$ on $\mathcal{V}$. This action is precisely the action of formula (4.2). One obtains the same type of commutative diagram as in the toy example:

$$\begin{array}{ccc} \mathcal{V} & \xrightarrow{Id} & \mathcal{V} \\ \downarrow & & \downarrow \\ N & \xrightarrow{p} & X \end{array} \qquad \text{(CE)}$$

A calculation shows that the fibers of $p : N_\Lambda \to X$ are compact complex tori of complex dimension m. This is equivalent to showing that every isotropy group under the action is a lattice isomorphic to $\mathbb{Z}^{2m}$. In fact this lattice can be explicitly calculated here is where one uses the rationality condition (**K**) in Definition 4.12. The lattice is constant in an open an dense set (in the Hausdorff or Zariski topology) but it could have special fibers that are finite quotients of the typical fibre (all the fibers are isogenous). In other words: the projection p corresponds to the quotient (i.e., the orbit space) of N by the holomorphic action of a compact complex torus of complex dimension m acting with finite isotropy groups. This implies that X has the structure structure of an orbifold such that $p : N \to X$ is a Seifert-Orlik fibration [OR72].

Remark 4.19 The pre-image under $p : N \to X$ of a singular point of X is necessarily a special fiber. However there could be above regular points special fibers. In fact, the locus on the base X having special fibers could be of codimension one but the singular locus of X as a normal projective toric manifold must have codimension at least two. The reason of this difference is that X is an orbifold in addition to being toric and this structure could have "fake" codimension one singularities. For instance in the examples 4.3.3 of if one replaces $\Lambda_2 = 1$ by $\Lambda_2 = p$, one constructs a Hopf surface with an elliptic fibration over $\mathbb{P}^1_{\mathbb{C}}$ having two singular points of orbifold type and one or two special fibers.

Theorem 4.13 has the following corollary:

Corollary 4.2 *Let N satisfy the conditions of Theorem 4.13. Then the algebraic reduction of N is a quasi-regular, projective, toric variety of dimension $n - 2m - 1$.*

As a particular case of the previous theorem one recovers the elliptic fibrations used by E. Calabi and B. Eckmann to provide the product of spheres $\mathbb{S}^{2p-1} \times \mathbb{S}^{2q-1}$ (for $p > 1$ et $q > 1$) with a complex structure. This generalization is given by the following

Definition 4.15 A generalized Calabi–Eckmann fibration is the fibration obtained by the previous theorem.

Since we know, fixing m and n, that the set of configurations satisfying condition **(K)** in Definition 4.12 is dense in the space of admissible configurations on obtains:

Corollary 4.3 *Every manifold N corresponding to an admissible configuration is a small deformation of a generalized Calabi–Eckmann fibration*

Remark 4.20 All the fibers are isogenous complex tori.

In the case where $X(\Delta) = \mathbb{P}^p_{\mathbb{C}} \times \mathbb{P}^q_{\mathbb{C}}$ with its manifold structure, then the fibration is the one in elliptic curves

$$\mathbb{S}^{2p+1} \times \mathbb{S}^{2q+1} \to \mathbb{C}P^p \times \mathbb{C}P^q$$

where $\mathbb{S}^{2p+1} \times \mathbb{S}^{2q+1}$ is endowed with a Calabi–Eckmann complex structure (see [CE53]). This explains the following definition.

Definition 4.16 We call such a Seifert bundle $N \to X(\Delta)$ a generalized Calabi–Eckmann fibration over $X(\Delta)$.

Corollary 4.4 *Let* Λ *be an admissible configuration satisfying condition* (**K**). *Let* $z \in S$ *and let*

$$J_z = \{i_1, \ldots, i_p\} = \{1 \leqslant i \leqslant n \quad | \quad z_i \neq 0\}$$

Then,

(1) The lattice in $\mathbb{C}^m$ *of the orbit through* z *is* $2\pi L^*_{(J_z)^c}$.
(2) The orbit through z *is an exceptional orbit if and only if* $L_0 \subsetneq L_{(J_z)^c}$. *In this case, it is a finite unramified quotient of the generic orbit of degree the index of* L_0 *in* $L_{(J_z)^c}$.

The construction in the preceding section is completely reversible. Let X be a quasi-regular toric projective variety (i.e., if it has singularities they are quotient singularities) then the construction by David Cox in [CO95] permits to realize X as the quotient of an open set X_{ss} by a linear algebraic action of $(\mathbb{C}^*)^p$ like the one given in formula (4.26) above. One can arrange this action in order to have p even and set $m = p/2$. Then one defines the configuration by the formulae:

$$\Lambda = \{\Lambda_j = a_j + ib_j \quad | \quad 1 \leqslant j \leqslant n\} \quad \text{(Realization of } \Lambda\text{)} \qquad \text{(F)}$$

where the natural numbers a_j b_j are the weights (like in formula (4.26)) of the algebraic action of $(\mathbb{C}^*)^{2m} = (\mathbb{C}^*)^m \times (\mathbb{C}^*)^m$ and one induces an action of $\mathbb{C}^m$ on X_{ss} via the inclusion i defined above. To achieve one uses the following technical Lemma found in [**MV04, Lemmas 2.12 et 4.9.**]:

Lemma 4.6 ([MV04, Lemmas 2.12 et 4.9]) *With the previous definition the configuration* Λ *is admissible and satisfies condition* (**K**) *in Definition 4.12. In addition the open set* $\mathcal{V}(\Lambda)$ *(see Definition 4.5) is equal to* X_{ss}.

One can show without difficulty that the variety X obtained by Cox construction is the generalized Calabi–Eckmann fibration associated to N_Λ via the commutative diagram (CE).

Therefore one the following theorem which is the reciprocal of Theorem 4.13:

Theorem 4.14 *Let* X *be a projective, quasi-regular, toric variety. Then there exists* $m > 0$ *and a manifold* N *corresponding to an admissible configuration which admits a generalized Calabi–Eckmann over* X *and whose fibres are complex torii of complex dimension* m.

Furthermore, if X *is nonsingular (smooth), one can choose* m *and* N *such that the fibration is a holomorphic principal fibration.*

Remark 4.21 The previous theorem motivated a possible definition of non-commutative toric varieties and its deformations (usual toric varieties are rigid). See [BZ15, KLMV14].

Example 4.4 ([MV04, Proposition I]; Hirzebruch Surfaces) Let $a \in \mathbb{N}$. Then the manifold N_Λ associated to the admissible configuration $\Lambda = \{\Lambda_1, \Lambda_2, \Lambda_3, \Lambda_4, \Lambda_5\}$ with

$$\Lambda_1 = \Lambda_4 = 1 \qquad \Lambda_2 = i \qquad \Lambda_3 = (2a^2+3a)+i(2a+1) \qquad \Lambda_5 = -2(a+1)-2i$$

is diffeomorphic to $\mathbb{S}^3 \times \mathbb{S}^3$ and it is a principal fibration in elliptic curves over the Hirzebruch surface $\mathbb{F}_a$.

The existence of such manifold was noticed by H. Maeda in [MA74].

4.9.2 Examples

In this subsection, we shall use the following facts (which are proven in [DV97]). Let $\Lambda = (\lambda_1, \dots, \lambda_5)$ be an admissible configuration with $m = 1$ and $n = 5$. Then, the classification of N_Λ *up to diffeomorphism* is completely determined by k, the number of indispensable points. We have

$$k = 0 \iff N \text{ is the quotient of } \#(5)(\mathbb{S}^3 \times \mathbb{S}^4) \text{ by a non-trivial action of } \mathbb{S}^1$$

$$k = 1 \iff N \text{ is diffeomorphic to } \mathbb{S}^3 \times \mathbb{S}^3$$

$$k = 2 \iff N \text{ is diffeomorphic to } (\mathbb{S}^5 \times \mathbb{S}^1)$$

where $\#(5)\mathbb{S}^3 \times \mathbb{S}^4$ denotes the connected sum of five copies of $\mathbb{S}^3 \times \mathbb{S}^4$. In all of the following examples, we shall give the fans in $\mathbb{R}^2$ with canonical basis (e_1, e_2) and lattice $\mathbb{Z}^2$, or in $\mathbb{R}$ with canonical basis e_1 and lattice $\mathbb{Z}$.

Remark 4.22 In the examples that follow we also use the very technical fact, proven in [MV04], that given the fan Δ of a toric variety with quotient singularities one can recover the configuration Λ satisfying condition (**K**) of Definition 4.12. In particular one can recover from the fan the number of equations m and the dimension n to have an admissible action of C^m on C^n.

Example 4.5 Consider the complete fan Δ generated by

$$w_1 = e_1 \qquad w_2 = e_2 \qquad w_3 = -e_1 - e_2$$

of the complex projective space $\mathbb{C}P^2$. There is a unique class of Kähler classes (in the sense of Theorem G in [MV04]), which is that of the (Chern class) of the anti-canonical divisor. Up to scaling and up to translation, the polytope Q is defined as

$$Q = \{u \in \mathbb{R}^2 \mid \langle w_1, u \rangle \geqslant -1, \ \langle w_2, u \rangle \geqslant -1, \ \langle w_3, u \rangle \geqslant -1\}.$$

By the methods of [MV04] (F) we can recover the configuration:

$$\lambda_1 = \lambda_2 = \lambda_3 = 1 \qquad \lambda_4 = -3 + i \qquad \lambda_5 = -i .$$

It is easy to check that this configuration has two indispensable points (λ_4 and λ_5) and thus the manifold N is diffeomorphic to $\mathbb{S}^5 \times \mathbb{S}^1$. We obtain finally the well-known (holomorphic) fibration $\mathbb{S}^5 \times \mathbb{S}^1 \to \mathbb{P}^2_{\mathbb{C}}$ and the pre-symplectic form of N scaled by 5 projects onto the (Chern class) of the anti-canonical divisor.

Notice that we may easily compute the modulus of the fiber in this case. From Corollary B in [MV04], we obtain the lattice and we obtain:

$$L_0^* = \frac{1}{5} \text{Vect}_{\mathbb{Z}}(-4 + i, 1 + i)$$

and this modulus is equal to

$$\frac{-4 + i}{1 + i} = \frac{-3}{2} + \frac{5}{2}i .$$

It is known [CE53] (see also [LN96]) that, for any choice of a modulus τ, there exists a complex structure on $\mathbb{S}^5 \times \mathbb{S}^1$ such that it fibers in elliptic curves of modulus τ over the complex projective space.

Fix $\tau = \alpha + i\beta$ with $\beta > 0$. A straightforward computation shows that the admissible configuration

$$\begin{pmatrix} \Re\lambda_i' \\ \Im\lambda_i' \end{pmatrix} = \begin{pmatrix} -\beta & \beta \\ 1 + \alpha & 4 - \alpha \end{pmatrix} \cdot \begin{pmatrix} \Re\lambda_i \\ \Im\lambda_i \end{pmatrix} \qquad 1 \leqslant i \leqslant 5$$

determines a complex threefold diffeomorphic to $\mathbb{S}^5 \times \mathbb{S}^1$ which fibers over $\mathbb{P}^2_{\mathbb{C}}$ with fiber an elliptic curve of modulus τ.

Example 4.6 Consider the complete fan Δ generated by

$$w_1 = e_1 \qquad w_2 = e_2 \qquad w_3 = -e_1 \qquad w_4 = -e_2$$

of $\mathbb{C}P^1 \times \mathbb{C}P^1$. The Kähler classes are

$$D_{\alpha,\beta} = \alpha(D_1 + D_3) + \beta(D_2 + D_4) \qquad\qquad \alpha > 0, \ \beta > 0$$

corresponding to the rectangles

$$Q_{\alpha,\beta} = \{(u_1, u_2) \in \mathbb{R}^2 \mid -\alpha \leqslant u_1 \leqslant \alpha, \ -\beta \leqslant u_2 \leqslant \beta\} .$$

We get the corresponding configuration:

$$\lambda_1 = \lambda_3 = 1 \qquad \lambda_2 = \lambda_4 = i \qquad \lambda_5 = -2\alpha - 2i\beta.$$

The corresponding manifold $N_{\alpha,\beta}$ is diffeomorphic to $\mathbb{S}^3 \times \mathbb{S}^3$ and we find the Calabi–Eckmann fibration $\mathbb{S}^3 \times \mathbb{S}^3 \to \mathbb{P}^1_{\mathbb{C}} \times \mathbb{P}^1_{\mathbb{C}}$. The pre-symplectic form of $N_{\alpha,\beta}$ projects onto a representative of the class of $D_{\alpha,\beta}$ (up to scaling).

Fix α and β. As in Example 4.5, for every choice of $\tau \in \mathbb{C}$ with Im $\tau > 0$, there exists a matrix A of $GL_2(\mathbb{R})$ such that the product of the previous configuration by A determines a LVM manifold diffeomorphic to $\mathbb{S}^3 \times \mathbb{S}^3$ which fibers over the product of projective lines with an elliptic curve of modulus τ as fiber (compare with [CE53]).

Alternatively, we may start from

$$\lambda_1 = \lambda_3 = 1 \qquad \lambda_2 = \lambda_4 = i \qquad \lambda_5 = -1 - i.$$

which is an admissible configuration such that the class of $\tilde{\omega}$ is $D_{1,1}$ (up to scaling) and perform a translation on $(\lambda_1, \dots, \lambda_5)$ to have another Kähler ray associated to $\tilde{\omega}$ (see Remarks 4.11 and 4.12 in [MV04]).

More precisely, assume that $\alpha < 1$ and $\beta < 1$ and let

$$b = \frac{1 - 2\alpha}{2\alpha + 2\beta + 1} + i\frac{1 - 2\beta}{2\alpha + 2\beta + 1}$$

then the class of $\tilde{\omega}$ of $N((\lambda_1, \dots, \lambda_5) + b)$ is $D_{\alpha,\beta}$.

Example 4.7 Consider the fan of $\mathbb{P}^1_{\mathbb{C}}$:

$$w_1 = e_1 \qquad w_2 = -e_1$$

There is a unique Kähler ray, that of $D = D_1 + D_2$. We choose $p_1 = p$ and $p_2 = q$ for p and q strictly positive integers, that is we want to recover all possible codimension one equivariant orbifold singularities on $\mathbb{P}^1_{\mathbb{C}}$. We take $n = 4$ and $m = 1$ and choose

One can show that that configuration is:

$$\lambda_1 = \frac{-3}{2p} + i\frac{1 - 2q}{2p} \qquad \lambda_2 = \frac{-1}{2q} - i\frac{2p + 1}{2q} \qquad \lambda_3 = 1 \qquad \lambda_4 = i .$$

It is easy to check that the corresponding manifold $N_{p,q}$ is diffeomorphic to $\mathbb{S}^3 \times \mathbb{S}^1$. It is the total space of a generalized Calabi–Eckmann fibration over $\mathbb{P}^1_{\mathbb{C}}$ with at most two singular points of order p and q.

Notice that if p and q are not coprime, then the orbifold structure cannot be obtained as a weighted projective space, i.e. cannot be obtained as quotient of $\mathbb{C}^2 \setminus \{(0,0)\}$ by an algebraic action of $\mathbb{C}^*$.

Example 4.8 Consider the complete fan Δ generated by

$$w_1 = e_1, \qquad w_2 = e_2, \qquad w_3 = -e_1, \qquad w_4 = -e_1 - e_2, \qquad w_5 = -e_2$$

of the toric del Pezzo surface X obtained as the equivariant blowing-up of $\mathbb{P}^2_\mathbb{C}$ in two points. Define Q as the pentagon associated to the anti-canonical divisor of X (which is ample for X is del Pezzo)

$$Q = \{u \in \mathbb{R}^2 \mid \langle w_1, u \rangle \geqslant -1, \ldots, \langle w_5, u \rangle \geqslant -1\}.$$

We want to construct a generalized Calabi–Eckmann fibration with elliptic curves as fibers. This implies that we must take $m = 1$ and $n = 5$, so we cannot add any indispensable point. As the sum of the w_i is not zero, we cannot take at the same time $p_1 = \ldots = p_5 = 1$. In other words, there does not exist any *non singular* generalized Calabi–Eckmann fibration over X with elliptic curves as fibers. However, if we allow exceptional fibers, the construction is possible keeping $m = 1$. For example, take

$$p_1 = 2 \qquad p_2 = 2 \qquad p_3 = 1 \qquad p_4 = 1 \qquad p_5 = 1.$$

This gives

$$v_1 = 2e_1, \qquad v_2 = 2e_2, \qquad v_3 = -e_1, \qquad v_4 = -e_1 - e_2, \qquad v_5 = -e_2,$$

$$\epsilon_1 = \epsilon_2 = \frac{2}{7}, \qquad \epsilon_3 = \epsilon_4 = \epsilon_5 = \frac{1}{7}$$

Notice that $\mathcal{L}$ is $\mathbb{Z}^2$. Taking a linear Gale transform of this, we obtain

$$\lambda_1 = 1, \qquad \lambda_2 = i, \qquad \lambda_3 = -2 - 4i, \qquad \lambda_4 = 4 + 4i \qquad \lambda_5 = -4 - 2i.$$

This gives a fibration in elliptic curves $N \to X$ where N is the quotient of the differentiable manifold $\#(5)\mathbb{S}^3 \times \mathbb{S}^4$ by a non trivial action of $\mathbb{S}^1$. The orbifold structure on X has two codimension one singular sets of index 2 and the form 7ω projects onto a representant of the Chern class of D.

Example 4.9 We consider the same toric variety X and the same polytope Q as in Example 4.8, but this time we want $p_1 = \ldots = p_5 = 1$, i.e. we want a non singular fibration. We are thus forced to increase m by one and take $m = 2$ and $n = 7$, which gives us two additional indispensable points. We take

$$v_1 = e_1, \, v_2 = e_2, \, v_3 = -e_1, \, v_4 = -e_1 - e_2,$$

$$v_5 = -e_2, \, v_6 = -v_1 - v_2 - v_3 - v_4 - v_5 = e_1 + e_2, \qquad v_7 = 0$$

and

$$\epsilon_1 = \ldots = \epsilon_5 = \epsilon_7 = \frac{1}{9}, \qquad \epsilon_6 = \frac{1}{3}$$

Notice that ϵ_6 is chosen so that v_6/ϵ_6 lies in the interior of $\mathcal{H}(v_1/\epsilon_1, \ldots, v_5/\epsilon_5)$. Making all the computations, we find that

$$(\Lambda_1, \ldots, \Lambda_7) = \begin{pmatrix} 1 & i & 0 & 0 & -1+i & -1 & 3-2i \\ 0 & 0 & 1 & i & 1 & 1+i & -5-4i \end{pmatrix}$$

defines a LVM manifold diffeomorphic to $(\#(5)\mathbb{S}^3 \times \mathbb{S}^4) \times \mathbb{S}^1$ (by application of Theorem 12 of [ME00]) which is the total space of a non singular principal holomorphic fibration over X with complex tori of dimension 2 as fibers. The form 9ω projects onto the anti-canonical divisor of X.

Example 4.10 Let $a \in \mathbb{N}$ and consider the complete fan Δ generated by

$$w_1 = e_1 \qquad w_2 = e_2 \qquad w_3 = -e_2 \qquad w_4 = -e_1 + ae_2$$

of the Hirzebruch surface F_a.

Let $D = D_1 + D_2 + D_3 + (a+1)D_4$. The divisor D is ample (see [FU93, p. 70]). We take $v_i = w_i$ for $1 \leqslant i \leqslant 4$ and add the vertex $v_5 = -v_1 - \ldots - v_4 = -ae_2$. We have $m = 1$ and $n = 5$. We choose

$$\epsilon_1 = \frac{1}{2a+5} \qquad \epsilon_2 = \frac{1}{2a+5} \qquad \epsilon_3 = \frac{1}{2a+5} \qquad \epsilon_4 = \frac{a+1}{2a+5} \qquad \epsilon_5 = \frac{a+1}{2a+5}$$

Taking a linear Gale transform of this, we obtain

$$\lambda_1 = \lambda_4 = 1 \qquad \lambda_2 = i \qquad \lambda_3 = (2a^2 + 3a) + i(2a+1) \qquad \lambda_5 = -2(a+1) - 2i$$

with only one indispensable point λ_5. We thus have the following proposition.

Proposition 4.9 *Let $a \in \mathbb{N}$. Consider the admissible configuration*

$$\lambda_1 = \lambda_4 = 1 \qquad \lambda_2 = i \qquad \lambda_3 = (2a^2 + 3a) + i(2a+1) \qquad \lambda_5 = -2(a+1) - 2i$$

Then, the corresponding LVM manifold N_a is diffeomorphic to $\mathbb{S}^3 \times \mathbb{S}^3$ and is a principal fiber bundle in elliptic curves over the Hirzebruch surface F_a (where F_0 is $\mathbb{C}P^1 \times \mathbb{P}^1_{\mathbb{C}}$). The scaling of the canonical Euler form of the bundle $M_1 \to N_a$ by $2a + 5$ projects onto a representant of the Chern class of the ample divisor $D = D_1 + D_2 + D_3 + (a+1)D_4$ on F_a.

Remark 4.23 The preceding example shows that, for special complex structures of Calabi–Eckmann type on $\mathbb{S}^3 \times \mathbb{S}^3$, there exists holomorphic principal actions of an elliptic curve whose quotient may be topologically different (since the Hirzebruch surfaces F_{2n} are all diffeomorphic to $\mathbb{S}^2 \times \mathbb{S}^2$ whereas the Hirzebruch surfaces F_{2n+1} are all diffeomorphic to the non-trivial $\mathbb{S}^2$-bundle over $\mathbb{S}^2$, and these two manifolds have different intersection form).

4.10   From Polytopes to Quadrics

This section and the following rely on the paper by Panov [PA10] and we also recommend [BP02] for this section. Let $\mathbb{R}^n$ be given the standard inner product $\langle \cdot, \cdot \rangle$ and consider convex polyhedrons P defined as intersections of m closed half-spaces:

$$\Pi_{(a_i, b_i)} = \{x \in \mathbb{R}^n \mid \langle a_i, x \rangle + b_i \geqslant 0\}, \quad for \quad i = 1, \ldots, m$$

with $a_i \in \mathbb{R}^n$, $b_i \in \mathbb{R}$. Assume that the hyperplanes defined by the equations $\langle a_i, x \rangle + b_i = 0$ are in general position, i. e. at least n of them meet at a single point. Assume further that $dim\, P = n$ and P is bounded (which implies that $m > n$). Then P is an n-dimensional compact simple polytope. Set

$$F_i = \{x \in P : \langle a_i, x \rangle + b_i = 0\} \qquad (F \text{ for } facet).$$

Since the hyperplanes are in general position F_i is either empty or a facet of P. If it is empty the linear equation is redundant and we can remove the corresponding inequality without changing P.

Let A_P be the $m \times n$ matrix of row vectors a_i, and b_P be the column m-vector of scalars $b_i \in \mathbb{R}$ ($i \in \{1, \cdots, m\}$). Then we can write:

$$P = \{x \in \mathbb{R}^n : A_P x + b_P \geqslant 0\}$$

and consider the affine map

$$i_P : \mathbb{R}^n \to \mathbb{R}^m,$$

$$i_P(x) = A_P x + b_P.$$

It embeds P into the first orthant

$$\mathbb{R}^m_{\geqslant 0} = \{(y_1, \cdots, y_m) \in \mathbb{R}^m \mid y_i > 0, \quad i \in \{1, \ldots, m\}\}.$$

We identify $\mathbb{C}^m$ (as a real vector space) with $\mathbb{R}^{2m}$ as usual using the map $z = (z_1, \ldots, z_m) \mapsto (x_1, y_1, \ldots, x_m, y_m)$, where $z_k = x_k + i y_k$ for $k = 1, \ldots, m$.

Consider the following commutative diagram where $\mathcal{Z}_P$ its obtained by pull-back and

$\mu : \mathbb{C}^m \to \mathbb{R}^m_{\geq 0}$ is given by $\mu(z_1, \ldots, z_m) = (|z_1|, \ldots, |z_m|)$:

$$
\begin{array}{ccc}
\mathcal{Z}_P & \xrightarrow{\;i_P^*\;} & \mathbb{C}^m \\
{\scriptstyle \pi}\downarrow & & \downarrow{\scriptstyle \mu} \\
P & \xrightarrow[\;i_P\;]{} & \mathbb{R}^m_{\geq 0}
\end{array}
$$

The map μ may be thought of as the quotient map for the coordinatewise action of the standard torus

$$\mathbb{T}^m = \{(z_1, \ldots, z_m) \in \mathbb{C}^m : |z_i| = 1 \ \text{for} \ 1 \leqslant i \leqslant m\}$$

on $\mathbb{C}^m$.

Therefore, $\mathbb{T}^m$ acts on $\mathcal{Z}_P$ with quotient P, and i_P^* is a $\mathbb{T}^m$-equivariant embedding.

The image of $\mathbb{R}^n$ under i_P is an n-dimensional affine plane in $\mathbb{R}^m$, which can be written as

$$
i_P(\mathbb{R}^n) = \{y \in \mathbb{R}^m : y = A_P(x) + b_P \ \text{for some} \ x \in \mathbb{R}^n\}
$$
$$
= \{y \in \mathbb{R}^m : \Gamma y = \Gamma b_P\},
$$

where $\Gamma = ((\gamma_{jk}))$ is an $(m - n) \times m$ matrix whose rows form a basis of linear relations between the vectors a_i. That is, Γ is of full rank and satisfies the identity $\Gamma A_P = 0$.

Then we obtain that $\mathcal{Z}_P$ embeds into $\mathbb{C}^m$ as the set of common zeros of $m - n$ real quadratic equations:

$$
i_P^*(\mathcal{Z}_P) = \left\{ z \in \mathbb{C}^m \ \Bigg| \ \sum_{k=1}^m \gamma_{jk}|z_k|^2 = \sum_{k=1}^m \gamma_{jk}b_k, \ \text{for} \ 1 \leqslant j \leqslant m - n \right\} \quad \textbf{(Quadratic $\star$)}
$$

$$
\text{(G)}
$$

The following properties of $\mathcal{Z}_P$ easily follow from its construction.

1. Given a point $z \in \mathcal{Z}_P$, the ith coordinate of $i_P^*(z) \in \mathbb{C}^m$ vanishes if and only if z projects onto a point $x \in P$ such that $x \in F_i$ for some facet F_i.
2. Adding a redundant inequality to results in multiplying $\mathcal{Z}_P$ by a circle.
3. $\mathcal{Z}_P$ is a smooth manifold of dimension $m + n$. The embedding $i_P^* : \mathcal{Z}_P \to \mathbb{C}^m$ has $\mathbb{T}^m$-equivariantly trivial normal bundle.

4.11 From Quadrics to Polytopes (Associated Polytope of LVM Manifolds)

Let N and

$$M_1 = \{z \in \mathbb{C}^n \mid \sum_{i=1}^{n} \Lambda_i |z_i|^2 = 0, \sum_{i=1}^{n} |z_i|^2 = 1\}$$

be as before in Definition 4.6.

Let us remark that the standard action of the torus $(\mathbb{S}^1)^n$ on $\mathbb{C}^n$

$$((\exp i\theta_1, \cdots, \exp i\theta_n), z) \longmapsto (\exp i\theta_1 \cdot z_1, \ldots, \exp i\theta_n \cdot z_n) \tag{H}$$

leaves M_1 invariant. The quotient of M_1 by this action can be identified, via the diffeomorphism $r \in \mathbb{R}_{>0}^+ \to r^2 \in \mathbb{R}_{>0}^+$, to

$$K = \{r \in (\mathbb{R}^+)^n \mid \sum_{i=1}^{n} r_i \Lambda_i = 0, \sum_{i=1}^{n} r_i = 1\} \tag{I}$$

Lemma 4.7 *The quotient K is a convex polytope of dimension $n - 2m - 1$ with $n - k$ facets.*

Proof By definition K is the intersection of the space A of solutions of an affine system with the closed sets $r_i \geqslant 0$. Each one of these closed sets defines an affine half-space $A \cap \{r_i \geqslant 0\}$ in the affine space A. In other words, K is the intersection of a finite number of affine half-spaces. Since this intersection is bounded (since M_1 is compact), one obtains indeed a convex polytope. The weak hyperbolicity condition implies that the affine system that defines K is of maximal rank. Hence, K is of dimension $n - 2m - 1$.

Let us consider in more detail the definition of K. The points $r \in K$ verifying $r_i > 0$ for all i are the points which belong to the interior of the convex polytope. They correspond to the points z de M_1 which also belong to $(\mathbb{C}^*)^n$, i.e. to the points of M_1 such that the orbit under the action (H) is isomorphic to $(\mathbb{S}^1)^n$. The points which belong to a hyperface are exactly the points r of K having all of its coordinates *except one* equal to zero. They correspond to the points z de M_1 which have a unique coordinate equal to zero, i.e. such that its orbit under the action (H) is isomorphic to $(\mathbb{S}^1)^{n-1}$. One obtains from the definition of K that there exist points of K having all coordinates different from zero except the ith coordinate if and only if 0 belongs to the convex envelope of the configuration formed by the Λ_j with j different from i; hence if and only if Λ_i is a point which can be eliminated keeping the conditions of Siegel and weak hyperbolicity. therefore one has $n - k$ hyperfaces. □

Definition 4.17 One calls the convex polytope $K = K_\Lambda$ corresponding to the admissible configuration Λ the *associated polytope*. The polytopes $\mathcal{H}(\Lambda)$ and K_Λ are related by the Gale transform.

One central idea is that the topology of the manifolds M_1, and therefore of the manifolds N, is codified by the combinatorial type of the polytope K. To make this idea more precise, it is interesting to push to the end the reasoning involved in the proof of the preceding Lemma. One had seen that

$$K_i = K \cap \{r_i = 0, \ r_j > 0 \text{ for } j \neq i\}$$

is nonempty, and therefore is a hyperface de K, if and only if

$$0 \in \mathcal{H}((\Lambda_j)_{j \neq i}).$$

Analogously, given I a subset of $\{1, \ldots, n\}$, the set

$$K_I = K \cap \{r_i = 0 \text{ for } i \in I, \ r_j > 0 \text{ for } j \notin I\}$$

is nonempty, and therefore it is a facet of K of codimension equal the cardinality of I, if and only if

$$0 \in \mathcal{H}((\Lambda_j)_{j \notin I})$$

One has therefore established a very important correspondence between two convex polytopes: the polytope K on one hand and the convex hull of the Λ_i's on the other hand.

This correspondence allows us to to prove the following result:

Remark 4.24 It follows from [MV04, Lemma 1.1] that

$$\text{rank}_{\mathbb{C}} \begin{pmatrix} \Lambda_1 & \ldots & \Lambda_n \\ 1 & \ldots & 1 \end{pmatrix} = m + 1.$$

Hence, up to a permutation, condition (4.21) is always fulfilled.

Definition 4.18 We say that N_Λ and $N_{\Lambda'}$ are *G-biholomorphic* if they are $(G(\Lambda), G(\Lambda')$-equivariantly biholomorphic.

Recall that by definition (D) the manifold N_Λ embeds in $\mathbb{P}^{n-1}$ as the C^∞ submanifold

$$N = \{[z] \in \mathbb{P}^{n-1} \ \mid \ \sum_{i=1}^{n} \Lambda |z_i|^2 = 0\}. \tag{4.27}$$

It is crucial to notice that this embedding is not arbitrary but has a clear geometric meaning. Indeed, it is proven in that action (4.1) induces a foliation of $\mathcal{S}_\Lambda$; that every leaf admits a unique point closest to the origin (for the Euclidean metric); and finally that N is the projectivization of the set of all these minima. This is a sort of non-algebraic Kempf-Ness Theorem. So we may say that this embedding is canonical.

The maximal compact subgroup $(\mathbb{S}^1)^n \subset (\mathbb{C}^*)^n$ acts on $\mathcal{S}_\Lambda$, and thus on N_Λ. This action is clear on the smooth model (4.27). Notice that it reduces to a $(\mathbb{S}^1)^{n-1}$ since we projectivized everything.

The quotient of N_Λ by this action is easily seen to be a simple convex polytope of dimension $n - 2m - 1$, cf. Up to scaling, it is canonically identified to

$$K_\Lambda := \{r \in (\mathbb{R}^+)^n \;\mid\; \sum_{i=1}^n \Lambda r_i = 0, \; \sum_{i=1}^n r_i = 1\}. \tag{4.28}$$

It is important to have a description of K_Λ as a convex polytope in $\mathbb{R}^{n-2m-1}$. This can be done as follows. Take a Gale diagram of Λ, that is a basis of solutions $(v_1, \ldots, v_n)$ over $\mathbb{R}$ of the system (S):

$$\begin{cases} \displaystyle\sum_{i=1}^n \Lambda_i x_i = 0 \\[2mm] \displaystyle\sum_{i=1}^n x_i = 0 \end{cases} \tag{S}$$

Take also a point ϵ in K_Λ. This gives a presentation of K_Λ as

$$\{x \in \mathbb{R}^{n-2m-1} \;\mid\; \langle x, v_i \rangle \geqslant -\epsilon_i \text{ for } i = 1, \ldots, n\} \tag{4.29}$$

This presentation is not unique. Indeed, taking into account that K_Λ is unique only up to scaling, we have

Lemma 4.8 *The projection* (4.29) *is unique up to action of the affine group of* $\mathbb{R}^{n-2m-1}$.

On the combinatorial side, K_Λ has the following property. A point $r \in K_\Lambda$ is a vertex if and only if the set I of indices i for which r_i is zero is maximal, that is has $n - 2m - 1$ elements. Moreover, we have

$$r \text{ is a vertex} \iff \mathcal{S}_\Lambda \cap \{z_i = 0 \text{ for } i \in I\} \neq \emptyset \iff 0 \in \mathcal{H}(\Lambda_{I^c}) \tag{4.30}$$

for I^c the complementary subset to I in $\{1, \ldots, n\}$. This gives a numbering of the faces of K_Λ by the corresponding set of indices of zero coordinates.

More precisely, we have

$$J \subset \{1, \ldots, n\} \text{ is a face of codimension Card } J$$

$$\Longleftrightarrow \mathcal{S}_\Lambda \cap \{z_i = 0 \text{ for } i \in J\} \neq \emptyset \Longleftrightarrow 0 \in \mathcal{H}(\Lambda_{J^c}) \tag{4.31}$$

In particular, K_Λ has $n - k$ facets. Observe moreover that the action (4.1) fixes $\mathcal{S}_\Lambda \cap \{z_i = 0 \text{ for } i \in J\}$, hence its quotient defines a submanifold N_J of N_Λ of codimension Card J.

4.12 Moment-Angle Manifolds

We will explain the link between the moment-angle manifolds in Definition 4.6 and the manifolds studied by Buchstaber and Panov in [BP02]. Let P be a simple convex polytope with the set $\mathcal{F} = \{F_1, \ldots, F_n\}$ of hyperfaces (i.e., codimension one faces). Let $T_i \simeq \mathbb{S}^1$ for $1 \leqslant i \leqslant n$ and let $T_\mathcal{F} = T_1 \times \cdots \times T_n \simeq (\mathbb{S}^1)^n = \mathbb{T}^n$ be the n torus with its standard group structure. For each hyperface F_i associate T_i, the circle corresponding to the ith coordinate of $\mathbb{T}^n$.

If G is a face of the polytope P let

$$T_G = \prod_{F_i \supset G} T_i \subset T_\mathcal{F}$$

For each point $q \in P$, let $G(q)$ be the unique face of P which contains q in its relative interior

Definition 4.19 The *moment-angle complex* $\mathcal{Z}_P$ associated to P is defined as

$$\mathcal{Z}_P = (T_\mathcal{F} \times P)/\sim$$

where the equivalence relation is: $(t_1, p) \sim (t_2, q)$ if and only if $p = q$ and $t_1 t_2^{-1} \in T_{G(q)}$.

The *moment-angle complex* $\mathcal{Z}_P$ depends only upon the combinatorial type of P and it admits a natural continuous action of the n-torus $T_\mathcal{F}$ having as quotient P. The fact that P is simple implies that $\mathcal{Z}_P$ is a topological manifold (see [BP02, Lemma 6.2]).

Consider now the moment-angle manifolds $M_1(\Lambda)$ defined by formula (4.6) and let K_Λ the associated polytope (4.17). One has the natural projection $\Pi : M_1(\Lambda) \to K_\Lambda$ The faces of codimension q of $K(\Lambda)$ correspond to the orbits of the of the points of $\mathcal{V}$ which have some precise q coordinates fixed. In other words the orbits above the relative interior of a codimension q face are isomorphic to $(\mathbb{S}^1)^{n-q}$. En poussant un peu plus loin cette description, on montre le lemme suivant.

Lemma 4.9 ([BM06, Lemma 0.15]) *Let N_Λ be an LVM manifold without indispensable points. Let K_Λ be its associated polytope. Then there exists an equivariant homeomorphism between $M_1(\Lambda)$ and the moment-angle variety Z_{K_Λ}.*

Equivariant homeomorphism means that the homeomorphism conjugates the action (H) of on $M_1(\Lambda)$ to the action of $T_{\mathcal{F}}$ on $\mathcal{Z}_P$.

Hence:

Corollary 4.5 *Let N_Λ et $N_{\Lambda'}$ two LVM manifolds without indispensable points. Then there exists an equivariant homeomorphism of the associated moment-angle varieties $M_1(\Lambda)$ and $M_1(\Lambda')$ if and only if the associated polytopes K_Λ et $K_{\Lambda'}$ are combinatorially equivalent. More generally, there exists an equivariant homeomorphism between $M_1(\Lambda)$ and $M_1(\Lambda')$ if and only K_Λ and $K_{\Lambda'}$ are combinatorially equivalent the number of indispensable points k and k', respectively, are equal.*

Proof The combinatorial equivalence between K_Λ and $K_{\Lambda'}$ implies the existence of an equivariant homeomorphism between $\mathcal{Z}_{K_\Lambda}$ and $\mathcal{Z}_{K_{\Lambda'}}$, and hence by Lemma 4.9 between entre $M_1(\Lambda)$ and $M_1(\Lambda')$. The proof for any number of indispensable points follows from the first result and Lemma 4.9.   □

It is more delicate to have the same result up to *equivariant diffeomorphism*, however one has the following theorem:

Theorem 4.15 ([BM06, Theorem 4.1]) *There is an equivalence between the following assertions:*

 (i) *The manifolds $M_1(\Lambda)$ and $M_1(\Lambda')$ are equivariantly diffeomorphic*
 (ii) *The corresponding associated polytopes K_Λ and $K_{\Lambda'}$ are combinatorially equivalent and the number of indispensable points k and k' are equal.*

4.13   Flips of Simple Polytopes and Elementary Surgeries on LVM Manifolds

The motivation of this section is to generalize the following result of Mac Gavran [MC79] adapted to our case.

Theorem 4.16 (Mac Gavran [MC79]) *Let Λ be an admissible configuration. Suppose that the associated polytope K_Λ is a polygon with p vertices. Then the moment-angle manifold $M_1(\Lambda)$ is diffeomorphic via an equivariant diffeomorphism to the connected sum of products of spheres;*

$$(\#_{j=1}^{p-3}(jC_{p-2}^{j+1})\mathbb{S}^{2+j} \times \mathbb{S}^{p-j}) \times (\mathbb{S}^1)^k$$

There are many cases of configurations for higher-dimensional polytopes where the manifolds $M_1(\Lambda)$ are similar to those of Mac Gavran i.e., the manifolds are products of manifolds of the type:

1. Odd dimensional spheres
2. Connected sums of products of spheres

Bosio and Meersseman [BM06] showed that some, but not all, moment-angle manifolds M_Λ are connected sums of products of spheres, and they conjectured that if the dual to the polytope is neighborly, then the manifold is such a connected sum. This conjecture was proven by Samuel Gitler and Santiago López de Medrano in [GL13].

Let us remember once more the results by de Medrano [DM88, DM89] on the classification of manifolds $M_1(\Lambda)$ when $m = 1$ given above in Sect. 4.3.5 given by Theorem 4.1.

When $m = 1$ the vectors are vectors Λ_i in $\mathbb{C} \simeq \mathbb{R}^2$ and S. López de Medrano shows that one can modify the configuration $\Lambda \in \mathbb{C}$ through a smooth homotopy Λ_t (just moving the vectors) that *satisfies the admissibility conditions of Siegel and weak hyperbolicity for all* $t \in [0, 1]$ such that $\Lambda_0 = \Lambda$ and Λ_1 is a regular polygon with an odd number of vertices $2l + 1$ and with multiplicities $n_1, \ldots, n_{2l+1}$. Thus, for instance, in Fig. 4.5, one can move from the pentagon at the left to the pentagon at the right to configurations with different multiplicities, for instance configurations with 4 vectors with multiplicities $n_1 = n_2 = 1$ and $n_3 = 3$, then 3 vectors of multiplicities $n_1 = 1, n_2 = n_3 = 2$, and finally 5 vectors of multiplicity 1. Ehresmann Lemma implies, that all manifolds belonging to the homotopy are diffeomorphic.

With these notations we recall Theorem 4.1 which was seen before:

Theorem 4.17 ([DM88, DM89]) *Let N be an LVM manifold $m = 1$ then M_1 is diffeomorphic to*

(i) The product of spheres $\mathbb{S}^{2n_1-1} \times \mathbb{S}^{2n_2-1} \times \mathbb{S}^{2n_3-1}$ *if* $l = 1$.
(ii) The connected sum

$$\#_{i=1}^{2l+1} \mathbb{S}^{2d_i-1} \times \mathbb{S}^{2n-2d_i-2}$$

if $l > 1$. *Where* $d_i = n_{[i]} + \ldots + n_{[i+l-1]}$ *and* $[a]$ *is the residue of the Euclidean division of a by* $2l + 1$.

Fig. 4.5 Chambers of a pentagon in $\mathbb{C}$ (the small cross is the origin in $\mathbb{C}$)

If $m > 1$ one has a higher dimensional polytope of n elements in $\mathbb{C}^m$ ($n > 2m$) and there is not a way to have a canonical model. One could consider a homotopy that takes the configuration to one with minimal number of vertices, but that is not enough to determine the polytope which is the convex hull of the points in the configuration. For this reason it is better to adapt the approach used by Mac Gavran in [MC79]. He considers simply connected manifolds of dimension $p + 2$ which admit a smooth action of a torus $(\mathbb{S}^1)^p$ which satisfies certain conditions, in particular one requires that the quotient under the action can be identified with a 2-dimensional convex polygon K with p vertices. If we write $M_1 \simeq (\mathbb{S}^1)^k \times M_0$, where $\simeq$ means "up to an equivariant diffeomorphism", then as in Lemma 4.4 one shows that the factor M_0 verifies the hypotheses of Mac Gavran. The proof of Mac Gavran Theorem is done by induction on the number p of vertices of K. If $p = 3$ one has a triangle and we know that M_1 is $\mathbb{S}^5 \times (\mathbb{S}^1)^k$, where M_0 is the sphere $\mathbb{S}^5$. To go from a polygon with p vertices to a polygon with $p + 1$ vertices one can do the following "surgery": remove an open neighborhood of a vertex and glue an interval. The reciprocal operation consists in collapsing to a point an edge. Now we recall that the faces of the associated polytope corresponding to the admissible sub-configurations of Λ (i.e., subsets of Λ) determine equivariant subvarieties of M_1 or M_0 where the quotient space identifies in a natural way with the given face. In other words to remove a neighborhood of a face means to remove an invariant (under the action of the torus) tubular neighborhood of the subvariety associated to the face in question in M_0. The invariant subvarieties have trivial tubular neighborhoods (by the Slice Theorem). Since we know that the subvarieties associated to a vertex is a torus and the subvarieties associated to an edge are the product of a torus with $\mathbb{S}^3$, one sees that if M_p denotes the manifold corresponding to a polygon K with p vertices, then to pass from M_p a M_{p+1} consists of applying an equivariant surgery

$$M_{p+1} = (M_p \times \mathbb{S}^1) \setminus ((\mathbb{S}^1)^{p-2} \times \mathbb{D}^4 \times \mathbb{S}^1) \cup ((\mathbb{S}^1)^{p-2} \times \mathbb{S}^3 \times \mathbb{D}^2)$$

Where $\mathbb{D}^s$ denotes the closed disk of dimension s. The work of Mac Gavran consists of understanding the meaning of these surgeries up to equivariant diffeomorphisms. To generalize this approach to higher dimensional polytopes K we need to generalize the notion of "surgery". and understand what is the construction one has to perform on the moment angle manifold M_1 associated to K. This is done using the following notion of cobordism between polytopes inspired by MacMullen [MA93] and Timorin [TI99].

Definition 4.20 Let P and Q be two simple convex polytopes of the same dimension p. One says that P and Q are obtained from each other by an *elementary cobordism* if there exists a simple convex polytope W of dimension $p + 1$ such that:

(i) P and Q are disjoint hyperfaces of W.
(ii) There exists a unique vertex v of W that does not belong neither to P or Q.

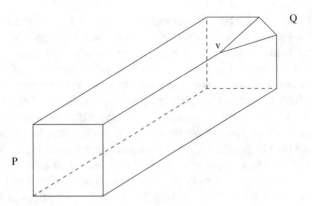

Fig. 4.6 Elementary cobordism between a square and a pentagon

Fig. 4.7 Flip de type (2, 2)

Let us recall that everything related to polytopes is up to combinatorial equivalence for instance Fig. 4.6 illustrates an elementary cobordism between a square and a pentagon.

Given a vertex v, since W is simple, there are exactly $q + 1$ edges that have v as an end point. Then, hypothesis (ii) these edges have the second end point either in P or in Q, then the *type* of the elementary cobordism is the pair (a, b) where a (respectively b) is the number of edges joining v to P (respectively Q). Of course $a + b = q + 1$,

Definition 4.21 One says that Q is obtained from P by a flip of type (a, b) if there exists an elementary cobordism of type (a, b) between P and Q.

Figure 4.7 shows an example of type (2, 2).

Let us consider the elementary cobordism W of dimension 4 between the 3-dimensional polyhedra P and Q and let us "cut" W with 3-dimensional hyperplanes parallel to P. Starting from P one sees that the edge $[AB]$ is contracted as one moves the cuts up to the point when the edge collapses to the vertex $v = A$ when the cut meets the vertex A. On the other hand if one makes cuts by hyperplanes parallel to Q the edge $[AB]$ is contracted to $v = A$. In some sense W is the *trace* of the cobordism. In other words Q is obtained P by removing a neighborhood of the edge $[AB]$ and gluing the "transverse" edge $[AB']$.

This description can be generalized for higher dimensional polytopes. A flip of type (a, b) is obtained by removing a simplicial face of dimension a and gluing the neighborhood of a simplicial face of "complementary" dimension b. Since the simplicial faces of dimension a correspond to products of a sphere of dimension $2a - 1$ by a torus [BM06, Proposition 3.6] an argument similar to that of Mac Gavran shows that if K' is obtained from K (of dimension q) by a flip of type (a, b), then the manifold $(M_0)'$ is obtained from M_0 (de dimension p) by an elementary *surgery of type* (a, b) then:

$$(M_0)' = (M_0 \times \mathbb{S}^1) \setminus ((\mathbb{S}^1)^{p-2b} \times \mathbb{D}^{2b} \times \mathbb{S}^1) \cup ((\mathbb{S}^1)^{p-2b} \times \mathbb{S}^{2b-1} \times \mathbb{D}^2)$$

if $a = 1$ and

$$(M_0)' = M_0 \setminus ((\mathbb{S}^1)^{p-2b-2a+1} \times \mathbb{D}^{2b} \times \mathbb{S}^{2a-1}) \cup ((\mathbb{S}^1)^{p-2b-2a+1} \times \mathbb{S}^{2b-1} \times \mathbb{D}^{2a})$$

if $a > 1$.

The proof of this fact is very delicate and technical since we must prove the equivariance of the constructions. All the details can be found in [BM06].

The essential difference with the case of polygons of Mac Gavran is that starting with an odd-dimensional sphere as M_0 after a finite number of elementary surgeries one does not end up with a manifold of the type connected sum of products of spheres or product of spheres. In fact in the next section one will describe the homology. However the previous considerations prove again that if two moment-angle manifolds of type M_1 of dimension p are combinatorially equivalent they are obtained from the sphere $\mathbb{S}^{2p-1}$ by the same sequence of elementary surgeries and therefore they are equivariantly diffeomorphic.

4.14 The Homology of LVM Manifolds

Recall that from Lemma we have that for a configuration Λ the associated moment angle manifold $M_1(\Lambda)$ factorizes as $M_1(\Lambda) =\simeq (\mathbb{S}^1)^k \times M_0(\Lambda)$ where k is the number of indispensable points and $M_0(\Lambda)$ is 2-connected. Since M_0 is a moment-angle manifold one can use the results of Buchstaber and Panov [BP02] to compute the homology and cohomology of these manifolds.

Theorem 4.18 ([BM06, Theorem 10.1]) *Let N_Λ be an LVM manifold and $M_0(\Lambda)$ the 2-connected factor as in Lemma 4.4. Let K be the associated polytope (quotient under the action of the torus). Let K^* be its dual which is therefore a convex*

Fig. 4.8 K^* is the octahedral
which is the dual of the cube
K, the subcomplex which
corresponds to the vertices
$\{1, 2, 3, 4\}$ is indicated in
boldface

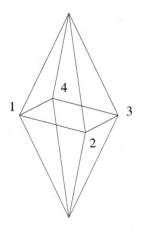

simplicial polytope. Let **F** *be its set of vertices. Then the homology of $M_0(\Lambda)$ with coefficients in $\mathbb{Z}$ is given by the formula:*

$$H_i(M_0(\Lambda), \mathbb{Z}) = \bigoplus_{\mathcal{I} \subset \mathbf{F}} \tilde{H}_{i - |\mathcal{I}| - 1}(K^*_{\mathcal{I}}, \mathbb{Z})$$

where $\tilde{H}_i$ denotes the reduced homology, $|\mathcal{I}|$ is the cardinality of $\mathcal{I}$ and $K^*_{\mathcal{I}}$ is the maximal simplicial subcomplex of K^* with vertices $\mathcal{I}$. (We remark that $H_i(M_0(\Lambda), \mathbb{Z}) = 0$ if $i < 0$).

Let us explain the meaning of *maximal simplicial subcomplex of K^* with vertices $\mathcal{I}$*. Given a q-tuple $(i_1, \ldots, i_q)$ in $\mathcal{I}$ it is a face of the simplicial subcomplex $K^*_{\mathcal{I}}$ if and only if a q-face of the simplicial complex K^*. For instance in Fig. 4.8 (K^* is the octahedral which the dual of the cube K) the subcomplex which corresponds to the vertices $\{1, 2, 3, 4\}$ is indicated in boldface.

Let us consider now the question of the level of complexity of the homology of the manifolds $M_0(\Lambda)$. By Theorem 4.18 the dual polytope K^* can be an arbitrary simplicial complex and the question of complexity becomes to ask which simplicial complexes can be *maximal* subcomplexes of a simplicial convex polytope. We claim that any finite simplicial complex can be a *maximal* sub-complexes of a simplicial convex polytope. In effect, let K_0 be any finite simplicial complex, we can always embed K_0 in a simplex $\mathfrak{S}^d$ of dimension d equal to the number of vertices of K_0 minus one. In general is not embedded in a maximal subcomplex For instance in Fig. 4.8, K_0 is the one-dimensional complex with is a circuit of four edges with vertices in boldface. It can be embedded in a tetrahedron $\mathfrak{S}^3$ as a circuit with 4 vertices but the maximal associated subcomplex is the tetrahedron itself so this embedded copy is not maximal but we can fix this by choosing a barycentric subdivision of the tetrahedron. In general it is enough to make barycentric subdivision of all the faces that belong to the maximal simplicial complex generated by K_0 to obtain an embedding which is maximal. *This is always possible* (see [BM06]).

Therefore one has:

Theorem 4.19 ([BM06, Théorèm 14.1]) *Let K_0 be any finite simplicial complex. Then there exists a 2-connected LVM manifold N such that its homology verifies:*

$$H_{i+q+1}(N, \mathbb{Z}) = \tilde{H}_i(K_0, \mathbb{Z}) \oplus \dots$$

for all i between 0 and the dimension of K_0.

Hence there exist an LVM manifold such that its homology has as a direct summand the homology of a given simplicial complex, in particular its homology can be as complex as one wishes. For instance, given a finite abelian group $\mathfrak{G}$ there exists a configuration Λ such that N_Λ has as subgroup $\mathfrak{G}$ in its group of torsion.

Remark 4.25 In [BM06, Theorem 10.1] one finds a formula describing the ring structure via the cup product of the cohomology of these manifolds.

Remark 4.26 More details and results about the homology of moment-angle manifolds using the fact that they have in many cases an open-book structure will be found in Sects. 4.17 and 4.17

4.15 Wall-Crossing

Let us consider again Fig. 4.5 before now considered as Fig. 4.9 with the purpose of illustrating the process of *wall-crossing*.

Consider in Fig. 4.9 different positions of the origin (marked as a cross) with respect to a configuration which is a regular pentagon and in the three positions the pentagon has been translated so that the origin is in different "chambers" bounded by the diagonals of the pentagon. We see that if the point marked with a cross moves from the figure on the left to the figure on the right then the figure in the left has two indispensable points, in the second there is one indispensable point and in the figure at right there are not indispensable points. The the manifolds from left to right are, respectively, $\mathbb{S}^5 \times \mathbb{S}^1 \times \mathbb{S}^1$, then $\mathbb{S}^3 \times \mathbb{S}^3 \times \mathbb{S}^1$ and finally $\#5(\mathbb{S}^3 \times \mathbb{S}^4)$.

Fig. 4.9 Wall-crossing from one chamber to another

As we mentioned before, these configurations are similar: one passes from one to the other translating Λ by a family Λ_t or if one wishes translating the origin. If we take the latter perspective and if we regard the translation of the origin as a homotopy along which 0 moves, we see that there is a moment in which 0 crosses at certain moment a "wall" $[\Lambda_i \Lambda_j]$ (in fact one crosses first a wall to go from left to the middle and the another to go from the middle to the right). The topology changes exactly after crossing the wall. In effect, if 0 does not encounters the wall the configuration is admissible and $M_1(\Lambda)$ does not change differentiably (again using Ehresmann Lemma). After crossing the wall the topology of M_1 changes drastically and after crossing the wall, by the same argument using Ehresmann Lemma, nothing happens for the rest of the homotopy.

This situation generalizes to every dimension.

Question *How does the topology of M_1 changes when we cross a wall?*

Let us see what happens in our example in Fig. 4.9 at the level of the associated polygons. At the left one has a triangle, in the middle a square and finally at right a pentagon. In other words one passes from the configuration at the left to the configuration in the middle by a surgery of type $(1, 2)$, then from the configuration in the middle to that in the right to a second surgery of type $(1, 2)$. This solves completely this particular case.

Some simple arguments of convex geometry allows us to see that everything is analogous in the general case. When one crosses a wall in a configuration $(\Lambda_1, \ldots, \Lambda_n)$, the wall is supported by $2m$ vectors Λ_i. This wall separates the convex envelope of Λ in two connected components one contains 0 *before* the other contains 0 *after*. The Λ_j's which do not belong to the wall divide in two parts: a belong to the part that contains 0 before crossing the wall and b to the part that contains 0 after crossing the wall. One of course has $a + b = n - 2m$, namely $a + b$ is equal to the dimension of the associated polytope plus one. We say that it is a wall-crossing of type (a, b).

With this notation we have:

Theorem 4.20 ([BM06, Theorem 5.4]) *Let Λ et Λ' be two admissible configurations. Suppose that Λ' is obtained from Λ through a wall crossing of type (a, b). Then*

(i) The polytope associated to K' is obtained from K by a flip of type (a, b).
(ii) The manifold $M_1(\Lambda')$ is obtained from $M_1(\Lambda)$ by an elementary surgery of type (a, b).

One can be more precise and characterize the face of the polytope where the "flip" happens (or equivalently the subvariety along which we perform an equivariant surgery in function of the wall).

4.16 LVMB Manifolds

Recall that

$$S := \{z \in \mathbb{C}^n \mid 0 \in \mathcal{H}(\Lambda_{I_z})\} \tag{4.32}$$

where

$$i \in I_z \iff z_i \neq 0. \tag{4.33}$$

Then N_Λ is the quotient of the projectivization $\mathbb{P}(S)$ by the holomorphic action

$$(T, [z]) \in \mathbb{C}^m \times \mathbb{P}(S) \longmapsto \left[z_i \exp\langle\Lambda_i, T\rangle\right]_{i=1,\dots,n} \tag{4.34}$$

where $\langle -, - \rangle$ denotes the inner product of $\mathbb{C}^m$, and not the hermitian one. It is a compact complex manifold of dimension $n - m - 1$, which is either a m-dimensional compact complex torus (for $n = 2m + 1$) or a non kähler manifold (for $n > 2m + 1$).

4.16.1 Bosio Manifolds

In [BO01], Frédéric Bosio gave a generalization of the previous construction. His idea was to relax the weak hyperbolicity and Siegel conditions for Λ and to look for all the subsets S of $\mathbb{C}^n$ such that action (4.1) is free and proper.

To be more precise, let $n \geqslant 2m + 1$ and let $\Lambda = (\Lambda_1, \dots, \Lambda_n)$ be a configuration of n vectors in $\mathbb{C}^m$. Let also $\mathcal{E}$ be a non-empty set of subsets of $\{1, \dots, n\}$ of cardinal $2m + 1$ and set

$$S = \{z \in \mathbb{C}^n \mid I_z \supset E \text{ for some } E \in \mathcal{E}\} \tag{4.35}$$

Assume that

(i) For all $E \in \mathcal{E}$, the affine hull of $(\Lambda_i)_{i \in E}$ is the whole $\mathbb{C}^m$.
(ii) For all couples $(E, E') \in \mathcal{E} \times \mathcal{E}$, the convex hulls $\mathcal{H}((\Lambda_i)_{i \in E})$ and $\mathcal{H}((\Lambda_i)_{i \in E'})$ have non-empty interior.
(iii) For all $E \in \mathcal{E}$ and for every $k \in \{1, \dots, n\}$, there exists some $k' \in E$ such that $E \setminus \{k'\} \cup \{k'\}$ belongs to $\mathcal{E}$.

Then, action (4.1) is free and proper [BO01, Théorème 1.4]. We still denote it by N_Λ although it also depends on the choice of S. As in the LVM case, it is a compact complex manifold of dimension $n - m - 1$, which is either a m-dimensional compact complex torus (for $n = 2m + 1$) or a non Kähler manifold (for $n > 2m + 1$).

Assume now that $(\Lambda_1, \dots, \Lambda_n)$ is an admissible configuration. Let

$$\mathcal{E} = \{I \subset \{1, \dots, n\} \mid 0 \in \mathcal{H}((\Lambda_i)_{i \in I})\} \tag{4.36}$$

Then (4.35) and (4.32) are equal, the previous three properties are satisfied and the LVMB manifold is exactly the corresponding LVM.

We say that Λ_i, or simply i, is an *indispensable point* if every point z of S satisfies $z_i \neq 0$. We denote by k the number of indispensable points.

4.16.1.1   The Associated Polytope of a LVM Manifold

In this section, N_Λ is a LVM manifold. Recall that the manifold N_Λ embeds in $\mathbb{P}^{n-1}$ as the C^∞ submanifold

$$N = \{[z] \in \mathbb{P}^{n-1} \mid \sum_{i=1}^{n} \Lambda |z_i|^2 = 0\}. \tag{4.37}$$

It is crucial to notice that this embedding is not arbitrary but has a clear geometric meaning. Indeed, it is proven in [ME98] that action (4.1) induces a foliation of S; that every leaf admits a unique point closest to the origin (for the Euclidean metric); and finally that (4.27) is the projectivization of the set of all these minima.[1] So we may say that this embedding is canonical.

The maximal compact subgroup $(\mathbb{S}^1)^n \subset (\mathbb{C}^*)^n$ acts on S, and thus on N_Λ. This action is clear on the smooth model (4.27). Notice that it reduces to a $(\mathbb{S}^1)^{n-1}$ since we projectivized everything.

The quotient of N_Λ by this action is easily seen to be a simple convex polytope of dimension $n - 2m - 1$, cf. [ME98] and [MV04]. Up to scaling, it is canonically identified to

$$K_\Lambda := \{r \in (\mathbb{R}^+)^n \mid \sum_{i=1}^{n} \Lambda r_i = 0, \ \sum_{i=1}^{n} r_i = 1\}. \tag{4.38}$$

It is important to have a description of K_Λ as a convex polytope in $\mathbb{R}^{n-2m-1}$. This can be done as follows. Take a Gale diagram of Λ, that is a basis of solutions $(v_1, \ldots, v_n)$ over $\mathbb{R}$ of the system

$$\begin{cases} \displaystyle\sum_{i=1}^{n} \Lambda_i x_i = 0 \\[2mm] \displaystyle\sum_{i=1}^{n} x_i = 0 \end{cases} \tag{4.39}$$

Take also a point ϵ in K_Λ. This gives a presentation of K_Λ as

$$\{x \in \mathbb{R}^{n-2m-1} \mid \langle x, v_i \rangle \geqslant -\epsilon_i \text{ for } i = 1, \ldots, n\} \tag{4.40}$$

---

[1]This is a sort of non-algebraic Kempf-Ness Theorem.

This presentation is not unique. Indeed, taking into account that K_Λ is unique only up to scaling, we have

Lemma 4.10 *The projection* (4.29) *is unique up to action of the affine group of* $\mathbb{R}^{n-2m-1}$.

On the combinatorial side, K_Λ has the following property. A point $r \in K_\Lambda$ is a vertex if and only if the set I of indices i for which r_i is zero is maximal, that is has $n - 2m - 1$ elements. Moreover, we have

$$r \text{ is a vertex} \iff S \cap \{z_i = 0 \text{ for } i \in I\} \neq \emptyset \iff 0 \in \mathcal{H}(\Lambda_{I^c}) \qquad (4.41)$$

for I^c the complementary subset to I in $\{1, \ldots, n\}$. This gives a numbering of the faces of K_Λ by the corresponding set of indices of zero coordinates. To be more precise, we have

$$J \subset \{1, \ldots, n\} \text{ is a face of codimension Card } J$$
$$\iff S \cap \{z_i = 0 \text{ for } i \in J\} \neq \emptyset \iff 0 \in \mathcal{H}(\Lambda_{J^c}) \qquad (4.42)$$

In particular, K_Λ has $n - k$ facets. Observe moreover that the action (4.1) fixes $S \cap \{z_i = 0 \text{ for } i \in J\}$, hence its quotient defines a submanifold N_J of N_Λ of codimension Card J.

Also, (4.31) implies that

$$S = \{z \in \mathbb{C}^n \mid I_z^c \text{ is a face of } K_\Lambda\} \qquad (4.43)$$

Remark 4.27 Other results related to LVMB manifolds were obtained by Battisti [BA13], Ishida [IS17, IS0] and Tambour [TA12].

4.17 Moment-Angle Manifolds and Intersection of Quadrics

Remark 4.28 This section is based on the papers [BLV17, BLV, GL13] and it borrows freely a lot from them. In order to be compatible with the notation in these papers, we use in this section sometimes different notations that the ones used in the previous sections, for instance the moment-angle manifolds $M_1(\Lambda)$ are called here $Z^{\mathbb{C}}(\Lambda)$ and the corresponding LVM manifolds N_Λ are denoted here $\mathcal{N}(\Lambda) = Z^{\mathbb{C}}(\Lambda)/\mathbb{S}^1$

The topology of generic intersections of quadrics in $\mathbb{R}^n$ of the form:

$$\sum_{i=1}^n \lambda_i x_i^2 = 0, \quad \sum_{i=1}^n x_i^2 = 1, \text{ where } \lambda_i \in \mathbb{R}^k, \, i = 1, \ldots, n$$

appears naturally in many instances and has been studied for many years. If $k = 2$ they are diffeomorphic to a triple product of spheres or to the connected sum of sphere products [GL05, DM89]; for $k > 2$ this is no longer the case [BBCG10, BM06]) but there are large families of them which are again connected sums of spheres products [GL13].

The generic condition, known as *weak hyperbolicity* and equivalent to regularity of the manifold, is the following:

> If $J \subset 1, \ldots, m$ has k or fewer elements then the origin is not in the
>
> convex hull of the λ_i with $i \in J$.

A crucial feature of these manifolds is that they admit natural group actions: all of them admit $\mathbb{Z}_2^n$ actions by changing the signs of the coordinates.

Their complex versions in $\mathbb{C}^n$, which we denote by $Z^{\mathbb{C}}$ or $Z^{\mathbb{C}}(\Lambda)$ (denoted by $M_1(\Lambda)$ in the previous sections),

$$\sum_{i=1}^n \lambda_i |z_i|^2 = 0, \qquad \sum_{i=1}^n |z_i|^2 = 1, \quad \text{where } \lambda_i \in \mathbb{C}^k, \ i = 1, \ldots, n$$

(now known as *moment-angle manifolds*) admit natural actions of the n-torus $\mathbb{T}^n$

$$((u_i, \ldots, u_n), (z_i, \ldots, z_n)) \mapsto (u_1 z_1, \ldots u_n z_n)$$

The quotient can be identified in both cases with the polytope $\mathcal{P}$ given by

$$\sum_{i=1}^n \lambda_i r_i = 0, \qquad \sum_{i=1}^n r_i = 1, \qquad r_i \geqslant 0.$$

that determines completely the varieties (so we can use the notations $Z(\mathcal{P})$ and $Z^{\mathbb{C}}(\mathcal{P})$ for them) as well as the actions. The weak hyperbolicity condition implies that $\mathcal{P}$ is a simple polytope and any simple polytope can be realized as such a quotient.

The facets of $\mathcal{P}$ are its non-empty intersections with the coordinate hyperplanes. If all such intersections are non-empty Z and $Z^{\mathbb{C}}$ fall under the general concept of *generalized moment-angle complexes* [BBCG10].

If we take the quotient of $Z^{\mathbb{C}}(\Lambda)$ by the scalar action of $\mathbb{S}^1$:

$$\mathcal{N}(\Lambda) = Z^{\mathbb{C}}(\Lambda)/\mathbb{S}^1,$$

we obtain a compact, smooth LVM manifold $\mathcal{N}(\Lambda) \subset \mathbb{P}_{\mathbb{C}}^{n-1}$.

When k is even, $\mathcal{N}(\Lambda)$ and $Z^{\mathbb{C}}(\Lambda) \times \mathbb{S}^1$ have natural complex structures and so does $Z^{\mathbb{C}}(\Lambda)$ itself when k is odd, but admit symplectic structures only in a few well-known cases [DV97, ME00].

An open book construction was used to describe the topology of Z for $k = 2$ in some cases not covered by Theorem 2 in [DM89]. In [GL13] it is a principal technique for studying the case $k > 2$. In Sect. I-1 we recall this construction, underlining the case of moment-angle manifolds:

If $\mathcal{P}$ is a simple convex polytope and F one of its facets, $Z^{\mathbb{C}}(\mathcal{P})$ admits an open

book decomposition with binding $Z^{\mathbb{C}}(F)$ and trivial monodromy.

When $k = 2$, the varieties Z and $Z^{\mathbb{C}}(\Lambda)$ can be put in a normal form given by an *odd cyclic partition* (see Sect. I-1) and they are diffeomorphic to a triple product of spheres or to the connected sum of sphere products (see [DM89, GL13]). Using the same normal form, we give a topological description of the leaves of their open book decompositions which is complete in the case of moment-angle manifolds:

The leaf of the open book decomposition of $Z^{\mathbb{C}}(\Lambda)$ is the interior of:

(a) *a product $\mathbb{S}^{2n_2-1} \times \mathbb{S}^{2n_3-1} \times \mathbb{D}^{2n_1-2}$,*
(b) *a connected sum along the boundary of products of the form $\mathbb{S}^p \times \mathbb{D}^{2n-p-4}$,*
(c) *in some cases, there may appear summands of the form:*

   a punctured product of spheres $\mathbb{S}^{2p-1} \times \mathbb{S}^{2n-2p-3} \backslash \mathbb{D}^{2n-4}$ or
   the exterior of an embedding $\mathbb{S}^{2q-1} \times \mathbb{S}^{2r-1} \subset \mathbb{S}^{2n-4}$.

The precise result (Theorem 4.22 in Sect. I-1) follows from Theorem 4.23 in Sect. I-4, a general theorem that gives the topological description of the *half* real varieties $Z_+ = Z \cap \{x_1 \geqslant 0\}$, and requires additional dimensional and connectivity hypotheses that should be avoidable using the methods of [GL05]. Some of the proofs follow directly from the result in [DM89], while other ones require the use of some parts of its proof. All these manifolds with boundary are also generalized moment-angle complexes. In Part II, using a recent deep result about contact forms due to Borman et al. [BEM15], we show that every odd dimensional moment-angle manifold admits a contact structure. This is surprising since even dimensional moment-angle manifolds admit symplectic structures only in a few well-known cases. We also show this for large families of more general odd-dimensional intersections of quadrics by a different argument.

Part I: Open Book Structures

I-1 Construction of the Open Books

Let Λ' be obtained from Λ by adding an extra λ_1 which we interpret as the coefficient of a new extra variable x_0, so we get the variety Z':

$$\lambda_1 \left(x_0^2 + x_1^2 \right) + \sum_{i>1} \lambda_i x_i^2 = 0, \qquad (x_0^2 + x_1^2) + \sum_{i>1} x_i^2 = 1.$$

Let Z_+ be the intersection of Z with the half space $x_1 \geqslant 0$. Z_+ admits an action of $\mathbb{Z}_2^{n-1}$ with quotient the same $\mathcal{P}$: Z_+ can be obtained by reflecting $\mathcal{P}$ on all the coordinate hyperplanes except $x_1 = 0$. Z_+ is a manifold with boundary Z_0 which is the intersection of Z with the subspace $x_1 = 0$.

Consider the action of $\mathbb{S}^1$ on Z' by rotation of the coordinates (x_0, x_1). This action fixes the points of Z_0 and all its other orbits cut Z_+ transversely in exactly one point. So Z' is the open book with binding Z_0, page Z_+ and trivial monodromy:

Theorem 4.21

(i) *Every manifold Z' is an open book with trivial monodromy, binding Z_0 and page Z_+.*

(ii) *If $\mathcal{P}$ is a simple convex polytope and F is one of its facets, there is an open book decomposition of $Z^{\mathbb{C}}(\mathcal{P})$ with binding $Z^{\mathbb{C}}(F)$ and trivial monodromy.*

(ii) follows because if we write the equations of $Z^{\mathbb{C}}(\mathcal{P})$ in real coordinates, we get terms $\lambda_i(x_i^2 + y_i^2)$ so each λ_i appears twice as a coefficient and $Z^{\mathbb{C}}(\mathcal{P})$ is a variety of the type Z' in several ways. It is then an open book with binding the manifold $Z_0^{\mathbb{C}}(\mathcal{P})$ obtained by taking $z_i = 0$.

When $k = 2$ it can be assumed Λ is one of the following normal forms (see [DM89]): Take $n = n_1 + \cdots + n_{2\ell+1}$ a partition of n into an odd number of positive integers. Consider the configuration Λ consisting of the vertices of a regular polygon with $(2\ell+1)$ vertices, where the i-th vertex in the cyclic order appears with multiplicity n_i.

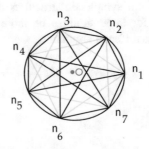

The topology of Z and $Z^{\mathbb{C}}(\Lambda)$ can be completely described in terms of the numbers $d_i = n_i + \cdots + n_{i+\ell-1}$, i.e., the sums of ℓ consecutive n_i in the cyclic order of the partition (see [DM89, GL05] and Sect. I-1):

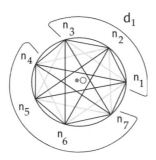

For $\ell = 1$: $Z = \mathbb{S}^{n_1-1} \times \mathbb{S}^{n_2-1} \times \mathbb{S}^{n_3-1}$, $Z^{\mathbb{C}} = \mathbb{S}^{2n_1-1} \times \mathbb{S}^{2n_2-1} \times \mathbb{S}^{2n_3-1}$.

For $\ell > 1$: $Z = \overset{2\ell+1}{\underset{j=1}{\#}} \left(\mathbb{S}^{d_i-1} \times \mathbb{S}^{n-d_i-2} \right)$, $Z^{\mathbb{C}} = \overset{2\ell+1}{\underset{j=1}{\#}} \left(\mathbb{S}^{2d_i-1} \times \mathbb{S}^{2n-2d_i-2} \right)$.

Now we have a similar description of the topology of the leaves in all moment-angle manifolds, where $\coprod$ denotes connected sum along the boundary and $\tilde{\mathcal{E}}^{2n-4}_{2n_2-1, 2n_4-1}$ is the exterior of $\mathbb{S}^{2n_2-1} \times \mathbb{S}^{2n_5-1}$ in $\mathbb{S}^{2n-4}$ (see Sect. I-3):

Theorem 4.22 *Let $k = 2$, and consider the manifold $Z^{\mathbb{C}}$ corresponding to the cyclic partition $n = n_1 + \cdots + n_{2\ell+1}$. Consider the open book decomposition of $Z^{\mathbb{C}}$ corresponding to the binding at $z_1 = 0$, as given by Theorem 4.21. Then the leaf of this decomposition is diffeomorphic to the interior of:*

(a) If $\ell = 1$, the product

$$\mathbb{S}^{2n_2-1} \times \mathbb{S}^{2n_3-1} \times \mathbb{D}^{2n_1-2}.$$

(b) If $\ell > 1$ and $n_1 > 1$, the connected sum along the boundary of $2\ell+1$ manifolds:

$$\coprod_{i=2}^{\ell+2} \left(\mathbb{S}^{2d_i-1} \times \mathbb{D}^{2n-2d_i-3} \right) \coprod \coprod_{i=\ell+3}^{1} \left(\mathbb{D}^{2d_i-2} \times \mathbb{S}^{2n-2d_i-2} \right).$$

(c) If $n_1 = 1$ and $\ell > 2$, the connected sum along the boundary of 2ℓ manifolds:

$$\coprod_{i=3}^{\ell+1} \left(\mathbb{S}^{2d_i-1} \times \mathbb{D}^{2n-2d_i-3} \right) \coprod \coprod_{i=\ell+3}^{1} \left(\mathbb{D}^{2d_i-2} \times \mathbb{S}^{2n-2d_i-2} \right)$$

$$\coprod \left(\mathbb{S}^{2d_2-1} \times \mathbb{S}^{2d_{\ell+2}-1} \setminus \mathbb{D}^{2n-4} \right).$$

(d) If $n_1 = 1$ and $\ell = 2$, the connected sum along the boundary of two manifolds:

$$\left(\mathbb{S}^{2d_2-1} \times \mathbb{S}^{2d_4-1} \backslash \mathbb{D}^{2n-4} \right) \coprod \tilde{\mathcal{E}}^{2n-4}_{2n_2-1, 2n_5-1}.$$

Theorem 4.22 will follow from its real version (see Theorem 4.23). It follows also that in cases (c) and (d) the product of the leaf with an open interval is diffeomorphic to the interior of a connected sum along the boundary of the type of case (b).

For $k > 2$, the topology of moment-angle manifolds and their leaves is much more complicated and it seems hopeless to give a complete description of them: they may have non-trivial triple Massey products [BA03], any amount of torsion in their homology [BM06] or may be a different kind of open books [GL13]. Nevertheless, it is plausible that a description of their leaves as above may be possible for large families of them in the spirit of [GL13].

The manifold $\mathcal{N}(\Lambda)$, defined in the introduction, also admits an open book decomposition, since the $\mathbb{S}^1$ action on the first coordinate commutes with the diagonal one.

Let

$$\pi_\Lambda : Z^{\mathbb{C}}(\Lambda) \to \mathcal{N}(\Lambda),$$

denote the canonical projection.

Consider now the open book decomposition of $Z^{\mathbb{C}}$ described above, corresponding to the variable z_1. If Λ_0 is obtained from Λ by removing λ_1 it is clear that the diagonal $\mathbb{S}^1$-action on $Z^{\mathbb{C}}$ has the property that each orbit intersects each page in a unique point and at all of its points this page is intersected tranversally by the orbits. This implies that the restriction of the canonical projection π_Λ to each page is a diffeomorphism onto its image $\mathcal{N}(\Lambda) - \mathcal{N}(\Lambda_0)$.

For k even we therefore obtain, since $\mathcal{N}(\Lambda) - \mathcal{N}(\Lambda_0)$ has a complex structure:

For k even, the page of the open book decomposition of $Z^{\mathbb{C}}(\Lambda)$ in Theorem 4.22 with binding $Z_0^{\mathbb{C}}(\Lambda_0)$ admits a natural complex structure which makes it biholomorphic to $\mathcal{N}(\Lambda) - \mathcal{N}(\Lambda_0)$.

For k odd, both the whole manifold and the binding admit natural complex structures.

So we have a very nice open book decomposition of every moment-angle manifold. Unfortunately, it does not have the right properties to help in the construction of contact structures on them.

The topology of these manifolds and of the leaves of their foliations is more complicated, even for $k = 2$, and only some cases have been described (see [DV97] for the simpler ones).

I-2. Homology of Intersections of Quadrics and Their Halves

We recall here the results of [DM89], whose proofs are equally valid for any intersection of quadrics and not only for $k = 2$.

Let $Z = Z(\Lambda) \subset \mathbb{R}^n$ as before, $\mathcal{P}$ its associated polytope and $F_1, \ldots, F_n$ the intersections of $\mathcal{P}$ with the coordinate hyperplanes $x_i = 0$ (some of which might be empty).

Let g_i be the reflection on the i-th coordinate hyperplane and for $J \subset \{1, \ldots, n\}$ let g_J be the composition of the g_i with $i \in J$. Let also F_J be the intersection of the F_i for $i \in J$.

The polytope $\mathcal{P}$, all its faces (the non-empty F_J) and all their combined reflections on the coordinate hyperplanes form a cell decomposition of Z. Then the elements $g_J(F_L)$ with non-empty F_L generate the chain groups $C_*(Z)$, where to avoid repetitions one has to ask $J \cap L = \emptyset$ (since g_i acts trivially on F_i).

A more useful basis is given as follows: let $h_i = 1 - g_i$ and h_J be the product of the h_i with $i \in J$. The elements $h_J(F_L)$ with $J \cap L = \emptyset$ are also a basis, with the advantage that $h_J C_*(Z)$ is a chain subcomplex of $C_*(Z)$ for every J and, since h_i annihilates F_i and all its subfaces, this subcomplex can be identified with the chain complex $C_*(\mathcal{P}, \mathcal{P}_J)$, where $\mathcal{P}_J$ is the union of all the F_i with $i \in J$. It follows that

$$H_*(Z) \approx \oplus_J H_*(\mathcal{P}, \mathcal{P}_J).$$

For the manifold Z_+ we start also with the faces of $\mathcal{P}$, but we cannot reflect them in the subspace $x_1 = 0$. This means we miss the classes $h_J(F_L)$ where $1 \in J$ and we get[2]

$$H_*(Z_+) \approx \oplus_{1 \notin J} H_*(\mathcal{P}, \mathcal{P}_J).$$

To compute the homology of $Z^{\mathbb{C}}(\Lambda)$ one can just take that of its real version (with each λ_i duplicated) or directly using instead of the basis $h_J(F_L)$ with $J \cap L = \emptyset$ the basis of (singular) cells $F_L \times T_J$ (with $J \cap L = \emptyset$) where T_J is the subtorus of T^n which is the product of its i-th factors with $i \in J$. This gives the splitting

$$H_i(Z^{\mathbb{C}}(\Lambda)) \approx \oplus_J H_{i-|J|}(\mathcal{P}, \mathcal{P}_J).$$

(See [BM06]).

These splittings have the same summands as the ones in [BBCG10] derived from the homotopy splitting of ΣZ. Even if it is not clear that they are the *same* splitting, having two such with different geometric interpretations is most valuable.

---

[2]The retraction $Z \to Z_+$ induces an epimorphism in homology and fundamental group.

I-3. The Space $\tilde{\mathcal{E}}^m_{p,q}$

Consider the standard embedding of $\mathbb{S}^p \times \mathbb{S}^q$ in $\mathbb{S}^m$, $m > p + q$ given by

$$\mathbb{S}^p \times \mathbb{S}^q \subset \mathbb{R}^{p+1} \times \mathbb{R}^{q+1} = \mathbb{R}^{p+q+2} \subset \mathbb{R}^{m+1}.$$

whose image lies in the m-sphere of radius $\sqrt{2}$.

We will denote by $\tilde{\mathcal{E}}^m_{p,q}$ the exterior of this embedding, i.e., the complement in $\mathbb{S}^m$ of the open tubular neighborhood $U = int\,(\mathbb{S}^p \times \mathbb{S}^q \times \mathbb{D}^{m-p-q}) \subset \mathbb{S}^m$. Observe that the boundary of $\tilde{\mathcal{E}}^m_{p,q}$ is $\mathbb{S}^p \times \mathbb{S}^q \times \mathbb{S}^{m-p-q-1}$ and that the classes $[\mathbb{S}^{m-p-q-1}]$, $[\mathbb{S}^p \times \mathbb{S}^{m-p-q-1}]$ and $[\mathbb{S}^q \times \mathbb{S}^{m-p-q-1}]$ are the ones bellow the top dimension that go to zero in the homology of U. By Alexander duality, the images of these classes freely generate the homology of $\tilde{\mathcal{E}}^m_{p,q}$.

Theorem A2.2 of [GL13] tells that, under adequate hypotheses (and probably always) $\tilde{\mathcal{E}}^m_{p,q} \times \mathbb{D}^1$ is diffeomorphic to a connected sum along the boundary of products of the type $\mathbb{S}^a \times \mathbb{D}^{m+1-a}$.

Under some conditions (and probably always), $\tilde{\mathcal{E}}^m_{p,q}$ is characterized by its boundary and its homology properties: Let X be a smooth compact manifold with boundary $\mathbb{S}^p \times \mathbb{S}^q \times \mathbb{S}^{m-p-q-1}$ and ι the inclusion $\partial X \subset X$.

Lemma *Assume that*

(i) *X and ∂X are simply connected.*
(ii) *the classes $\iota_*[\mathbb{S}^{m-p-q-1}]$, $\iota_*[\mathbb{S}^p \times \mathbb{S}^{m-p-q-1}]$ and $\iota_*[\mathbb{S}^q \times \mathbb{S}^{m-p-q-1}]$ freely generate the homology of X.*

Then X is diffeomorphic to $\tilde{\mathcal{E}}^m_{p,q}$.

Proof Observe that condition (i) implies that $p, q, m-p-q-1 \geq 2$ so $dim(X) = m \geq 7$. Consider the following subset of ∂X:

$$K = \mathbb{S}^p \times * \times \mathbb{S}^{m-p-q-1} \cup * \times \mathbb{S}^q \times \mathbb{S}^{m-p-q-1}$$

and embed K into the interior of X as $K \times \{1/2\}$ with respect to a collar neighborhood $\partial X \times [0, 1)$ of ∂X. Finally, let V be a smooth regular neighborhood [HI62] of $K \times \{1/2\}$ in $\partial X \times [0, 1)$.

Now, the inclusion $V \subset X$ induces an isomorphism in homology. Since the codimension of K in X is equal to $1 + min(p, q) \geq 3$, $X \setminus int(V)$ is simply connected and therefore an h-cobordism, so X is diffeomorphic to V.

Since $\tilde{\mathcal{E}}^m_{p,q}$ verifies the same properties as X, the above construction *with the same V* shows that $\tilde{\mathcal{E}}^m_{p,q}$ is also diffeomorphic to V and the Lemma is proved. $\square$

I-4. Topology of Z and Z+ When k = 2

For $k = 2$ and $\ell = 1$ a simple computation shows that

$$Z_+ = \mathbb{D}^{n_1-1} \times \mathbb{S}^{n_2-1} \times \mathbb{S}^{n_3-1}.$$

For the case $\ell > 1$ we recall here the main steps in the proof of the result about the topology of Z in [DM89], underlining those that are needed to determine the topology of Z_+. For the cyclic partition $n = n_1 + \cdots + n_{2\ell+1}$ we will denote by J_i the set of indices corresponding to the n_i copies of the i-th vertex of the polygon in the normal form. Let also $D_i = J_i \cup \cdots \cup J_{i+\ell-1}$ and $\tilde{D}_i$ its complement.

It is shown in [DM89] that for $k = 2$, the pairs $(\mathcal{P}, \mathcal{P}_J)$ with non-trivial homology are those where J consists of ℓ or $\ell + 1$ consecutive classes, that is, those where J is some D_i or $\tilde{D}_i$. In those cases there is just one dimension where the homology is non-trivial and it is infinite cyclic.

In the case of D_i that homology group is in dimension $d_i - 1$ where $d_i = n_i + \cdots + n_{i+\ell-1}$ is the length of D_i. A generator is given by the face F_{L_i} where

$$L_i = \tilde{D}_i \setminus (\{j_{i-1}\} \cup \{j_{i+\ell}\})$$

and $j_{i-1} \in J_{i-1}$, $j_{i+\ell} \in J_{i+\ell}$ are any two indices in the extreme classes of $\tilde{D}_i$ (in other words, those contiguous to D_i).

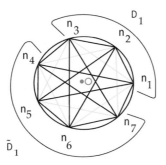

F_{L_i} is non empty of dimension $d_i - 1$. It is not in $\mathcal{P}_{D_i}$, but its boundary is. Therefore it represents a homology class in $H_{d_i-1}(\mathcal{P}, \mathcal{P}_{D_i})$, which defines a generator $h_{D_i} F_{L_i}$ of $H_{d_i-1}(Z)$. Since F_{L_i} has exactly d_i facets it is a $(d_i - 1)$-simplex so when reflected in all the coordinate subspaces containing those facets we obtain a sphere, which clearly represents $h_{D_i} F_{L_i} \in H_{d_i-1}(Z)$.

The class corresponding to $\tilde{D}_i$ is in dimension $n - d_i - 2$ and is represented by the face $F_{\tilde{L}_i}$, where $\tilde{L}_i = D_i \setminus \{j\}$ and j is any index in one of the extreme classes of D_i. It represents a generator of $H_{n-d_i-2}(Z)$, but now it is a product of spheres. For $\ell = 1$ this cannot be avoided, but for $\ell > 1$, with a good choice of j and a surgery, it can be represented by a sphere (this also follows from [GL13]). We will not make use of this class in what follows.

The final result is that, if $\ell > 1$, all the homology of Z below the top dimension can be represented by embedded spheres with trivial normal bundle.

Let Z'_+ be the manifold with boundary obtained by setting $x_0 \geq 0$ in Z' (as defined in Sect. I-1). Then Z'_+ can be deformed down to Z_+ by folding gradually the half-plane $x_0 \geq 0$, x_1 onto the ray $x_1 \geq 0$. This shows that the inclusion $Z \subset Z'_+$ induces an epimorphism in homology so one can represent all the classes in a basis of $H_*(Z'_+)$ by embedded spheres with trivial normal bundle. Those spheres can be assumed to be disjoint since they all come from the boundary Z and can be placed at different levels of a collar neighborhood. Finally, one forms a manifold Q with boundary by joining disjoint tubular neighborhoods of those spheres by a minimal set of tubes and then the inclusion $Q \subset Z'_+$ induces an isomorphism in homology. If Z is simply connected and of dimension at least 5, then Z'_+ minus the interior of Q is an h-cobordism and therefore Z is diffeomorphic to the boundary of Q which is a connected sum of spheres products. Knowing its homology we can tell the dimensions of those spheres:

If $\ell > 1$ and Z is simply connected of dimension at least 5, then:

$$Z = \overset{2\ell+1}{\underset{j=1}{\#}} \left(\mathbb{S}^{d_j-1} \times \mathbb{S}^{n-d_j-2} \right).$$

For the moment-angle manifold $Z^{\mathbb{C}}$ this formula gives, without any restrictions

$$Z^{\mathbb{C}} = \overset{2\ell+1}{\underset{j=1}{\#}} \left(\mathbb{S}^{2d_j-1} \times \mathbb{S}^{2n-2d_j-2} \right).$$

(In [GL71] this has recently been proved without any restrictions also on Z).

The topology of Z'_+ is implicit in the above proof: Z'_+ is diffeomorphic to Q and therefore it is a connected sum along the boundary of manifolds of the form $\mathbb{S}^p \times \mathbb{D}^{n-3-p}$. Since any Z with $n_1 > 1$ is such a Z' we have:

If Z_0 is simply connected of dimension at least 5, and $\ell > 1$, $n_1 > 1$ then:

$$Z_+ = \coprod_{i=2}^{\ell+2} \left(\mathbb{S}^{d_i-1} \times \mathbb{D}^{n-d_i-2} \right) \coprod \coprod_{i=\ell+3}^{1} \left(\mathbb{D}^{d_i-1} \times \mathbb{S}^{n-d_i-2} \right).$$

The classes D_i and $\tilde{D}_i$ that now give no homology are the ones that contain n_1.

The case $n_1 = 1$ is different. When $n_1 > 1$ the inclusion $Z_0 \subset Z_+$ induces an epimorphism in homology (since it is of the type $Z \subset Z'_+$). This is not the case for $n_1 = 1$: for the partition $5 = 1 + 1 + 1 + 1 + 1$, the polytope $\mathcal{P}$ is a pentagon and an Euler characteristic computation (from a cell decomposition formed by $\mathcal{P}$ and its reflections) shows that Z is the surface of genus 5. Now Z_0 has partition $4 = 1 + 2 + 1$ and consists of four copies of $\mathbb{S}^1$. From this, Z_+ must be a torus minus four disks that can be seen as the connected sum of a sphere minus four disks (all whose homology comes from the boundary) and a torus that carries the homology not coming from the boundary.

In general, when $n_1 = 1$ Z_0 is given by a normal form with $2\ell - 1$ classes, has $4\ell - 2$ homology generators below the top dimension, only half of which survive in Z_+. But Z_+ has $2\ell + 1$ homology generators, so two of them do not come from its boundary and actually form a handle.

To be more precise, the removal of the element $1 \in J_1$ allows the opposite classes $J_{\ell+1}$ and $J_{\ell+2}$ to be joined into one without breaking the weak hyperbolicity condition.

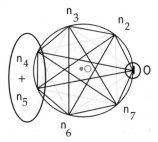

Therefore Z_0 has fewer such classes and $D_2 = J_2 \cup \cdots \cup J_{\ell+1}$, which gives a generator of $H_*(Z_+)$, does not give anything in $H_*(Z_0)$ because there *it is not a union of classes* (it lacks the elements of $J_{\ell+2}$ to be so).

The two classes in $H_*(Z_+)$ missing in $H_*(Z_0)$ are thus those corresponding to $J = D_2$ and $J = D_{\ell+2}$; all the others contain both $J_{\ell+1}$ and $J_{\ell+2}$ and thus live in $H_*(Z_0)$.

As shown above, these two classes are represented by embedded spheres in Z_+ with trivial normal bundle built from the cells F_{L_2} and $F_{L_{\ell+2}}$ by reflection. Now $F_{L_2} \cap F_{L_{\ell+2}}$ is a single vertex v, all coordinates except $x_1, x_{j_{\ell+1}}, x_{j_{\ell+2}}$ being 0.

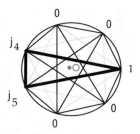

The corresponding spheres are obtained by reflecting in the hyperplanes corresponding to elements in D_2 and $D_{\ell+2}$, respectively. Since these sets are disjoint, the only point of intersection is the point v.

Now, a neighborhood of the vertex v in $\mathcal{P}$ looks like the first orthant of $\mathbb{R}^{n-3}$ where the faces F_{L_2} and $F_{L_{\ell+2}}$ correspond to complementary subspaces. When reflected in all the coordinates hyperplanes of $\mathbb{R}^{n-3}$, one obtains a neighborhood of v

in Z_+ where those subspaces produce neighborhoods of the two spheres. Therefore the spheres intersect transversely in that point.

A regular neighborhood of the union of those spheres is diffeomorphic to their product minus a disk:

$$\mathbb{S}^{d_2-1} \times \mathbb{S}^{d_\ell+2-1} \backslash \mathbb{D}^{n-3}.$$

Joining its boundary with the boundary of Z_+ we see that Z_+ is the connected sum along the boundary of two manifolds:

$$Z_+ = \mathbb{S}^{d_2-1} \times \mathbb{S}^{d_\ell+2-1} \backslash \mathbb{D}^{n-3} \coprod X$$

where $\partial X = Z_0$ and X is simply connected.

Now, all the homology of X comes from its boundary which again is Z_0, since all those classes actually live in the homology of Z and are the ones corresponding to the classes D_i and $\tilde{D}_i$ that do not contain n_1. Those classes also exist in the homology of Z_0 and are given by the same generators, so this part of the homology of Z_0 embeds isomorphically onto the homology of X.

If $\ell > 2$, Z_0 is a connected sum of sphere products, so the homology classes of X can be represented again by disjoint products $\mathbb{S}^p \times \mathbb{D}^{n-p-3}$ and finally we construct the analog of the manifold with boundary Q and the h-Cobordism Theorem gives:

If Z is simply connected of dimension at least 6, and $n_1 = 1$, $\ell > 2$ then

$$Z_+ = \coprod_{i=3}^{\ell+1} \left(\mathbb{S}^{d_i-1} \times \mathbb{D}^{n-d_i-2} \right) \coprod \coprod_{i=\ell+3}^{1} \left(\mathbb{D}^{d_i-1} \times \mathbb{S}^{n-d_i-2} \right)$$

$$\coprod \left(\mathbb{S}^{d_2-1} \times \mathbb{S}^{d_\ell+2-1} \backslash \mathbb{D}^{n-3} \right).$$

The homology classes of Z_+ are those corresponding to D_2, D_4 (not coming from the boundary) and to D_3, $\tilde{D}_1$, $\tilde{D}_5$. Clearly the last ones come from the classes $[\mathbb{S}^{n_3+n_4-1}]$, $[\mathbb{S}^{n_2-1} \times \mathbb{S}^{n_3+n_4-1}]$ and $[\mathbb{S}^{n_5-1} \times \mathbb{S}^{n_3+n_4-1}]$ in the boundary. This means that X satisfies the hypotheses of the lemma in Sect. I-3 with $p = n_2-1$, $q = n_5-1$ and $m = n - 3$, so we can conclude that X is diffeomorphic to $\tilde{\mathcal{E}}^{n-3}_{n_2-1,n_5-1}$. We have proved all the cases of the

Theorem 4.23 *Let $k = 2$, and consider the manifold Z corresponding to the cyclic decomposition $n = n_1 + \cdots + n_{2\ell+1}$ and the half manifold $Z_+ = Z \cap \{x_1 \geqslant 0\}$. When $\ell > 1$ assume Z and $Z_0 = Z \cap \{x_1 = 0\}$ are simply connected and the dimension of Z is at least 6. Then Z_+ diffeomorphic to:*

(a) If $\ell = 1$, the product

$$\mathbb{S}^{n_2-1} \times \mathbb{S}^{n_3-1} \times \mathbb{D}^{n_1-1}.$$

(b) If $\ell > 1$ and $n_1 > 1$, the connected sum along the boundary of $2\ell+1$ manifolds:

$$\coprod_{i=2}^{\ell+2} \left(\mathbb{S}^{d_i-1} \times \mathbb{D}^{n-d_i-2} \right) \coprod_{i=\ell+3}^{1} \left(\mathbb{D}^{d_i-1} \times \mathbb{S}^{n-d_i-2} \right).$$

(c) If $n_1 = 1$ and $\ell > 2$, the connected sum along the boundary of 2ℓ manifolds:

$$\coprod_{i=3}^{\ell+1} \left(\mathbb{S}^{d_i-1} \times \mathbb{D}^{n-d_i-2} \right) \coprod_{i=\ell+3}^{1} \left(\mathbb{D}^{d_i-1} \times \mathbb{S}^{n-d_i-2} \right)$$

$$\coprod \left(\mathbb{S}^{d_2-1} \times \mathbb{S}^{d_{\ell+2}-1} \setminus \mathbb{D}^{n-3} \right).$$

(d) If $n_1 = 1$ and $\ell = 2$, the connected sum along the boundary of two manifolds:

$$\left(\mathbb{S}^{d_2-1} \times \mathbb{S}^{d_4-1} \setminus \mathbb{D}^{n-3} \right) \coprod \tilde{\mathcal{E}}^{n-3}_{n_2-1, n_5-1}.$$

When $n_1 = 1$ and $\ell = 2$ we have the additional complication that restricting to $x_1 = 0$ takes us from the pentagonal Z_+ *to the* triangular Z_0, *which is not a connected sum but a product of three spheres and not all of its homology below the middle dimension is spherical.*

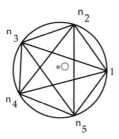

Theorem 4.23 immediately describes, under the same hypotheses, the topology of the page of the open book decomposition of Z' given by Theorem 4.21, since this page is precisely the interior of Z_+.

Theorem 4.22 about the page of the open book decomposition of the moment-angle manifold $Z^{\mathbb{C}}$ follows also, since this page is Z_+ for Z the (real) intersection of quadrics corresponding to the partition $2n - 1 = (2n_1 - 1) + (2n_2) + \cdots + (2n_{2\ell+1})$. In this case all the extra hypotheses of Theorem 4.23 hold automatically.

Theorem 4.23 applies also to the topological description of some *smoothings* of the cones on our intersections of quadrics. In this case the normal form is not sufficient to describe all possibilities as it was in [DM88] where actually only the sums d_i were needed to describe the topology or in the present work where additional information about n_1 only is required.

Part II: Some Contact Structures on Moment-Angle Manifolds

The even dimensional moment-angle manifolds and the **LVM**-manifolds have the characteristic that, except for a few, well-determined cases, do not admit symplectic structures. We will show that the odd-dimensional moment-angle manifolds (and large families of intersections of quadrics) admit contact structures.

Theorem 4.24 *If k is even, $Z^{\mathbb{C}}(\Lambda)$ is a contact manifold.*

First we show that $Z^{\mathbb{C}}(\Lambda)$ is an almost-contact manifold. Recall that a $(2n + 1)$-dimensional manifold $\mathcal{M}$ is called *almost contact* if its tangent bundle admits a reduction to $\mathbf{U}(n) \times \mathbb{R}$. This is seen as follows: consider the fibration $\pi : Z^{\mathbb{C}}(\Lambda) \to \mathcal{N}(\Lambda)$ with fibre the circle, given by taking the quotient by the diagonal action. Since $\mathcal{N}(\Lambda)$ is a complex manifold, the foliation defined by the diagonal circle action is transversally holomorphic. Therefore, $Z^{\mathbb{C}}(\Lambda)$ has an atlas modeled on $\mathbb{C}^{n-2} \times \mathbb{R}$ with changes of coordinates of the charts of the form

$$((z_1, \cdots, z_{n-2}), t) \mapsto (h(z_1, \cdots, z_{n-2}, t), g(z_1, \cdots, z_{n-2}, t)),$$

where $h : U \to \mathbb{C}^{n-2}$ and $g : U \to \mathbb{R}$ where U is an open set in $\mathbb{C}^{n-2} \times \mathbb{R}$ and, for each fixed t the function $(z_1, \cdots, z_{n-2}) \mapsto h(z_1, \cdots, z_{n-2}, t)$ is a biholomorphism onto an open set of $\mathbb{C}^{n-2} \times \{t\}$. This means that the differential, in the given coordinates, is represented by a matrix of the form

$$\left[\begin{array}{c|c} A & * \\ \hline 0 \ldots 0 & r \end{array} \right]$$

where $*$ denotes a column $(n - 2)$-real vector and $A \in \mathbf{GL}(n - 2, \mathbb{C})$. The set of matrices of the above type form a subgroup of $\mathbf{GL}(2n - 3, \mathbb{R})$. By Gram-Schmidt this group retracts onto $\mathbf{U}(n - 2) \times \mathbb{R}$.

Now it follows from [BEM15] that $Z^{\mathbb{C}}(\Lambda)$ is a contact manifold and the Theorem is proved.

In [BV14] it is given a different construction, in some sense more explicit, of contact structures, not on moment-angle manifolds but on certain non-diagonal generalizations of moment-angle manifolds of the type that has been studied by Gómez Gutiérrez and Santiago López de Medrano in [GL71]. It consists in the construction of a positive confoliation which is constructive and uses the heat flow method described in [AW00].

The argument used there applies however for many other intersections of quadrics that are not moment angle manifolds, for which the proof of the previous Theorem need not apply:

Theorem 4.25 *There are infinitely many infinite families of odd-dimensional generic intersections of quadrics that admit contact structures.*

First consider the odd-dimensional intersections of quadrics that are connected sums of spheres products:

An odd dimensional product $\mathbb{S}^m \times \mathbb{S}^n$ of two spheres admits a contact structure by the following argument: let n even and m odd, and $n, m > 2$. Without loss of generality, we suppose that $m > n$ (the other case is analogous) then $\mathbb{S}^m$ is an open book with binding $\mathbb{S}^{m-2}$ and page $\mathbb{R}^{m-1}$. Hence $\mathbb{S}^n \times \mathbb{S}^m$ is an open book with binding $\mathbb{S}^{m-2} \times \mathbb{S}^n$ and page $\mathbb{R}^{m-1} \times \mathbb{S}^n$. This page is parallelizable since $\mathbb{R} \times \mathbb{S}^n$ already is so. Then, since $m + n - 1$ is even the page has an almost complex structure. Furthermore, by hypothesis, $2n \leqslant n + m$ hence by a theorem of Eliashberg [EL90] the page is Stein and is the interior of an compact manifold with contact boundary $\mathbb{S}^{m-2} \times \mathbb{S}^n$. Hence by a theorem of Giroux [GI] $\mathbb{S}^n \times \mathbb{S}^m$ is a contact manifold.

Now, it was shown by Meckert [ME82] and more generally by Weinstein [WE91] (see also [EL90]) that the connected sum of contact manifolds of the same dimension is a contact manifold. Therefore all odd dimensional connected sums of sphere products admit contact structures.

Additionally, it was proved by Bourgeois in [BO02] (see also Theorem 10 in [GI]) that if a closed manifold $\mathcal{M}$ admits a contact structure, then so does $\mathcal{M} \times \mathbb{T}$. Therefore, all moment-angle manifolds of the form $Z \times \mathbb{T}^{2m}$, where Z is a connected sum of sphere products, admit contact structures. For every case where Z is a connected sum of sphere products we have an infinite family obtained by applying construction $Z \mapsto Z'$ an infinite number of times and in the different coordinates (as well as other operations). The basic cases from which to start these infinite families constitute also a large set and the estimates on their number in each dimension keep growing. Adding to those varieties their products with tori we obtain an even larger set of cases where an odd-dimensional Z admits a contact structure.

Another interesting fact is that most of them (including moment-angle manifolds) also have an open book decomposition. However, for these open book decompositions there does not exist a contact form which is supported in the open book decomposition like in Giroux's Theorem because the pages are not Weinstein manifolds (i.e manifolds of dimension $2n$ with a Morse function with indices of critical points lesser or equal to n). It is possible however that the pages of the book decomposition admit Liouville structures in which case one could apply the techniques of McDuff [MC91] and Seidel [SE11] to obtain contact structures.

Acknowledgements I would like to thank Yadira Barreto, Santiago López de Medrano, Ernesto Lupercio and Laurent Meersseman for their suggestions, lecture notes, and sharing their thoughts about the different aspects of the subject of these notes with me during several years. I would like to thank also the referee for pointing several typos and important mathematical details.

This work was partially supported by project IN106817, PAPIIT, DGAPA, Universidad Nacional Autónoma de México.

References

[AK01] Yu. Abe, K. Kopfermann, *Toroidal Groups: Line Bundles, Cohomology and Quasi-Abelian Varieties*. Lecture Notes in Mathematics, vol. 1759 (Springer, Berlin, 2001), viii+133pp.

[AW00] S.J. Altschuler, L.F. Wu, On deforming confoliations. J. Differ. Geom. **54**, 75–97 (2000)

[BBCG10] A. Bahri, M. Bendersky, F.R. Cohen, S. Gitler, The polyhedral product functor: a method of decomposition for moment-angle complexes, arrangements and related spaces. Adv. Math. **225**(3), 1634–1668 (2010)

[BV14] Y. Barreto, A. Verjovsky, Moment-angle manifolds, intersection of quadrics and higher dimensional contact manifolds. Moscow Math. J. **14**(4), 669–696 (2014)

[BLV] Y. Barreto, S. López de Medrano, A. Verjovsky, Open book structures on moment-angle manifolds $Z^C(\Lambda)$ and higher dimensional contact manifolds. arXiv:1303.2671

[BLV17] Y. Barreto, S. López de Medrano, A. Verjovsky, Some open book and contact structures on moment-angle manifolds. Bol. Soc. Mat. Mex. **23**(1), 423–437 (2017)

[BHPV04] W. Barth, K. Hulek, C. Peters, A. Van de Ven, *Compact Complex Surfaces* (Springer, Berlin, 2004)

[BA03] I.V. Baskakov, Massey triple products in the cohomology of moment-angle complexes. Russ. Math. Surv. **58**, 1039–1041 (2003)

[BP01] F. Battaglia, E. Prato, Generalized toric varieties for simple nonrational convex polytopes. Int. Math. Res. Not. **24**, 1315–1337 (2001)

[BA01] F. Battaglia, E. Prato, Simple nonrational convex polytopes via symplectic geometry. Topology **40**, 961–975 (2001)

[BP01] F. Battaglia, E. Prato Generalized toric varieties for simple nonrational convex polytopes. Int. Math. Res. Not. **24**, 1315–1337 (2001)

[BZ15] F. Battaglia, D. Zaffran, Foliations modeling nonrational simplicial toric varieties. Int. Math. Res. Not. IMRN **2015**(22), 11785–11815 (2015)

[BA13] L. Battisti, LVMB manifolds and quotients of toric varieties. Math. Z. **275**(1–2), 549–568 (2013)

[BEM15] M.S. Borman, Y. Eliashberg, E. Murphy, Existence and classification of overtwisted contact structures in all dimensions. Acta Math. **215**, 281–361 (2015)

[BO01] F. Bosio, Variétés complexes compactes: une généralisation de la construction de Meersseman et López de Medrano-Verjovsky. Ann. Inst. Fourier **51**(5), 1259–1297 (2001)

[BM06] F. Bosio, L. Meersseman, Real quadrics in $\mathbb{C}^n$, complex manifolds and convex polytopes. Acta Math. **197**, 53–127 (2006)

[BO02] F. Bourgeois, Odd dimensional tori are contact manifolds. Int. Math. Res. Not. **30**, 1571–1574 (2002)

[BP02] V.M. Buchstaber, T.E. Panov, *Torus Actions and Their Applications in Topology and Combinatorics* (AMS, Providence, 2002)

[CE53] E. Calabi, B. Eckmann, A class of compact, complex manifolds which are not algebraic. Ann. Math. **58**, 494–500 (1953)

[CO85] A. Connes, Non-commutative differential geometry. Publ. Math. l'IHES **62**(1), 41–144 (1985)

[CO94] A. Connes, *Noncommutative Geometry* (Academic Press, San Diego, 1994) xiv+661pp.

[CO95] D. Cox, The homogeneous coordinate ring of a toric variety. J. Algebraic Geom. **4**, 17–50 (1995)

[CZ07] S. Cupit-Foutu, D. Zaffran, Non-Kähler manifolds and GIT quotients. Math. Z. **257**, 783–797 (2007)

[DJ91] M. Davis, T. Januszkiewicz Convex polytopes, coxeter orbifolds and torus actions. Duke Math. J. **62**(2), 417–451 (1991)

[DG17] J.P. Demailly, H. Gaussier, Algebraic embeddings of smooth almost complex structures. J. Eur. Math. Soc. **19**(11), 3391–3419 (2017)

[DM88] S.L. de Medrano, *The Space of Siegel Leaves of a Holomorphic Vector Field*. Lecture Notes in Mathematics, vol. 1345 (Springer, Berlin, 1988), pp. 233–245

[DM89] S.L. de Medrano, *The Topology of the Intersection of Quadrics in $\mathbb{R}^n$*. Lecture Notes in Mathematics, vol. 1370 (Springer, Berlin, 1989), pp. 280–292

[DM14] S.L. de Medrano, Singularities of homogeneous quadratic mappings. Rev. R. Acad. Cienc. Exactas Fís. Nat. Ser. A Math. **108**(1), 95–112 (2014)

[DM17] S.L. de Medrano, Samuel Gitler and the topology of intersections of quadrics. Bol. Soc. Mat. Mex. **23**(1), 5–21 (2017)

[DV97] S.L. de Medrano, A. Verjovsky, A new family of complex, compact, non-symplectic manifolds. Bull. Braz. Math. Soc. **28**(2), 253–269 (1997)

[EL90] Y. Eliashberg, Topological characterization of Stein manifolds of dimension > 2. Int. J. Math. **1**(1), 29–46 (1990)

[FU93] W. Fulton, *Introduction to Toric Varieties* (Princeton University Press, Princeton, 1993)

[GHS83] J. Girbau, A. Haefliger, D. Sundararaman, On deformations of transversely holomorphic foliations. J. Reine Angew. Math. **345**, 122–147 (1983)

[GI] E. Giroux, Geometrie de contact: de la dimension trois vers les dimensions superieures. Proc. Int. Congress Math. **II**, 405–414 (2002)

[GL13] S. Gitler, S. López de Medrano, Intersections of quadrics, moment-angle manifolds and connected sums. Geom. Topol. **17**(3), 1497–1534 (2013)

[GL71] R. Goldstein, L. Lininger, *A Classification of 6-Manifolds with Free S^1-Action*. Lecture Notes in Mathematics, vol. 298 (Springer, Berlin, 1971), pp. 316–323

[GL05] V. Gómez Gutiérrez, S. Lépez de Medrano, *Stably Parallelizable Compact Manifolds are Complete Intersections of Quadrics*. Publicaciones Preliminares del Instituto de Matemáticas (UNAM, México, 2004)

[GL14] G.V. Gómez, S. López de Medrano, Topology of the intersections of quadrics II. Bol. Soc. Mat. Mex. **20**(2), 237–255 (2014)

[GH78] P. Griffiths, J. Harris, *Principles of Algebraic Geometry*. Pure and Applied Mathematics (Wiley, New York, 1978), xii+813pp.

[HA85] A. Haefliger, Deformations of transversely holomorphic flows on spheres and deformations of Hopf manifolds. Compos. Math. **55**, 241–251 (1985)

[HI62] M.W. Hirsch, Smooth regular neighborhoods. Ann. Math. **76**(3), 524–530 (1962)

[IS0] H. Ishida, Towards transverse toric geometry. arXiv:1807.10449

[IS17] H. Ishida, Torus invariant transverse Kähler foliations. Trans. Am. Math. Soc. **369**(7), 5137–5155 (2017)

[KA86] M. Kato, A. Yamada, Examples of simply connected compact complex 3-folds II. Tokyo J. Math. **9**, 1–28 (1986)

[KLMV14] L. Katzarkov, E. Lupercio, L. Meersseman, A. Verjovsky, The definition of a noncommutative toric variety, in *Algebraic Topology: Applications and New Directions*. Contemporary Mathematics, vol. 620 (American Mathematical Society, Providence, 2014), pp. 223–250

[KO64] K. Kopfermann, Maximale Untergruppen Abelscher komplexer Liescher Gruppen. Schr. Math. Inst. Univ. Münster **29**, iii+72pp. (1964)

[LM02] F. Lescure, L. Meersseman, Compactifications équivariantes non kählériennes d'un groupe algébrique multiplicatif. Ann. Inst. Fourier **52**, 255–273 (2002)

[LN96] J.J. Loeb, M. Nicolau, Holomorphic flows and complex structures on products of odd-dimensional spheres. Math. Ann. **306**, 781–817 (1996)

[LN99] J.J. Loeb, M. Nicolau, On the complex geometry of a class of non-Kählerian manifolds. Israel J. Math. **110**, 371–379 (1999)

[MA93] P. MacMullen, On simple polytopes. Invent. Math. **113**, 419–444 (1993)

[MA74] H. Maeda, Some complex structures on the product of spheres. J. Fac. Sci. Univ. Tokyo **21**, 161–165 (1974)

[MC91] D. McDuff, Symplectic manifolds with contact type boundaries. Invent. Math. **103**(3), 651–671 (1991)

[MC79] D. McGavran, Adjacent connected sums and torus actions. Trans. Am. Math. Soc. **251**, 235–254 (1979)

[ME82] C. Meckert, Forme de contact sur la somme connexe de deux variétés de contact de dimension impare. Ann. L'Institut Fourier **32**(3), 251–260 (1982)

[ME98] L. Meersseman, Un nouveau procédé de construction géométrique de variétés compactes, complexes, non algébriques, en dimension quelconque. Ph.D. Thesis, Lille (1998)

[ME00] L. Meersseman, A new geometric construction of compact complex manifolds in any dimension. Math. Ann. **317**, 79–115 (2000)

[MV04] L. Meersseman, A. Verjovsky, Holomorphic principal bundles over projective toric varieties. J. für die Reine und Angewandte Math. **572**, 57–96 (2004)

[MV08] L. Meersseman, A. Verjovsky, *Sur les variétés LV-M* (French). [On LV-M manifolds] Singularities II. Contemporary Mathematics, vol. 475 (American Mathematical Society, Providence, 2008), pp. 111–134

[MI65] J. Milnor, *Lectures on the H-Cobordism Theorem* (Princeton University Press, Princeton, 1965)

[60] [MO66] A. Morimoto, On the classification of non compact complex abelian Lie groups. Trans. Am. Math. Soc. **123**, 200–228 (1966)

[OR72] P. Orlik, *Seifert Manifolds*. Lecture Notes in Mathematics, vol. 291 (Springer, Berlin, 1972)

[PA10] T. Panov, *Moment-Angle Manifolds and Complexes*. Lecture notes KAIST'2010. Taras Panov. Trends in Mathematics - New Series. ICMS, KAIST, vol. 12, no. 1 (2010), pp. 43–69

[PR01] E. Prato, Simple non-rational convex polytopes via symplectic geometry. Topology **40**(5), 961–975 (2001)

[SC61] G. Scheja, Riemannsche Hebbarkeitssätze für Cohomologieklassen. Math. Ann. **144**, 345–360 (1961)

[SE11] P. Seidel, Simple examples of distinct Liouville type symplectic structures. J. Topol. Anal. **3**(1), 1–5 (2011)

[ST96] P.R. Stanley, *Combinatorics and Commutative Algebra*. Progress in Mathematics, 2nd edn., vol. 41 (Birkhäuser, Boston, 1996)

[ST51] N. Steenrod, *The Topology of Fibre Bundles* (Princeton University Press, Princeton, 1951)

[ST83] S. Sternberg, *Lectures on Differential Geometry*, 2nd edn. With an appendix by Sternberg and Victor W. Guillemin (Chelsea Publishing, New York, 1983)

[TA12] J. Tambour, LVMB manifolds and simplicial spheres. Ann. Inst. Fourier **62**(4), 1289–1317 (2012)

[TI99] V.A. Timorin, An analogue of the Hodge-Riemann relations for simple convex polytopes. Russ. Math. Surv. **54**, 381–426 (1999)

[WA80] C.T.C. Wall, Stability, pencils and polytopes. Bull. Lond. Math. Soc. **12**, 401–421 (1980)

[WE91] A. Weinstein, Contact surgery and symplectic handlebodies. Hokkaido Math. J. **620**(2), 241–251 (1991)

[WE73] R.O. Wells Jr., *Differential Analysis on Complex Manifolds* (Upper Saddle River, Prentice Hall, 1973)

LECTURE NOTES IN MATHEMATICS

 Springer

Editors in Chief: J.-M. Morel, B. Teissier;

Editorial Policy

1. Lecture Notes aim to report new developments in all areas of mathematics and their applications – quickly, informally and at a high level. Mathematical texts analysing new developments in modelling and numerical simulation are welcome.

   Manuscripts should be reasonably self-contained and rounded off. Thus they may, and often will, present not only results of the author but also related work by other people. They may be based on specialised lecture courses. Furthermore, the manuscripts should provide sufficient motivation, examples and applications. This clearly distinguishes Lecture Notes from journal articles or technical reports which normally are very concise. Articles intended for a journal but too long to be accepted by most journals, usually do not have this "lecture notes" character. For similar reasons it is unusual for doctoral theses to be accepted for the Lecture Notes series, though habilitation theses may be appropriate.

2. Besides monographs, multi-author manuscripts resulting from SUMMER SCHOOLS or similar INTENSIVE COURSES are welcome, provided their objective was held to present an active mathematical topic to an audience at the beginning or intermediate graduate level (a list of participants should be provided).

   The resulting manuscript should not be just a collection of course notes, but should require advance planning and coordination among the main lecturers. The subject matter should dictate the structure of the book. This structure should be motivated and explained in a scientific introduction, and the notation, references, index and formulation of results should be, if possible, unified by the editors. Each contribution should have an abstract and an introduction referring to the other contributions. In other words, more preparatory work must go into a multi-authored volume than simply assembling a disparate collection of papers, communicated at the event.

3. Manuscripts should be submitted either online at www.editorialmanager.com/lnm to Springer's mathematics editorial in Heidelberg, or electronically to one of the series editors. Authors should be aware that incomplete or insufficiently close-to-final manuscripts almost always result in longer refereeing times and nevertheless unclear referees' recommendations, making further refereeing of a final draft necessary. The strict minimum amount of material that will be considered should include a detailed outline describing the planned contents of each chapter, a bibliography and several sample chapters. Parallel submission of a manuscript to another publisher while under consideration for LNM is not acceptable and can lead to rejection.

4. In general, **monographs** will be sent out to at least 2 external referees for evaluation.

   A final decision to publish can be made only on the basis of the complete manuscript, however a refereeing process leading to a preliminary decision can be based on a pre-final or incomplete manuscript.

   Volume Editors of **multi-author works** are expected to arrange for the refereeing, to the usual scientific standards, of the individual contributions. If the resulting reports can be forwarded to the LNM Editorial Board, this is very helpful. If no reports are forwarded or if other questions remain unclear in respect of homogeneity etc, the series editors may wish to consult external referees for an overall evaluation of the volume.

5. Manuscripts should in general be submitted in English. Final manuscripts should contain at least 100 pages of mathematical text and should always include

   – a table of contents;
   – an informative introduction, with adequate motivation and perhaps some historical remarks: it should be accessible to a reader not intimately familiar with the topic treated;
   – a subject index: as a rule this is genuinely helpful for the reader.
   – For evaluation purposes, manuscripts should be submitted as pdf files.

6. Careful preparation of the manuscripts will help keep production time short besides ensuring satisfactory appearance of the finished book in print and online. After acceptance of the manuscript authors will be asked to prepare the final LaTeX source files (see LaTeX templates online: https://www.springer.com/gb/authors-editors/book-authors-editors/manuscriptpreparation/5636) plus the corresponding pdf- or zipped ps-file. The LaTeX source files are essential for producing the full-text online version of the book, see http://link.springer.com/bookseries/304 for the existing online volumes of LNM). The technical production of a Lecture Notes volume takes approximately 12 weeks. Additional instructions, if necessary, are available on request from lnm@springer.com.

7. Authors receive a total of 30 free copies of their volume and free access to their book on SpringerLink, but no royalties. They are entitled to a discount of 33.3 % on the price of Springer books purchased for their personal use, if ordering directly from Springer.

8. Commitment to publish is made by a *Publishing Agreement*; contributing authors of multiauthor books are requested to sign a *Consent to Publish form*. Springer-Verlag registers the copyright for each volume. Authors are free to reuse material contained in their LNM volumes in later publications: a brief written (or e-mail) request for formal permission is sufficient.

Addresses:
Professor Jean-Michel Morel, CMLA, École Normale Supérieure de Cachan, France
E-mail: moreljeanmichel@gmail.com

Professor Bernard Teissier, Equipe Géométrie et Dynamique,
Institut de Mathématiques de Jussieu – Paris Rive Gauche, Paris, France
E-mail: bernard.teissier@imj-prg.fr

Springer: Ute McCrory, Mathematics, Heidelberg, Germany,
E-mail: lnm@springer.com

Printed in the United States
By Bookmasters